Environmental Division

Core Programming Topic at the 2012 AIChE Spring Meeting & 8th Global Congress on Process Safety

Houston, Texas, USA
1-5 April 2012

ISBN: 978-1-62276-556-0

Printed from e-media with permission by:

Curran Associates, Inc.
57 Morehouse Lane
Red Hook, NY 12571

Some format issues inherent in the e-media version may also appear in this print version.

Printed by Curran Associates, Inc. (2012)

For permission requests, please contact AIChE
at the address below.

AIChE
3 Park Avenue
New York, NY 10016-5991

Phone: (203) 702-7660
Fax: (203) 775-5177

www.aiche.org

Additional copies of this publication are available from:

Curran Associates, Inc.
57 Morehouse Lane
Red Hook, NY 12571 USA
Phone: 845-758-0400
Fax: 845-758-2634
Email: curran@proceedings.com
Web: www.proceedings.com

TABLE OF CONTENTS

(4a) Equations of Change: How Patent Law Has Changed, and What Changes Are to Come

Jeffrey L. Wendt, The Wendt Firm PC, The Woodlands, TX

Patent law, particularly in the United States, has undergone many changes in the last 20 years, both in the courts and in the patent statute. My focus will be on what those changes have been, what changes are likely to come, and how these changes impact practicing chemical engineers. For example, 20 years ago the so-called "doctrine of equivalents" was widely used to ensnare infringers of patents. Now, the doctrine is almost an afterthought. My thoughts will touch on whether this is good or bad. Another change has been the term of patents. Before 1995, an issued patent had a term of 17 years from issue, even if the patent application had been filed many years before. Now, patents are granted for 20 years from the earliest filing date claimed as priority. Finally, I will touch on how a change to a "first to file" patent regime will make life different for engineer/inventors than the present "first to invent" system, particularly for small entities.

(4b) What Is Prior Art in the "First-to-File" Patent System Under the America Invents Act of 2011

Mark Rygiel, Sterne Kessler, Washington, DC and Anbar Khal, Sterne, Kessler, Goldstein & Fox PLLC, Washington, DC

On September 16, 2011, the Leahy-Smith America Invents Act was signed into law by President Obama. This law brings sweeping change to the current U.S. patent system, including transforming the current first-to-invent system into a first-inventor-to-file system. The first-to-file provisions of the Act take effect on March 16, 2013. These provisions substantially overhaul what acts are considered to be "prior art" and what acts are excepted. This presentation will provide an in-depth review of the new scope of prior art under the America Invents Act of 2011, focusing on what forms of disclosure prior to the effective filing date of a patent application will prevent the applicant from obtaining a patent in the United States. Case studies will be presented to illustrate the new prior art provisions, their exceptions, and how the new provisions compare to the existing prior art standards. The presentation will touch on how these new prior art provisions compare to the prior art rules of other developed countries having first-to-file systems.

Avoiding Common Mistakes in Licensing Agreements and Other Technology Agreements

By: Anthony P. Venturino, Esq. , Partner of Novak, Druce + Quigg, LLP Washington, DC

AIChE Spring Meeting - April 2, 2012 Houston, TX

The views presented here are those of the author and not attributable to his firm or its clients.

I. Technology agreements Can't live with them, can't live without them

Whether you consider technology agreements a help or a hindrance we need them to control the flow of technology. They set the ground rules so everyone involved knows how to handle technology being transferred and/or invented so there are no unwelcome surprises.

Thus, chemical engineers and managers need to know enough to know when they need an agreement or, how to perform if they are operating under an agreement.

In the course of performing their jobs chemical engineers and managers encounter many kinds of agreements having technology or other forms of intellectual property as a critical component. A few such agreements are as follows:

- Confidentiality Agreements
- Vendor Supplier Agreements
- Joint Development/Commercialization Agreements
- Exclusive Licenses / Non-exclusive Licenses
- Co-promotion Agreements
- Employee Agreements
- Settlement Agreements
- Joint Venture Agreements
- Contract R&D Agreements
- Consulting Agreements

II. Things can go wrong before, during and after the agreement

Here are some examples of where things can go wrong:

- Evaluation of the technology
- Selection of partners
- Conduct of the negotiations
- The agreement itself
- Implementing and servicing the agreement

Classic Licensee mistakes include:

- Royalties paid when licensed patent had expired
- Payments made after royalty terms had been satisfied
- Payments made under a trademark license when the trademark is not used

- Patents covered in a license no longer cover products that are currently made, yet royalty is paid against the patent coverage
- Royalties paid at the wrong rate, i.e. one higher or lower than provided Source:

Licensing – A Strategy for Profits, Edward P. White (1990)

Licensors also make mistakes. Here are some classics:

- Failure to take action if royalty payments cease when they should not, e.g., prior to expiration of the patents.
- Failure to monitor royalty payments, e.g. sudden reductions or failure to meet minimum royalties that could trigger a change from exclusive to nonexclusive or termination of the licenses.
- Failure to stay in technical contact with the licensee, with the result of licensee not interested in paying royalties, in maintaining quality, or keeping good records.

Licensing – A Strategy for Profits, Edward P. White (1990)

Before doing the deal.

- Know your business goals, strengths and weaknesses
- Know the available protection for the technology
- Have a realistic idea of the value of the technology
- Be open to creative technology agreements
- Do not let paranoia get in the way of a good deal
- Watch out for foreseeable legal and technical problems
- Admit when you are in over your head and get help
- Admit when you are in over your head and get help from a lawyer or other licensing professional (did I just say that?)

III. <u>Avoiding Common Traps</u>

A. <u>Define ownership and control of joint inventions of improvements</u>

Jointly developed invention often occur when smart people from different companies work together, for example while performing a joint R & D agreement or a variety of other agreements between companies. Joint inventions are jointly owned by US law unless there is a contract stating otherwise. This can cause problems.

If possible, have one party own the invention and, if necessary, grant the other party an exclusive field of use license. The following clause does this:

> In the event a joint invention arises during performance of this Agreement Company A owns the joint invention and controls patent prosecution and Company B gets an exclusive field of use license to the joint invention in the field of _______________.

B. Maintain good records of what the agreement covers or when it expires

The company, its legal department or a law firm should docket the agreement in a database. The docketing record should include: the type of agreement, the parties, the technology covered by the agreement, the agreement deadlines, the expiration date, the contact person and backup contact person, and have a miscellaneous portion for anything unusual.

Preferably an institutional person, for example, a docket clerk or contracts administrator, should be responsible for tracking notices in the docketing system and then relay the information to the business people. The more automatic you can make the timely conveying of such information, the better.

C. Do not accidentally create an agreement

The first time you negotiate with a party, especially if the other party sees this agreement as a one shot deal, be careful not to make records during the negotiations which unexpectedly bind you. Do not let an agreement to agree inadvertently become an agreement.

Summaries of the negotiations signed by both parties can lead to binding agreements even when a clause that might be important to you is missing. Let's take a hypothetical scenario where two parties bring their business people and lawyers to a meeting to negotiate a potential IP license. Mid-afternoon the attorney for the IP owner (potential licensor) had to leave to catch a plane. The potential licensor's business person (let's call him "Bill") had a later plane. The potential licensee kindly offered to take Bill to the airport later so he would not have to waste time in a boring airport. While Bill was by himself with the potential licensee's attorney and business people one of the business people suggested they draft and sign a summary of the day's discussions. They did this and then Bill caught his plane. The next day Bill's company was asked to sign an Assignment document to reflect the assignment they agreed to yesterday. Bill's response was "What assignment?" After a lawsuit the parties settled. The moral of the story is train the business people not to sign anything.

A deal can also inadvertently arise if the parties act like there is an agreement can create an agreement. This sometimes arises where the parties are in a rush to do joint development. While the deal is only a draft they start working together. Sometimes the unsigned draft becomes the *de facto* agreement.

The opposite can also happen. Sometimes you think you have an agreement but in reality you do not, or at least not all the clauses apply.

Take the case of *DataTreasury v. Wells Fargo,* 522 F.3d 1368 (Fed. Cir. 2008). In 2003 WMR e-Pin licensed a patent to Wells Fargo Services Corp. a subsidiary of Wells Fargo & Co. The license included an arbitration clause. In 2006 WMR e-Pin assigned 4 patents to DataTreasury. At least one of the 4 patents was a patent of the license. Afterwards DataTreasury sued Wells Fargo & Co. for patent infringement in court. Wells Fargo & Co. moved to dismiss or stay pending arbitration. Under 5th Circuit law, as applied by the

Federal Circuit, "to be enforceable, an arbitration clause must be in writing and signed by the party invoking it. *DataTreasury* at 1373, quoting *Westmorland v. Sadoux*, 299 F.3d 462, 465 (5th Cir. 2002). Neither Wells Fargo & Co. nor DataTreasury signed the original patent license. Thus, the court held these parties were not bound by the arbitration clause.

D. Remember important clauses

Clause checklists are important tools when reviewing a license to see if all the necessary clauses are present. However, use the right checklist for the type of agreement and the type of IP. Do not assume the clauses in your standard patent license belong in your new trademark license and *vice versa*. Unlike patents which protect the patent holder, trademarks protect the public by assuring the public a good labeled with a particular trademark will have the characteristics they expect. For example, a PIRELLI brand tire has certain handling properties. Trademark licenses without quality safeguards are termed naked licenses and invalidate the trademark rights.

A short checklist is provided in TABLE 1:

TABLE 1
Introductory Clauses
Definitions
Grant of Rights
Consideration/Compensation
Diligence
Product Quality Control
Advertising
3rd Party Infringement Suits
3rd Party Product Liability Suit
Licensor/ Licensee Suits
Termination
Continuing Obligations

A detailed set of clause checklists for a licensing agreement is available upon request from the author.

E. Maintain control of the technology after making the deal

Proprietary technology is a competitive advantage. If the licensor loses control of the technology he loses control of the competitive advantage. Thus, the licensor should determine the level of control he is willing to lose. Maybe the licensor thinks he can permit the licensee to sublicense because he trusts the judgment of licensee's president. What if the president leaves? Thus, the agreement preferably precludes sublicensing or assigning to competitors.

Another common way to lose control is for the licensee to be bought by the licensor's competitor. The Agreement should specify it may be assigned by ONE of the following:

1. Anyone;
2. Goes with the business;
3. Goes with the company;
4. Licensor maintains right of first refusal;
5. Licensor maintains right of first refusal which will not be unreasonably withheld;
6. No one.

F. Promote good communication after making the deal

Problems arise from misunderstandings, lack of thoroughness, carelessness, or failure to administer the agreement at all, i.e., no one in authority is responsible for administration.

The lawyer can structure the agreement to include mechanisms for communication between the licensor and licensee so little problems and misunderstandings do not become big ones. You cannot force the business people to communicate but the mechanisms provide fertile soil for communication. For example, the agreement should include a contact person and back-up contact person. The agreement can also mandate times for periodic meetings.

G. You need the equivalent of a pre-nuptial agreement

Agreements, like romance, can end badly. As with a newlywed couple, the enthusiastic partners to a new business relationship think nothing will go wrong. They can and they do.

Include termination provisions in the agreement to give each party a reasonable way to terminate the agreement. This includes planning for unexpected events, for example, bankruptcy or purchase of the other party. This also includes providing ways to terminate or renegotiate if the economics of the deal no longer make the current agreement viable.

IV. Organize A Licensing Department – A final word

Some companies only occasionally do a license or may be at the beginning of setting up a license project group. In these cases, the company might ask his or her lawyer for advice on how to organize the licensing project group.

The license project group will need the support of a top executive. Without this the license project group will generally not get the resources it needs.

Recognize the different functions to be performed. Then, divide or share these functions depending on available resources, business people, consultants or lawyers:

- Profit Center Group - optimizes the licensing in and licensing out to make money for the company
- Support Group - coordinates technical, legal and accounting aspects of tech transfer
- Administrative Group -handles accounting, record keeping and day to day details of maintaining the license.

V. Conclusion

Technology agreements help people work together while managing the intellectual property of their efforts. These agreements let everyone involved know their rights and responsibilities. Knowing your goal and having the appropriate agreement to help get to the goal and understanding the agreement can avoid nasty surprises. If you know your goals, technology and agreements you will be in a better position to turn that technology into profit.

Innovative Cost-Effective Control Device for Wastewater VOC Emissions (As Required by NESHAPs [BWON, HON, MON, MACT] and Other Regulations)

Dr. Carl E. Adams, Jr., PE, Senior Author[1] *
Lial F. Tischler[2]
Andrew W. Edwards, PE[3]

1 ENVIRON International Corporation, Nashville, TN
2 Tischler/Kocurek
3 ENVIRON International Corporation, Houston, TX
*Email: cadams@environcorp.com

ABSTRACT

The USEPA National Emission Standards for Hazardous Air Pollutants (NESHAP) compliance regulations require 95 percent or greater removal of regulated Hazardous Air Pollutants (HAPs) in off-gas from storage vessels, process vents, and wastewater equipment. Previous to the effort discussed herein, the only Control Technologies, routinely acceptable to the USEPA, were Vapor Phase Adsorption and Thermal Incineration, including Flares. Both control devices are capital and operationally expensive and generate high carbon footprints. However, the rule allows implementation of an Alternative Control Device. This paper describes innovative protocols, developed by the authors and Marathon Petroleum Corporation (MPC) in Garyville, Louisiana, to certify a cost-effective Alternative Control Device that has a minimal carbon footprint. The authors have developed bench-scale and full-scale confirmation protocols, which delineate more realistic and reliable benzene biodegradation rates (f_{bio}) than the EPA default rates. The biorates, thus determined, are more representative of full-scale conditions than the typical USEPA approach. Modification of typical, conventional protocols negated the required use of VOC biodegradation rate default values. This benzene-biodegradation system was approved in writing by the State of Louisiana and accepted by the USEPA - Research Triangle, North Carolina. It is premised that any properly designed and operated activated sludge system can be qualified as an Alternative Control Device when developed with the protocols herein.

KEY WORDS

Biodegradable VOCs, BWON, MACT, VOC Control, Innovative VOC Control Technology

INTRODUCTION

Case History: Marathon Petroleum Corporation - Garyville, Louisiana Refinery (MPC-Garyville Refinery)

ENVIRON International Corporation (ENVIRON) was selected by Marathon-Garyville Refinery to develop protocols for certification of an innovative Alternative Control Device for benzene biodestruction, using the existing refinery activated sludge system to replace a much more expensive activated carbon system (see Figure 1).

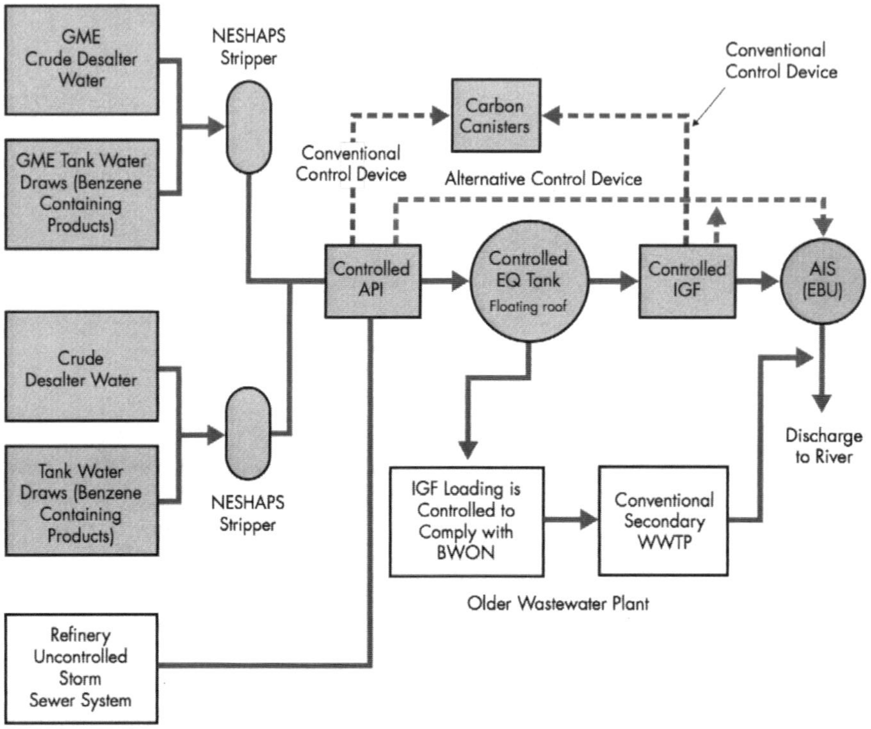

Figure 1. MPC-Garyville Original BWON Plan (Activated Carbon) and Revised Implementation of Alternative Control Device

The project was part of the major facility expansion increasing refining capacity from 185,000 to 550,000 bpd. Under BWON (§342(e)(2)(i)), this refinery requires:

- Total benzene mass in all uncontrolled aqueous wastes less than 6 megagrams/year
- (MG/yr) benzene and
- All organic emissions from wastewaters are controlled.

MPC-Garyville had already installed an activated carbon system when the the refinery discussed a payback on the carbon, using biodestruction in the existing biosystem. Figure 1 illustrates the potential use of carbon on all sources, resulting in 34,200 lbs/yr (15,550 kgs/yr). Also seen in Figure 1 is the proposed rerouting of the gases to the AIS biosystem as an Alternative Control Device. A rendering of the AIS configuration at MPC-Garyville is presented in Figure 2.

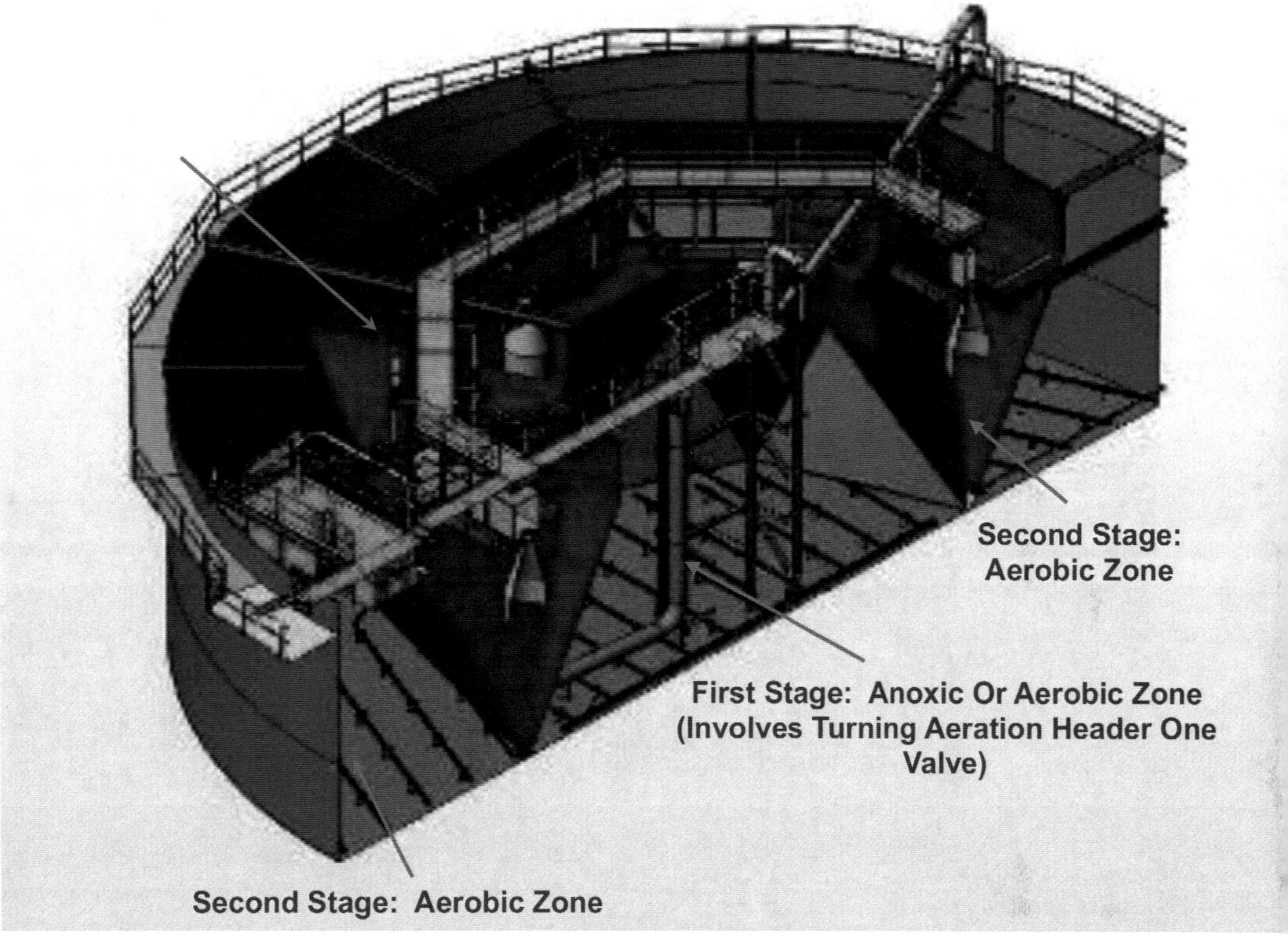

Figure 2. AIS Bioreactor Rendering

The full-scale system, during and after completion of construction, is shown in Figure 3.

To verify and implement the protocols, efforts were precisely implemented and correlated with expert Toxchem+ modeling:

1. Accurate benzene mass balance
2. Innovative assembly and utilization of BOX test
3. On-line continuous benzene monitoring of BOX test off-gas
4. Modeling, using Toxchem+ of impact of significant operational parameters at determinedbenzene biodegradation rate
5. Verification of accurate data fit to EPA criteria
6. Full-scale confirmation with Core Column simulation

Figure 3. Full-Scale AIS System at MPC-Garyville

Project Drivers

Regulatory Drivers - Under BWON regulations, an Alternative Control Device must be used to reduce benzene emissions from these sources by 98 percent. Current approved controls are Thermal Oxidizers and Vapor-Phase Activated Carbon, as cited below. However, the regulations also allow for an Alternative Control Device, provided that certain performance and operating guidelines are followed. Specifically, the rule is:

- *Title 40: Protection of Environment: 40 CFR § 61.340 presents two basic Control Devices that are acceptable, pursuant to specific design constraints:*
  - *An* ***enclosed combustion device*** *(e.g., vapor incinerator, boiler, or process heater)*
  - *A* ***vapor recovery system*** *(e.g., a carbon adsorption system or a condenser)*

- *A flare*

- *Title 40: Protection of Environment: 40 CFR § 61.340 also states* ***"other" Control Devices*** *can be used provided that certain conditions are met.*
  (iv) A control device other than those described in paragraphs (a)(2) (i) through (iii) of this section may be used provided that the following conditions are met:
  (A) The device shall recover or control the organic emissions vented to it with an efficiency of 95 weight percent or greater, or shall recover or control the benzene emissions vented to it with an efficiency of 98 weight percent or greater.
  (B) The owner or operator shall develop test data and design information that documents the control will achieve an emission control efficiency of either 95 percent or greater for organic compounds or 98 percent or greater for benzene.
  (C) The owner or operator shall identify:

    1) *The critical operating parameters that affect the emission control performance of the device;*
    2) *The range of values of these operating parameters that ensure the emission control efficiency specified in paragraph (a)(2)(iv)(A) of this is maintained during operation of the device; and*
    3) *How these operating parameters will be monitored to ensure the proper operation and maintenance of the device.*

Economic and Sustainability Drivers - The approach, presented herein, is based on utilizing existing biotreatment facilities for biodegradation of benzene and other regulated VOCs, rather than more expensive activated carbon and thermal oxidation. If the existing biosystem fits the criteria of an Enhanced Biodegradation Unit, the incremental capital costs for implementation will be very low, simply piping and proper conveyance from the sources to the biotreatment facility. Table 1 presents the economics of the common Control Devices and the preliminary ACD projections for MPC-Garyville. The incremental operating costs of the ACD are negligible due to the low quantity of VOCs being bio-oxidized.

Table 1. Costs Comparison for Major VOC Control Devices at MPC-Garyville Refinery

Process Technology	**Cost-Effective Impact**	
	Capital Cost ($)	**Annual Operating Cost ($)**
Thermal Oxidizer	$1,200,000	$300,000
Granular Activated Carbon (6 carbon canisters on each of two API separators,22	$240,000	$240,000

change-outs/yr per API)		
Biological (piping, fans and connection to blowers)	$500,000	Minimal

This methodology will eliminate the need for activated carbon or thermal oxidation, both of which have large carbon footprints and are intensive energy users. Table 2 presents the impact of energy and carbon footprint on the currently-approved control devices by USEPA.

Table 2. Green House Gas Emission and Energy Consumption Comparison for Major VOC Control Devices at MPC-Garyville Refinery

Process Technology	**Annual Impact**	
	Energy Consumption, Million BTUs per year	**CO_2 Emissions, tons CO_2 per year**
Thermal Oxidizer (calculated)	868	49
Granular Activated Carbon (in operation)	192	10
Biological (no additional energy required or CO_2 generated, due to minimal organics being treated)	Minimal	Minimal

Alternative Control Device (ACD)

Ideally, the ACD should be an existing biotreatment facility, and, more specifically, an activated sludge system. These systems are exposed to benzene routinely in a refinery and the biomass should be highly acclimated to benzene biodegradation. Thus, it is logical to assume that a properly designed and operated activated sludge system could easily assimilate the small quantity of benzene emitted from miscellaneous gaseous sources in the vicinity of the WWTP.

In order to achieve the required 98 percent benzene biodestruction, the ACD must be configured to thoroughly mix and adequately aerate and disperse and transfer the benzene from the gas phase to the liquid phase for microbial assimilation. This objective is evaluated by utilizing an Enhanced Biodegradation Unit and properly developing a site-specific predictive model, which allows evaluation of significant parameters to determine the most effective operation of the system. Additionally, each system must be qualified using site-specific criteria.

Modeling Techniques

The development of site-specific predictive models for wastewater processes involves a choice of two models, i.e., USEPA's WATER 9 or Toxchem+™. Toxchem+ is preferred in the case of biodegradable VOC gaseous emissions for several reasons:

- The Toxchem+ model allows introduction of target pollutants in gas streams as well as in aqueous wastewaters.
- Toxchem+ is approved in 40 CFR Part 63 Appendix C for emissions modeling.
- Toxchem+ assumes non-equilibrium conditions for vapor phase-liquid phase transfer.
- Benzene biodegradation rate is the most important element of model.
- Toxchem+ is very sensitive to modelers' expertise.

The mass balance for an air bubble rising through a well-mixed column of water is

$$\frac{dm}{dt} = \frac{dC_gV}{dt} = K_LaV\left(C_1 - \frac{C_g}{H}\right) \quad [1]$$

where:

m = compound mass (mg),
C_g = gas-phase compound concentration (mg/m^3),
V = local liquid volume (m3),
t = time (min),
K_La = overall mass transfer coefficient (1/min),
C_l = liquid-phase compound concentration (mg/m^3),

and

H = Henry's constant ($m^3{}_{liguid}/m^3{}_{gas}$)
Integrating Equation 1 between the limits of Cg = Cg,0

and

Cg = Cg and t = 0 and t = t yields

$$-Hln\left[\frac{C_{1-\frac{C_g}{H}}}{C_{1-\frac{C_{go}}{H}}}\right] \quad [2]$$

Where
Cg,0 = initial gas-phase compound concentration (mg/m3). Time also can be expressed

As

$$t = \frac{V}{Q_g} \quad [3]$$

Where:

Qg = gas flow rate (m3/min), and Equation 2 can be rearranged into the form:

$$C_g = HC_1 - (HC_1 - C_{go})exp\left(\frac{-K_L aV}{Q_g H}\right)$$

[4]

Toxchem+ uses the initial chemical concentrations of a gas stream (Cg,0) to calculate treatment removal efficiencies. For example, it uses the following mass balance for an uncovered diffused-air activated sludge system (neglecting sorption):

$$QC_{l\,-1} + Q_g C_{gO} = QC1 + V_k v C_l + VX_m k_h C_l + Q_g C_g \qquad [5]$$

Substituting Equation 4 into Equation 5 yields:

$$QCl - 1 + QgCg,O = QCl + VXmkbCl + VkvCl + Qg[HCl - (HCl - Cg,0)(1 - f) \qquad [6]$$

Where

Q = liquid flow rate (m3/min),
Cl-1 = influent liquid-phase compound concentration (mg/min),
Cl = effluent liquid-phase compound concentration (mg/min),
Xm = concentration of mixed liquor volatile suspended solids (mg/m^3),
Kb = aerobic biodegradation rate coefficient (m^3/mg•min),
Kv = surface volatilization mass transfer coefficient (1/min), and

$$f = 1 - \exp\left(\frac{-K_L aV}{Q_g H}\right)$$

When the initial gas-phase concentration is zero, Equation 6 reduces to

$$QCl - 1 = QCl + VXmkbCl + VkvCl + VksCl \qquad [7]$$

Where, under MPC's demonstration testing conditions, Equation 6 can be reduced to:

$$QgCg,O = VXmkbCl + VkvCl + QgCg,O\,(1 - f) \qquad [8]$$

where:

m = compound mass (mg),
C_g = gas-phase compound concentration (mg/m^3),
V = local liquid volume (m3),
t = time (min),
$K_L a$ = overall mass transfer coefficient (1/min),
C_l = liquid-phase compound concentration (mg/m^3),
H = Henry's constant (m^3_{liguid}/m^3_{gas})
Cg,0 = initial gas-phase compound concentration (mg/m3).
Q = liquid flow rate (m3/min),

Cl-1 = influent liquid-phase compound concentration (mg/min),
Cl = effluent liquid-phase compound concentration (mg/min),
Xm = concentration of mixed liquor volatile suspended solids (mg/m^3),
Kb = aerobic biodegradation rate coefficient (m^3/mg•min),
Kv = surface volatilization mass transfer coefficient (1/min),

Project Approach

Due to the unique nature of this project and the request to use a modification of USEPA's typical methodology to demonstrate a reliable and realistic benzene biodegradation rate, iIt was decided to coordinate and interface with both the USEPA Office of Air Quality Planning and Standards - Research Triangle Park, NC and the State of Louisiana Department of Environmental Quality (LDEQ). Several meetings were held to discuss protocols in both modeling and BOX and full-scale testing. The basic project approach was as follows:

1. A preliminary feasibility estimate of performance and economics of utilizing the existing activated sludge system as an Alternative Control Device.
2. Confirmation that the activated sludge process (AIS) at Garyville is thoroughly mixed and aerated properly an Enhanced Biodegradation Unit (EBU).
3. Development of a site-specific Toxchem+ model for accurately predicting benzene biodegradation.
4. Development of a site-specific Henry's Law coefficient and benzene biodegradation rate.
5. Full scale demonstration of 98 percent benzene biodegradation.

Items 1 and 2 have been discussed above. Items 3, 4 and 5 are discussed below.

Development of Site-Specific Benzene Biodegradation Model

Several basic data needs are required for developing the site-specific predictive model:

1. Major Variables
    a. Benzene Biodegradation Rate
    b. Air Flow
    c. Biomass Concentrations
    d. Potential Benzene Injection Locations into AIS
    e. Benzene Loadings
2. Other Significant Variables
    a. Air Distribution in Zones
    b. Depth of BioReactor
    c. Aeration Tank Surface Area
    d. Temperature
    e. Hydraulic Flow Rate & COD Loading
3. Sizing, dimensions and configuration of bio-unit
4. Maximum operating criteria of bio-unit
5. Mass balance of benzene in emissions and water

These data inputs are summarized below for the MPC-Garyville refinery:

- Influent Wastewater Conditions
  - Influent to API and IGF = 2,100 gpm
  - Influent to AIS-1 = 1,100 gpm
  - Benzene in water to API = 0.35 mg/L
  - Benzene in water to IGF = 0.35 mg/L (worst case)
  - Benzene in water to BioReactor = 0.1 mg/L
  - COD = 556 mg/L, 7,900 lbs/day
  - BOD = 278 mg/L, 3,000 lbs/day
  - TKN = 45 mg/L, 500 lbs/day

- Design aeration process air flows

Parameter	Units	Conditions	
		Average	Maximum
Average air flow	scfm	2,200	3,200
Aeration turnover time	minutes	14	9
Air flow per volume	scfm per 1,000 cu ft	22	33
Air flow per area	scfm / sq ft	0.44	0.66

- Off-Gas Emissions Conditions
  - API Off-Gas Flow = 65 scfm
  - API Off-Gas Benzene Concentration = 0.1 ppmv
  - IGF Off-Gas Flow = 3.8 scfm
  - IGF Off-Gas Benzene Concentration = 0.1 ppmv

- Design aeration process air flows

- Design Average Conditions
  - Average air flow = 2,200 scfm
  - Aeration turnover time = 14 minutes
  - Air flow per volume = 21.7 scfm/1000 ft^3
  - Air flow per area = 0.44 scfm/ft^2

- Design Maximum Conditions
  - Maximum air flow = 3,200 scfm
  - Aeration turnover time = 8.9 minutes
  - Air flow per volume = 32.5 scfm/1000 ft^3
  - Air flow per area = 0.66 scfm/ft^2

- Design Average Wastewater Characteristics
  - Flow = 1,100 gpm
  - COD = 556 mg/L, 7,900 lbs/day
  - BOD = 278 mg/L, 3,000 lbs/day
  - TKN = 45 mg/L, 500 lbs/day

- Design Parameters
  - F/M_{COD} = 0.30 lbs COD applied/lb MLVSS-day (0.40 max)
  - $F/M_{Nitrogen}$ = 0.025 lbs TKN applied/lb MLVSS-day (0.033)
  - Operating warm weather temperature = 35 °C (closed-circuit cooling tower installed for optimum control and no VOC emissions from cooler)

From these data, it is possible to develop the site-specific model to predict percent benzene removal:

- Select design parameters,
- Assume benzene biodegradation rate, and
- Calculate percent benzene removal at assumed biodegradation rate.

Based on the raw data and system operating variables, the specific modeling variables and the assumed benzene biodegradation rate, it is possible to evaluate influences of parameters on the system performance regarding benzene biodegradation and stripping. Modeling results are then generated using assumed benzene biodegradation rates. The design mass balance of benzene at the MPC-Garyville refinery is given in Figure 4.

Parameter	**Liquid Streams**			**Gaseous Streams**						
	units	**1**	**2**	**units**	**5**	**6**	**7**	**8**	**3**	**4**
Flow	gpm	1,110		scfm	68.8	3,131	3,200		65	3.8
	m^3 per hr	250		m^3/hr	117.5	5,348	5,465		111	6.5
Benzene	mg/L	0.142		Ug/L	2.07		0.0446		0.32	32
	mg/hr	35,500		Mg/hr	244		244		34	208

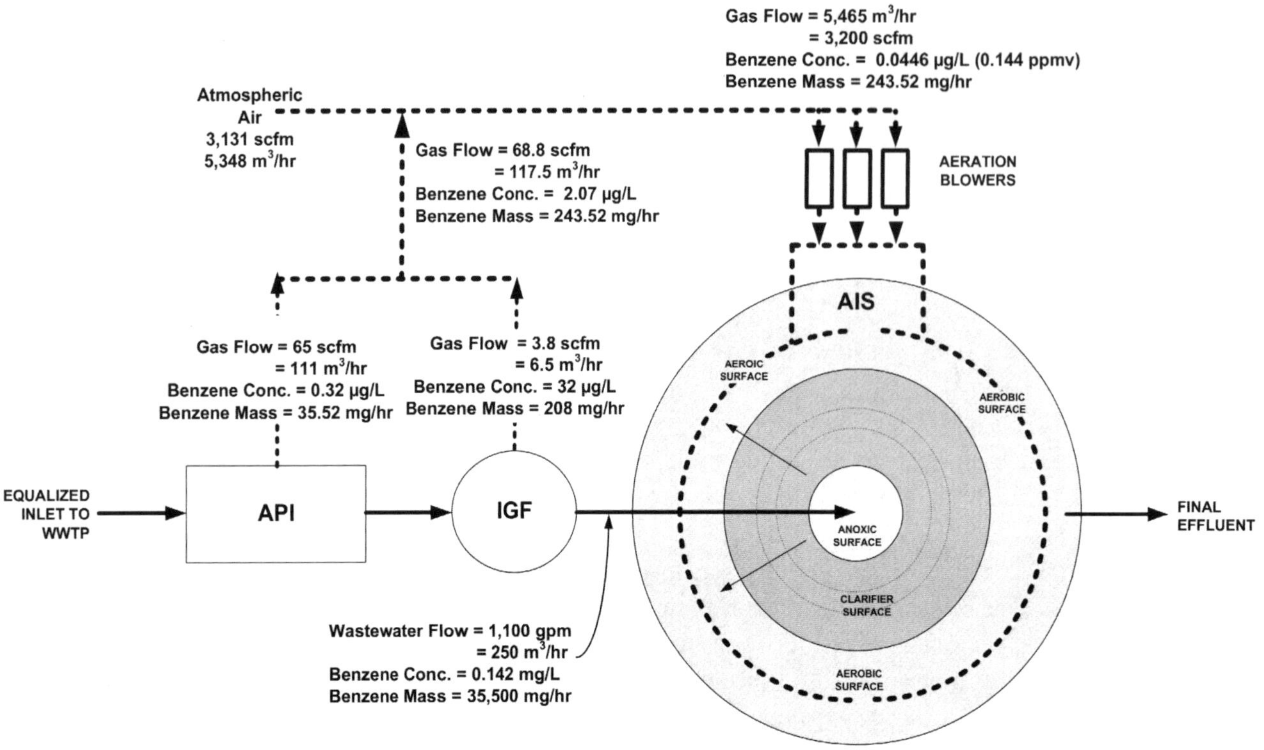

Figure 4. Benzene Mass Design Mass Balance

The preliminary modeling results, graphically presented in Figure 5, are examined for limitations of the system and desired operational flexibility. As can be observed, the benzene destruction efficiency can be improved by lowering the airflow rate (reduces stripping) and/or increasing the biomass concentration (increase mass rate of benzene biodestruction). However, the WWTP operators did not want to vary from current operational protocols; thus, it was decided that a benzene biodestruction rate of greater than 6.5 would be required (see Figure 5) in order to maintain current operational criteria, i.e., airflow of 3,200 scfm and MLSS of 3,000 mg/L.

The results in Figure 5 can be summarized as:

- A minimum benzene biodegradation rate of 6.5 L/g MLVSS-hr is required to achieve 98 percent gaseous benzene removal at current operating conditions which are stated here.
- If rate is < 6 L/gm-hr, system will require lower air flow rate and/or higher MLVSS concentrations, i.e., air flow <2,800 scfm and MLVSS > 5,200 mg/L.
- Neither option is desirable for operational protocol at site.
- Operators wished to remain at 3,200 scfm and 3,000 gm/L MLVSS.

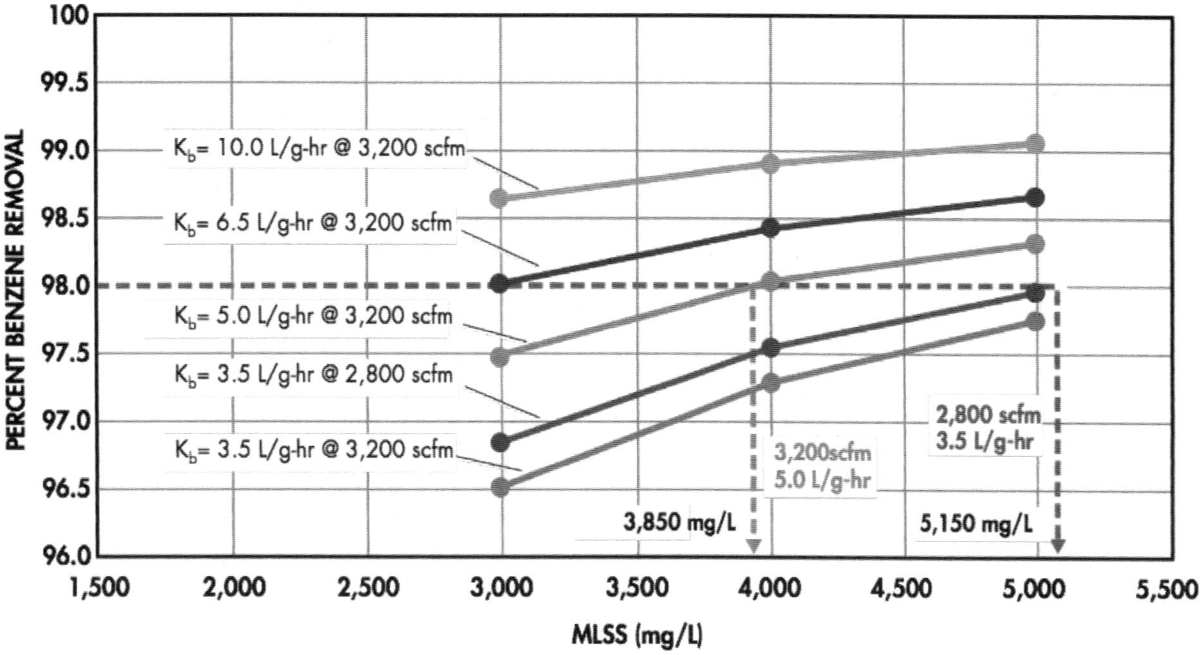

Figure 5. Preliminary Modeling Results using Assumed Benzene Biodegradation Rates and Toxichem+

Development of Site-Specific Henry's Law and Benzene Biodegradation Rate

The USEPA offers five experimental methods of determining the fraction or rate of benzene biodegradation as shown in Figure 6. Due to the very low inlet and assumed outlet benzene concentrations, the laboratory BOX Test method was chosen.

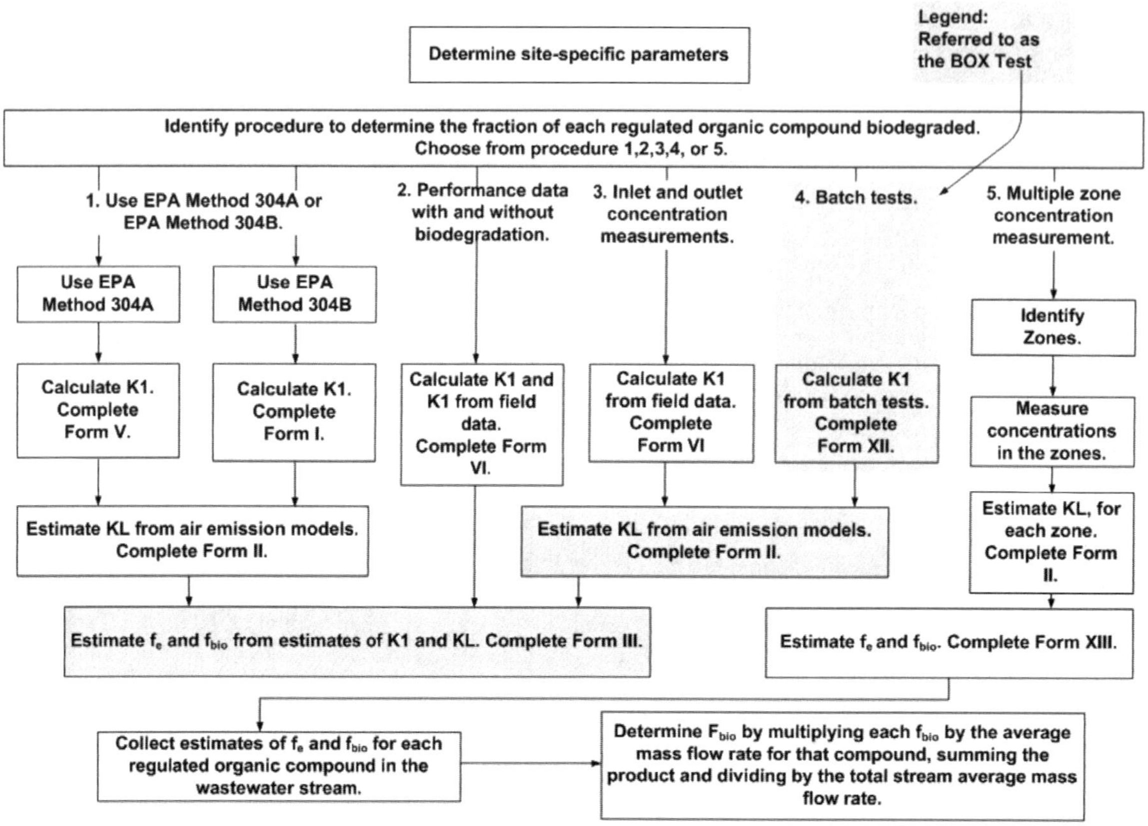

Figure 6. Alternative Experimental Methods for Determining Fraction of Organic Biodegraded (F_{bio}) in a Biological Treatment Unit

The authors felt that the typical BOX Test, using a 2-L reactor, was too small in volume (generating a very high air/liquid rate in the BOX reactor) and more realistic and reliable rates could be achieved in a larger volume bench reactor Thus, a modified version of the typical BOX apparatus was constructed and tested, then proposed to the USEPA as shown in Figure 7.

This test setup was considered more realistic of actual field conditions and complied with Procedure 4, 40 CFR 63, Appendix C. In this study the total off-gas from the reactor was monitored continuously with a gas chromatograph (GC) rather than measure the residual water benzene concentrations. In this way, the actual benzene present in the emission off-gas would be continuously monitored rather than using calculations based on aqueous grab samples collected at selected time intervals during the test (the BOX method in Appendix C specifies that either method, off-gas or water samples, is acceptable.

An online PhotoVac Voyager GC, shown in Figure 8, was utilized to monitor the BOX reactor exhaust gas. The laboratory setup for the BOX Test apparatus is also shown in Figure 8.

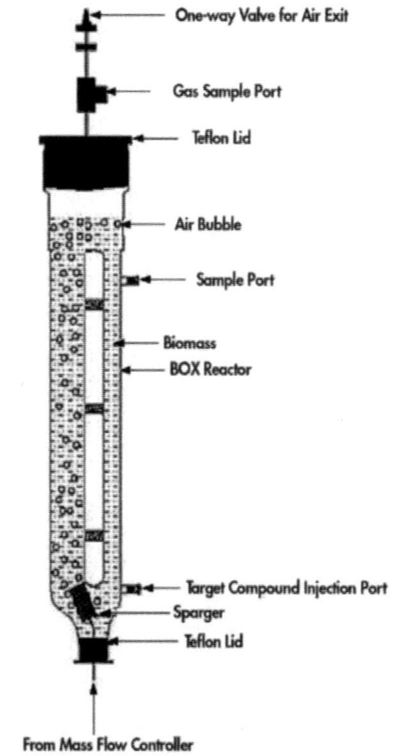

Figure 7. Typically used box test apparatus (2-L Volume)

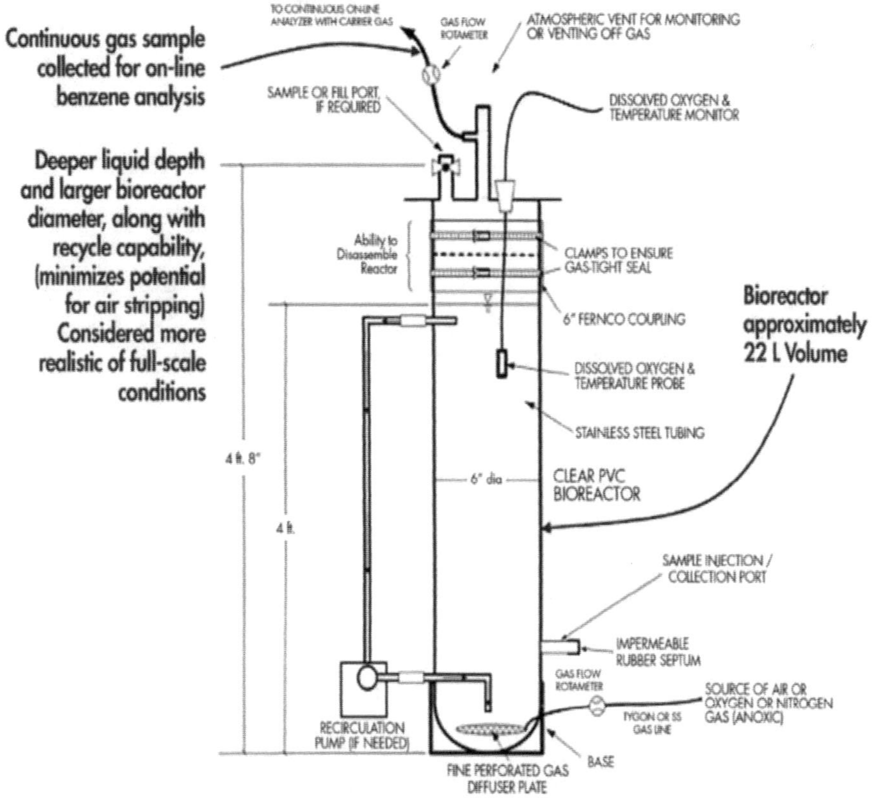

Figure 8. Recommended Box Test Apparatus (57-L Volume)

In order to conduct the BOX Test determination of site-specific Henry's Law coefficient and benzene biodegradation rate, the first exercise is to determine the H_L coefficient using filtered (0.45 micron) effluent from the site-specific operating activated sludge system. The on-site testing program was a cooperative effort with MPC and the authors. Site-specific conditions followed regulatory protocols, i.e.:

- Filtered bioeffluent effluent (0.45 micron filter paper)
- Bioreactor MLVSS < 4 hours removed
- Temperature maintained within 1 ^{o}C
- Dissolved Oxygen > 2.0 mg/L

The emission benzene results are given in Figure 9.

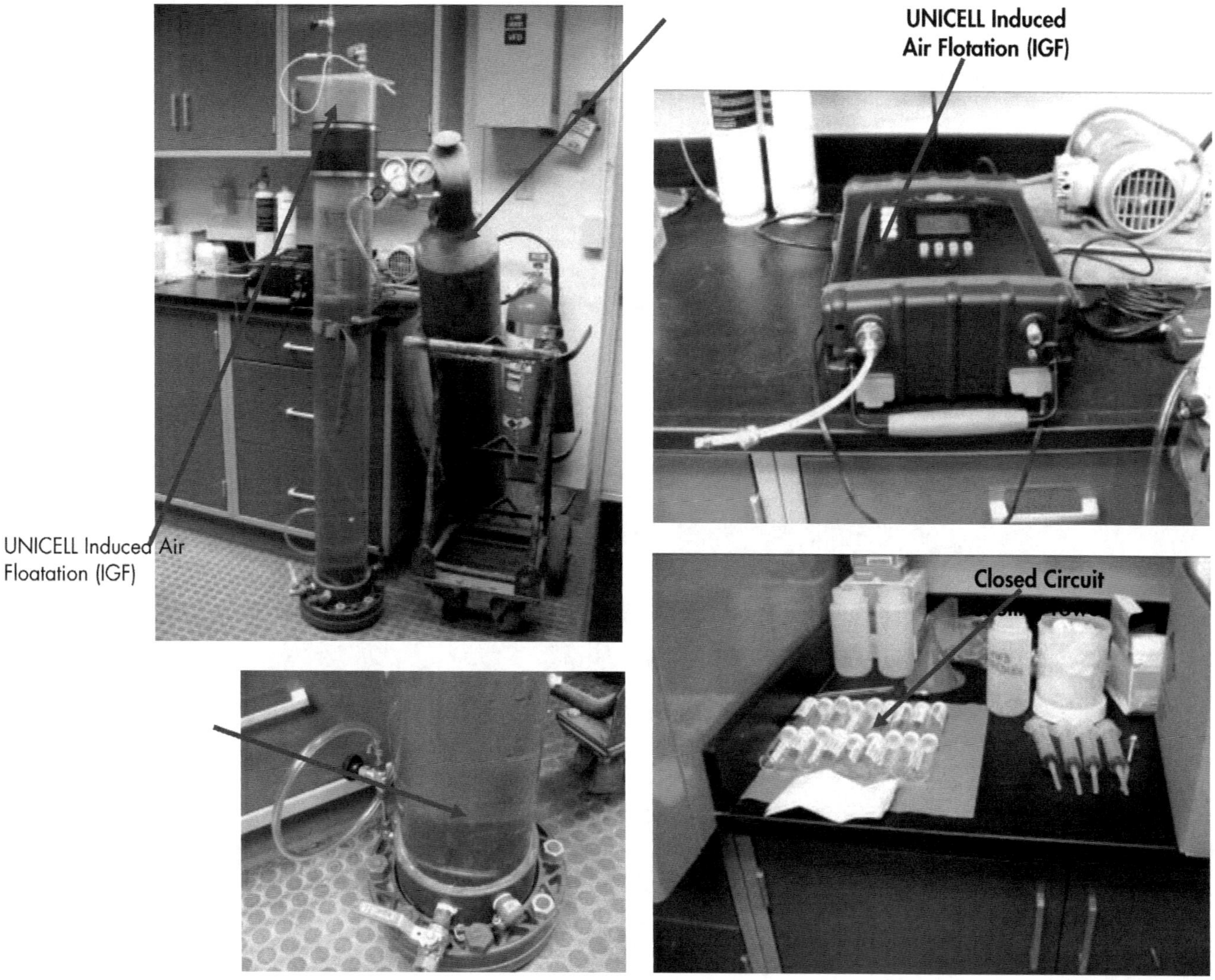

Figure 9. Box Test Setup Recommended in Figure 8

It is observed that the benzene (Figure 10) is completely stripped in 350 to 400 minutes in the absence of biomass. With the rapid biodegradation of benzene (Figure 9, lower curve), the benzene is completely removed by 30 minutes. The overlay of the stripping and bio-removal data in Figure 9 illustrates the highly biodegradable nature of benzene under proper conditions.

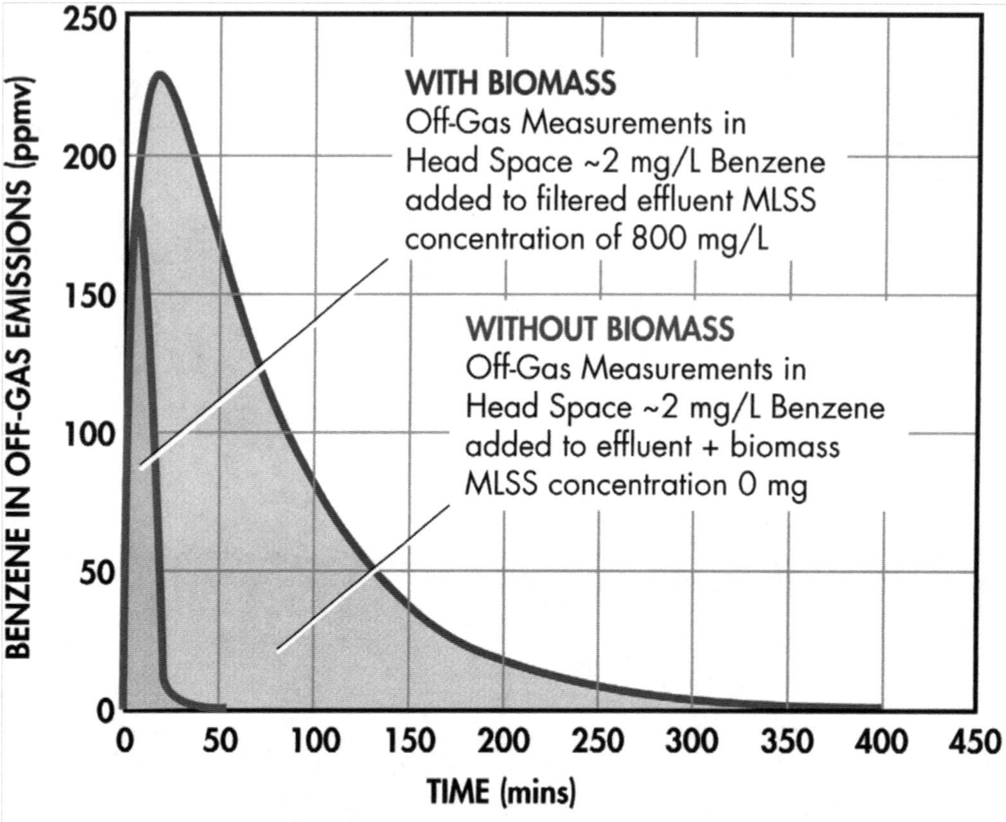

Figure 10. Combined Stripping and Biodegradation Benzene Emissions

For the data to be valid, the USEPA requires that the data provide a "smooth fit" as demonstrated in Figure 11. With this validation, it is acceptable to follow the USEPA protocol to calculate the benzene biodegradation rate, based on the BOX Test data.

Thus, the following conclusions were taken from the BOX Test and, subsequent, calculation procedures:

- The K_1 rate (site-specific biodegradation rate) is:
  - *29.3 L / g VSS-hr* at 25.8 °C
  - This rate must be corrected to 20 °C for input into the Toxchem+ model
- The site-specific biodegradation rate, corrected to 20 °C, is:
  - *22.6 L / g VSS-hr* @ 20 °C
- The Toxchem+ model will adjust this rate to the selected temperature for full-scale operating conditions.
- Subsequent modeling runs evaluating potential points of gaseous injection were done but are not reported herein. The optimum injection point for this system was directly into the inlet to the blowers.

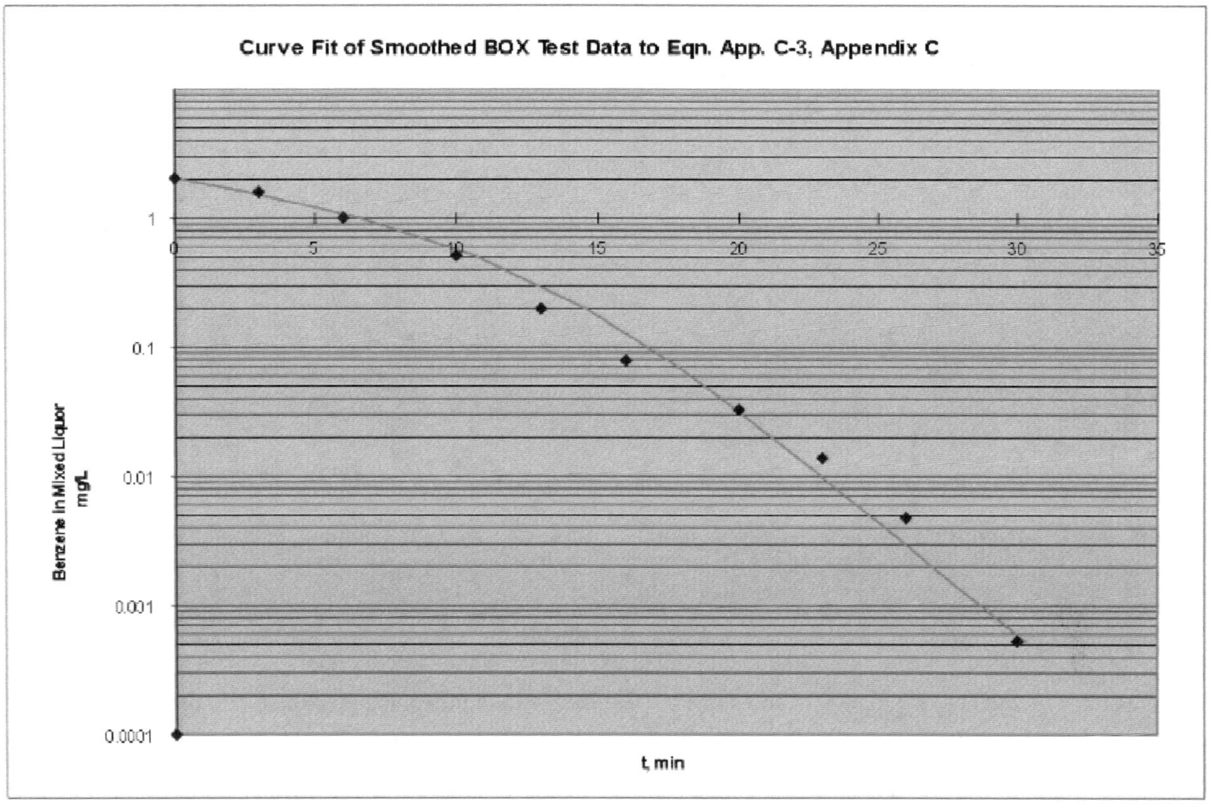

Figure 11. Smoothed Data Fit as per USEPA Criteria for F_{bio} Calculation

The model was calibrated with the site-specific biodegradation rate (see Figure 12, upper Blue values)

- Final modeling at the selected injection location (inlet to aeration blowers) and at the temperature-corrected (20 °C) benzene biodegradation rate, has confidently predicted 99+% removal for the benzene in the IGF + API off-gas and sumps, without modifications to the aeration system, i.e., aeration rate, biomass concentration, F/M, etc.
- IGF & API benzene off-gases are assumed to be injected into the inlet of the aeration blowers at 3,200 scfm total air flow to bioreactor
- The modeled benzene removals, as a function of MLVSS and air flow rate (scfm) at 30 °C (annual average at MPC-Garyville), are shown in Figure 12.
- The final site-specific modeling is presented by the upper blue curve at the top of the graph.
- Note that, at the actual site-specific benzene biodegradation rate determined for the bioreactor, there is little variation in benzene removal over a range of MLVSS concentrations from 3,000 mg/L to 5,000 mg/L. This means that the benzene removal from the gas stream is very operationally robust.

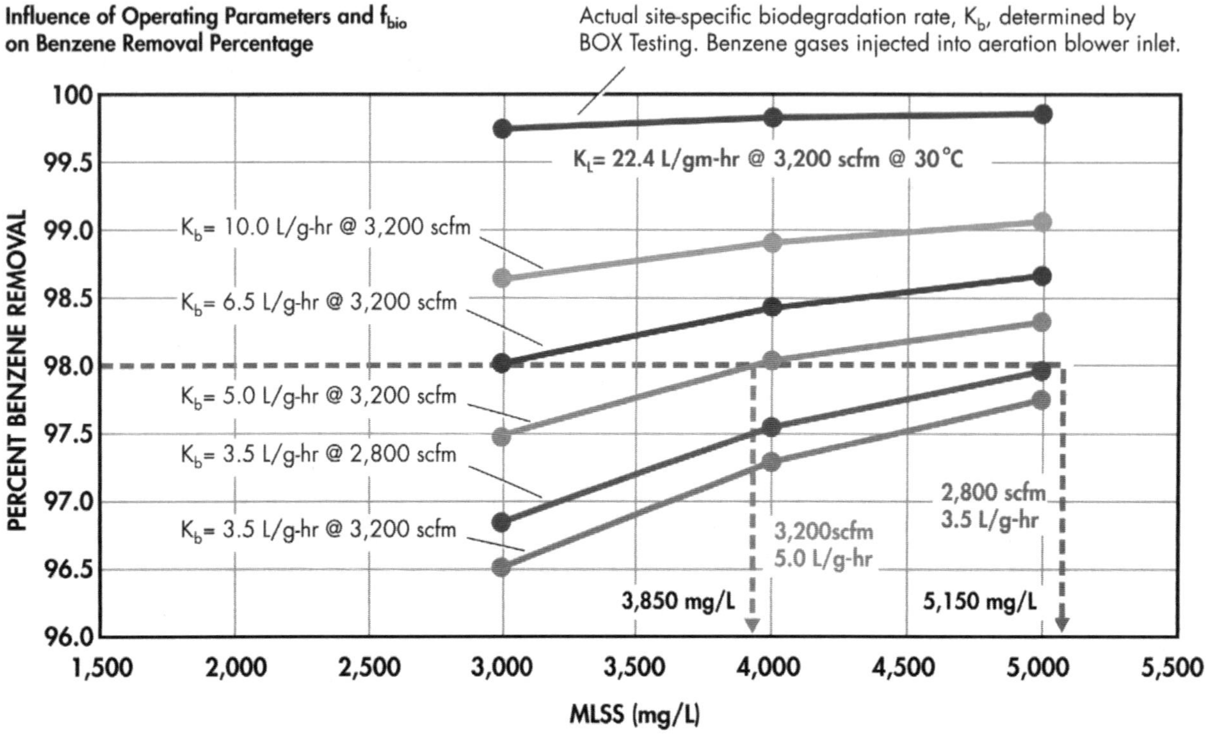

Figure 12. Benzene Removal as Function of Biomass (Biodegradation Site-Specific Rate is Upper Curve = 22.6 L / G-Hr @ 20 ^{O}C)

Full-Scale Confirmation of 98% Benzene Biodegradation

Although full-scale verification of performance is optional before implementing capital improvements, it is prudent to understand the capacity and robustness of the site biosystem to benzene biodegradation before initiating the changes.

Thus, the authors developed and recommended an *External Core Column* (Figure 13 representing a "core sample" of the full-scale bioreactor and operating in parallel to the full-scale bioreactor. The column is same operating depth as full-scale bioreactor and is supplied with proper pumps, compressors, benzene calibration cylinders, etc. (see details on Figure 14). The major advantage of the *Core Column* is that it is possible to control all operating parameters so that performance under maximum stress conditions can be evaluated. The *Core Column* was developed and received approval from both USEPA and the State of Louisiana.

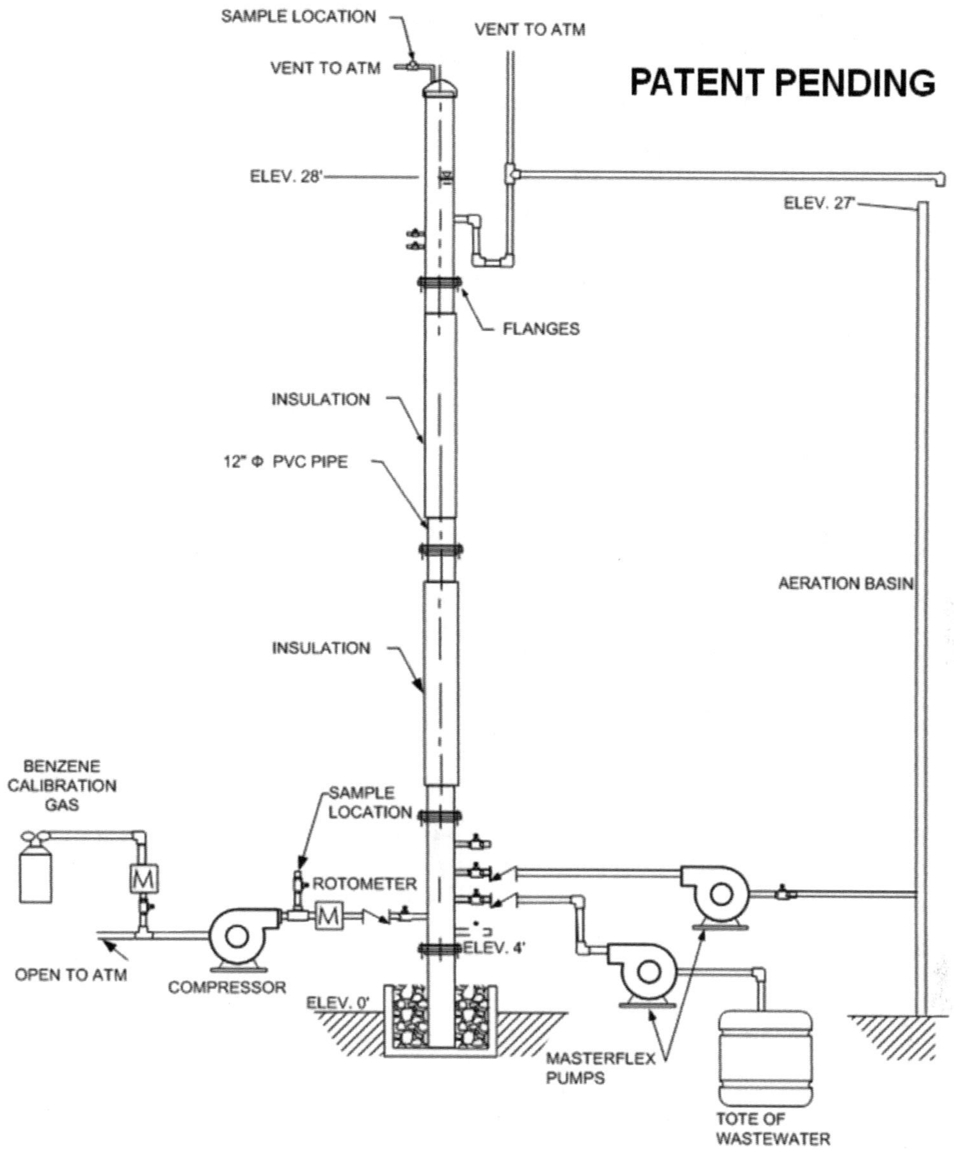

Figure13. Core Column Simulation Recommended by ENVIRON

Observations of the Full-Scale Performance Testing

The operating conditions and adjustments to the pilot Core Column were identical to the full-scale system, except the inlet benzene concentrations to the pilot aeration blower were varied. The full-scale performance investigation required about one week in extremely hot and humid conditions. Five runs were made, each run requiring about 8-10 hours. Each consecutive run subjected the biomass to increasing benzene loadings, both in the gaseous and aqueous influent loading phases. The study plan was designed to increase vapor phase loadings of benzene until breakthrough of benzene in the reactor exhaust was measured. The data from these runs are given in Table 5 and plotted in Figure 14.

Figure 14. Photo of Core Column Simulation Setup in Field

Run 2 was deemed unreliable due to overheating of the PhotoVac Voyager GC. Otherwise, the runs were successful, and it was concluded that >98% benzene biodegradation could be achieved at loadings up to 16 times design by the acclimated biomass under the techniques of gas dispersion in the AIS system. The GC vapor phase benzene detection limit was <2 ppbv and the aqueous phase detection limit was <1 ppbv (μg/L). There was breakthrough of the vapor phase benzene emissions at an inlet benzene concentration of 482 ppbv, which is almost 35 times the maximum design inlet concentration. Even at this extraordinary loading, the effluent emission was only 13 ppbv for 97.2 percent biodegradation.

Table 5. Benzene Analytical Results Of Full-Scale Confirmation								
Parameter	**Units**	**Run 1**	**Run 3**		**Run 4**		**Run 5**	
			A	**B**	**A**	**B**	**A**	**B**
Inlet Benzene Gas Concentration (Design = 14 ppbv)	ppbv	21	121	153	153	482	182	226
Outlet Off-Gas Benzene Concentration	ppbv	<2	<2	<2	<2	13.3	<2	<2
Percent Benzene Removal	%	NA	>98.35	>98.69	>98.72	97.24	>98.80	>99.11
Influent Aqueous Benzene	μg/L	109	129	129	137	137	245	245
Effluent Aqueous Benzene	μg/L	<1	<1	<1	<1	<1	<1	<1
Percent Benzene Removal	%	>99.09	>99.22	>99.22	>99.27	>99.27	>99.59	>99.59
Inlet Raw Wastewater Flow	gpm	0.21	0.21	0.21	0.21	0.21	0.35	0.35
Biomass Recycle from Clarifier	gpm	0.35	0.35	0.35	0.35	0.35	0.35	0.35
Internal Anoxic Recycle	gpm	0	0	0	1.1	1.1	1.1	1.1

A presentation of percent benzene removals vs. emission concentration is presented in Figure 15.

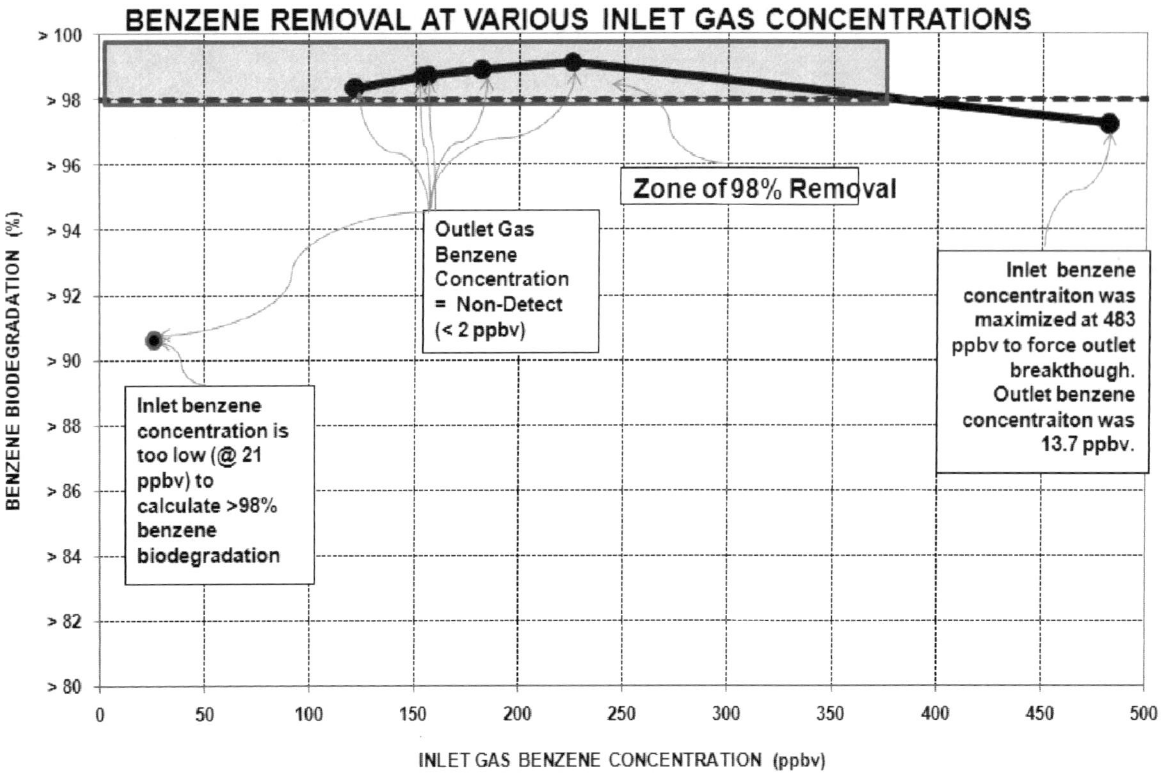

Figure 15. Benzene Analytical Results of Full-Scale Confirmation

Conclusions

The information provided herein summarizes the BOX Test and Full-Scale Confirmation testing objectives and achievement of benzene removal goals under maximum stress conditions at the MAP-Garyville site. The methodology, protocols and apparatus developed and employed by the authors have been approved by the state and federal agencies. The following information confirms compliance with 40 CFR § 61.340:

- ENVIRON has developed bench-scale BOX Test batch and Core Column Simulation full-scale confirmation protocols and parameters, which delineate more realistic and reliable benzene biodegradation rates (f_{bio}) than the EPA default rates.
- The biorates, thus determined, are more representative of full-scale conditions than the typical USEPA approach.
- This effort has required modifying the recommended protocols and approach by the USEPA and obtaining their approval of data collection, testing methodology, and appropriate modeling techniques.
- It is premised that any properly designed and operated activated sludge system can be qualified as an Alternative Control Device when certified with the protocols herein.
- The approach, presented herein, is an environmental-friendly, sustainable VOC Control Device
  - Negligible additional energy usage
  - Minimal carbon footprint

- The site-specific benzene biodegradation rate is 29.3 L / g VSS-hr @ 26 °C (22.6 L/gm-hr, corrected to 20 °C).

Thus, a properly designed activated sludge system provides excellent configuration and flexibility to achieve benzene removals >99+% even under benzene loadings >16 times projected maximum operating inlet benzene concentrations.

(7b) Automotive Emissions and Regulations Overview of U.S., European, Asian, and Global Standards

Aaron Matthews, Product Development, Navistar- Engine Group, Melrose Park, IL

Over the yearsThe U.S. EPA and CARB have instituted increasingly stringent emissions standards. Both of these institutions have jurisdiction over mobile and non-mobile sources of emissions. The one of increasing interest for consumers is how these regulations affect the fuel efficiency of automobiles. Additionally there are contrasts between critical components of emissions in various parts of the world. As these sometimes competing standards come into play there is often a battle between trading fuel efficiency for reduced emissions. This talk will feature a brief overview of the emissions types and their trade-offs for fuel efficiency, also a review of increasing emissions regulations, and a look into how these regulations are tested.

(7c) LCA As a Sustainability Decision-Making Tool for Regulations and Investment

Sarah Widder, Energy and Efficiency Division, Pacific Northwest National Laboratory, Richland, WA

Life-cycle analysis (LCA) is a unique tool which provides a framework to quantify impacts from a specific process or technology from resource extraction to disposal, or "cradle to grave." LCA can be employed to assess the "sustainability" of a technology by identifying the life-cycle impacts on environmental, economic, and social indicators. These sustainability scores can be used to compare technologies or identify areas for improvement and further research. The comprehensive sustainability impacts of a technology should inform the regulatory and financial support a technology receives. This talk will use examples to indentify how LCA can serve as a useful tool to assess the sustainability of technologies and how the results can be used in policy development or future research portfolios.

(7d) Sustainability Initiatives At Siemens

Stephanie Nadeau, Siemens Corporation, Washington, DC

Stephanie Nadeau will discuss Siemens and their internal sustainability programming, including their range of sustainability goals and integration of environmental considerations in product design. She will also cover many of the Siemens technology solutions deployed around the world that help Siemens customers with their own sustainability goals and efficient operations.

A Review of Regulatory & Engineering Issues Dealing with Flares Based on Forthcoming EPA Flare Regulations

Bruce C. Davis, DuPont Engineering, Houston, TX

Introduction

The EPA is in the process of developing a new set of regulations dealing with flares. These regulations will be part of the new Uniform Standards which deal with Storage Vessels and Transfer Operations, Equipment Leaks, and Closed Vent Systems and Control Devices; and Revisions to the National Uniform Emission Standards General Provisions. These proposed rules were signed by the EPA Administrator on Feb. 24, 2012. The new flare rules will be part of the closed vent system and control device rules. In the proposed rules signed in February, the EPA does not propose new flare regulations at this time. As of February, 2012, the EPA is continuing to gather data, review flare research papers and test reports, and investigate operating conditions that may influence the performance of a flare. Based on this information, the EPA may, in the future, propose to add new flare requirements to the Uniform Standards. The Uniform Standard rule proposal currently refers to the existing flare requirements in 40 CFR 63.11(b) of subpart A.

This paper discusses some of the issues dealing with EPA efforts to develop flare regulations.

Discussion

Existing Regulations

The existing flare rules are provided at 40 CFR 60.18(b) and 63.11(b). These rules require:

- A pilot flame
- A sufficiently low exit velocity
- A minimum heating value (e.g. 300 btu/scf for assisted flares and 200 btu/scf for non assisted flares)
- Smokeless operation

In addition regulations at 40 CFR 60.11(d), 60.18(d), 60.482-10(e), 63.6(e), 63.11(b)(1), 63.172(e) require:

- Adhere to the flare's design documents
- Use Good Air Pollution Control Practices

Flares are also subject to New Source Review rules when changes or projects are implemented in plants. The existing rules deal with four categories of flare operations.

These are steam assisted flares, air assisted flares, non-assisted flares and hydrogen rich flares.

Recent EPA Flare Enforcement and Flare Test Results

Recent EPA flare enforcement activities have centered on the use of good air pollution control practice. This concept is not defined in the rules and is somewhat subjective. The latest EPA presentation summarizing their enforcement efforts and their efforts dealing with flare regulations is provided in reference 1. EPA's flare enforcement activities have been ongoing and the current round of enforcement activities started in 2005. The inspections conducted by EPA have found problems dealing with over steaming for steam assisted flares, over aeration for air assisted flares, and issues dealing with low BTU flared gases.

Some of these inspections and investigations have resulted in flare tests of operating flare systems. These tests have been conducted using remote sensing techniques using passive fourier transform infrared spectroscopy (PFTIR) technology. This technique has been underdevelopment since the early 2000's. Recent testing on a 36 inch flare at low flows was sponsored by the Tx Commission on Environmental Quality (TCEQ). In this test work, the PFTIR and other remote sensing techniques were compared with direct sampled data in a single blind study. The results compared reasonably well. The PFTIR technique has been used at all of the recent flare testing done on operating flares. The TCEQ test results and all the recent PFTIR flare test reports can be found at the TCEQ flare stakeholders web page at: http://www.tceq.texas.gov/airquality/stationary-rules/stakeholder/flare_stakeholder.html. Links to the historic EPA test work, the CMA/EPA test work and the DuPont hydrogen flare test work done in the 80's can be found at this web link. The test work done in the 80's was used as the foundation for the current flare regulations.

The historic flare studies did look at flare operations at high and low steam or air to fuel ratios. The existing regulations did not include specific limits on steam or air assist ratios. The regulations simply limit the amount of smoking from flares. The historic studies did look at low BTU flaring and specific limits were adopted (300 BTU/scf for assisted flares and 200 BTU/scf for unassisted flares). The existing rules also establish exit velocity limits as a function of heat content.

International Flare Consortium Data

A literature review related to emissions from flares was published in April of 2010. (Reference 2) The work was sponsored by the International Flare Consortium (IFC). The work was conducted at the CANMET Energy Center, Natural Resources Canada and was led by Dr. Peter Gogolek. The work reviews the literature through 2004. The International Flare Consortium was formed in 2006 and sponsored the flare test work and the literature review performed at the CANMET Energy Center. The project was conducted in 2006, 2007, 2009 and 2010. The Consortium Members were: BP, ExxonMobil, Saudi Aramco, Chevron, John Zink Company LLC, Shell, DuPont, Nova Chem, and Total. The Consortium agreed to release the Literature search in April, 2010 and agreed to release the study results in December of 2011.

These are specific research areas that formed the general objectives of the IFC Flare Test Program. The specific program objectives were:

1. The first objective of the IFC Flare Test Program was to produce a method of predicting flare efficiency from operating variables: flare gas composition and flow rate; steam-assist rate; and wind speed. The method developed will rely upon original experimental work as well as published data.

2. The second objective was to measure the emissions of: NOx; the most important HRVOCs (ethylene, propylene, 1-3 butadiene, 1-butene and combined cis/trans-2-butene) and BTEX (benzene, toluene, ethylbenzene, and the xylenes); and to attempt to predict emission factors for these compounds based on the same set of operating variables.

3. The third objective was to identify optimal operating conditions and identify the operating envelope for flares.

The results confirm the most recent full scale flare test data results. The work was conducted on small flares (1 – 6" flares.). The bulk of the test work was done using a 3 " flare. The results using the 3" flare were thought to be scalable to larger flares based on fluid dynamic scaling. Some flare parameters don't scale based on fluid dynamics but scale based on geometry. The test work was conducted in a wind tunnel and one of the objectives was to examine the effects of wind speed. Wind effects were observed for some of the 3" flare tests. The effects for pipe flares were found to scale based on fluid dynamics. When wind stabilization devices were used, differences, in wind effects were suppressed. Flare tips incorporate wind stabilization devices, so the IFC work is not necessarily representative of full scale flare performance.

The main observation that develops when the IFC test data results are compared to full scale test data is that the wind effects observed for the smaller flares tested in the IFC work are not found in the full scale test work. Wind effects were not found in the TCEQ test work based on an analysis of replicate tests, where the only difference in the tests was wind speed. Wind effects have also not been observed in full scale flare tests. (References 3 & 4).

EPA Thinking on Regulatory Changes

The main issue that develops from EPA's enforcement and inspection activities and the recent flare test programs is related to over steaming. Current rules focus on steam control to limit smoking but the rules don't focus on steam control to deal with efficient combustion. Proper steam to fuel ratios need to be maintained over the full operating range of the flare. This is particularly important at low flows. The EPA has analyzed the available test work and is considering two potential new flare efficiency metrics. The first is the net heating value in the combustion zone. This is the heating value of the flared gas and the pilot gas and the added assist gas combined. This term is called the combustion zone net heating value. Available test data seems to collapse to a suggested value of 300 BTU/scf. This differs from the existing limit which is the heating value of the flared gas only and does not include the pilot gas nor the steam or air assist gas. An alternate parameter is based on a measure of the combustible material volume in the combustion zone. This parameter would be compared to the combustible volume at the LEL of the mixture. These values would need to be based on measurement to the extent possible. For smaller flares, the measurement could be based on a sampling program or a calorimeter measurement. For larger flares, the EPA thinking is that continuous measurements would be made.

EPA's thinking on basic practice that would possibly apply to smaller flares is:

- Meet Current Standards
    - Smoke, Exit Velocity, NHV
- •Train Operators
- •Adhere to O&M Manual
- •Additional Equipment and Monitoring
    - Vent gas flow monitor (e.g., GE Panametrics)
- –Air Assist
    - Effective air flow control
- –Steam Assist
    - Effective steam measurement and control
    - Reduce minimum steam and center steam
    - Steam/Vent Gas ratios less than 2 or 3

EPA's thinking on better than basic practice that would possibly apply to larger flares is:

Steam Flares

- Gas Chromatograph, heating value analyzer, video camera
- Automate control of supplemental gas and steam
- Requirement to operate at 99% combustion efficiency
- Monitor for wind effects
- Ensure NHVcz/combustible gas volume is sufficient to meet the minimum needed

Air Flares

- Monitoring Vent Gas Flow
- Variable Frequency Drive
- Modulate air commensurate with vent gas

In addition to the performance scheme outline above EPA is also considering requiring:

- Flare Minimization
    - Waste gas mapping
- Root cause analysis/corrective action
- Flaring Cap
- Appropriately-sized flare gas recovery

Conclusions

The results of the recent EPA enforcement and inspection activity and the results of recent flare testing show possible problems due to steam or air assist management. Over steaming or over aeration can result in inefficient performance. The EPA's regulatory thinking is outlined above. The challenge EPA is facing right now and the reason the flare regulations have been delayed is an ongoing review of the cost effectiveness of the proposed regulatory scheme. Problems have been found due to air or steam assist management, but significant excess emissions occur only if poor operation occurs for conditions of high VOC load (flow) for long time periods. Poor performance for short periods of time at low and high flows does not result in significant excess emissions. In order to craft cost effective regulations, the EPA needs to target the regulations to address problems where significant excess emissions are occurring. For those operations that are managing air and steam assist properly, a new, severe set of regulatory requirements is not cost effective and will not result in a significant emission reduction.

The recent full scale test work shows no demonstrated effects from winds on flare performance. The full scale test data does not warrant regulatory changes to deal with winds.

The data cited in this paper and the problems noted only deal with air assisted and steam assisted flares. There is no new data dealing with unassisted or hydrogen rich, unassisted flaring. The available data indicates that current regulations dealing with air and steam assisted flares need to be modified to regulate steam and air assist management. There is no new data that allows one to conclude that new regulations are needed for unassisted flaring or hydrogen rich, non-assisted flaring.

References

1. Dickens, Brian and Foley, Patrick (2012), "U. S. EPA Flare Activity Overview", presented at the LDAR, BWON, Tanks & Flares Conference, Feb 22, 2012, Austin, TX.

2. Gogolek, P, Aldred, A, Schwartz, R, Seebold, J & Pohl, J. "Emissions from Flares – A Survey of the Literature" prepared for the International Flare Consortium by Canmet Energy, Natural Resources Canada, April, 2010 available at: Publications | Natural Resources Canada's CanmetENERGY | Leadership in ecoInnovation

3. Evans Scott. "Insights from Passive FTIR Flare Performance Testing" presented at AFRC Industrial Flares Colloquium, • Houston, Texas • September 18-21, 2011 & at the the LDAR, BWON, Tanks & Flares Conference, Feb 22, 2012, Austin, TX.

4. Torres,, Vincent M., Herndon, Scott, and Allen, David T. "Industrial flare performance at low flow conditions: Part 2.Steam- and Air-Assisted Flares" submitted for publication to the Journal of Industrial & Engineering Chemistry Research, Nov, 2011.

(23b) Air Compliance Options:T/A Decontamination, Tank Cleaning & Maintenance, and Flaring Alternatives for Low Flow...What Are They, Do They Work, and At What Cost

Lisha Salathiel, Global Vapor Control, Houston, TX and Paul Komperda, United Laboratories International, LLC, Houston, TX

As the regulations continue to be in a state of flux and the pressure to capture and/or destroy VOC's and H2S increases, industry is being presented with an ever varying list of options to allow compliance. From the old standby of carbon to new 'chemistry' capture methods, these options are sometimes confusing as to whether they work and it is rarely clear just what the cost will be. Just what the options are (both new and old), where & how they work, and what the general costs are for each application will be presented for review and discussion.

(23c) Developing An EMS to Track Compliance with MSS Activities

Maria Gou, Zephyr Environmental Corp., Austin, TX and **Thuy Mai**, Safety, Health, Environmental Dept., DuPont, La Porte, TX

Texas now requires New Source Review (NSR) permit holders to amend their NSR air permits to include emissions from routine and planned Maintenance, Startup, and Shutdown (MSS) actvities. MSS activities include emissions from shutting down or starting up a chemical process unit. They also include emissions from degassing columns, process vessels, pumps, and storage tanks.

This presentation will cover the development os a computerized Environmental Management System to allow a permit holder to document MSS activities and the emissions that result so that compliance with the permit emission limits can be shown to an agency inspector.

(23d) Selective Catalytic Reduction Reactors and Their Uses

Bryan Pourchot, Air Pollution Control Catalyst & Technology , Haldor Topsoe, Inc., Houston, TX

Selective Catalytic Reduction Reactors and Their Uses

Bryan Pourchot

Sales & Service Engineer

Haldor Topsoe, Inc.

The burning of fossil fuels and other materials create the means for a civilization and economy to flourish. With this burning comes the consequence of releasing large amounts of compounds which can effect global climate change and the air quality of the civilization it supports. The compounds to be discussed during this presentation, and a technology to combat them, comprise NO_x. NO_xrepresents the compounds NO and NO_2. NO_xcan react with other chemicals in the air to create nitric acid, ozone, and other chemicals which can damage lung tissue and cause a variety of other health problems. NO_xcan also destroy ozone in the stratosphere, increasing the amount of UV light let down to the Earth's surface causing, among other ailments, skin cancer.

One way of destroying $NO_x$is by the implementation of a Selective Catalytic Reduction (SCR) System at the exhaust side of processes which are responsible for the creation of NO_x. There are numerous applications for SCRs from things as small as the exhaust on a car engine to things as large as a coal power plant. A SCR will convert NO_x, through a reaction with ammonia (NH_3) or another reducing agent, into the benign compounds of diatomic nitrogen (N_2) and water (H_2O). A SCR is capable of achieving very low levels of NO_xemissions, on the nature of $\leq$ 1ppm, coming from a multitude of processes while expelling a minimum of unreacted reducing agent out of the exhaust, also on the nature of $\leq$ 1ppm. This allows many of the processes to comply with the ever challenging requirements of the Environmental Protection Agency (EPA) on $NO_x$and ammonia emissions.

This presentation will discuss the chemistry behind an SCR, the structure of some common SCR reactors, the types of processes that use or can use a SCR, and some of the design considerations and issues inherent to the process.

(23e) The BP Texas City Refinery Explosion and the Involvement of Four Chemical Engineers

Jim Tarr, Stone Lions Environmental Corporation, Rolling Hills Estates, CA

The BP Texas City Refinery Explosion and the Involvement of Four Chemical Engineers

Their names are Glenn Bolton, Lorena Cruz-Alexander, Rafael Herrera, Daniel Hogan, Jimmy Hunnings, Morris King, Larry Linsenbardt, Arthur Ramos, Ryan Rodriguez, James Rowe, Linda Rowe, Kimberly Smith, Susan Taylor, Larry Thomas, and Eugene White. They died as a result of an explosion and fire that occurred at the BP Refinery in Texas City, Texas on March 23, 2005.

Simply stated, the incident at the BP Refinery occurred because the individuals involved in a process unit start-up forgot the first lesson taught in chemical engineering, the lesson about material balances: in minus out plus generation equals accumulation. In other words, BP personnel forgot to provide a means to remove material from a distillation column known as the raffinate splitter as they filled the column with process feed material. The result was that several pressure relief valves opened and discharged vapor and liquid to a "blowdown drum" that quickly overfilled, causing a volatile vapor/liquid mixture to spill from a one hundred and thirteen foot tall stack into the process area below.

The level of inattention exhibited by the BP personnel in charge of the raffinate splitter start-up defies the imagination. It is doubtful that a similar problem can be documented in the history of oil refinery operations anywhere in the world.

Notably, BP has admitted that they made ridiculous mistakes that day which caused people to die. They also paid a heavy monetary price.

The Texas air pollution control agency on the other hand has completely ignored their complicity in this matter. The unfortunate truth is that the Texas agency bears a collateral responsibility for the fifteen deaths that occurred at the BP Refinery on March 23, 2005.

As strange as it seems, it appears that the regulatory agency didn't even understand that the blowdown drum existed for the thirty years that it was operated illegally by BP in spite of a series of related regulatory actions that took place over the course of three decades.

This paper will explore the role of four chemical engineers who played a role in certain legal aspects of the circumstances related to the refinery explosion; two regulatory agency permit engineers, an expert witness for BP, and an expert witness for several people injured by the explosion.

About the Author: Jim Tarr, President of Stone Lions Environmental Corporations, holds two degrees in chemical engineering with over 35 years experience doing technical work in the legal arena.

All Sulphur Plants are Not Created Equal

Angela F. Slavens, P.E.
Business Development Director, Sulphur Technology
WorleyParsons
Monrovia, CA, USA

EXTENDED ABSTRACT

Sulphur recovery from natural resources is required to meet environmental regulations on sulphur content in final products, as well as SO_x emissions restrictions imposed on processing facilities. While sulphur recovery facilities in oil refineries, gas plants, gasification facilities and various chemical plants may appear to be very similar, there exist key differences in each type of processing facility that create unique, and often critical, design concerns for the sulphur plant. Factors such as feed gas strength (H_2S composition), type and concentration of impurities, feed gas stability, and availability of utilities alter the design and operational factors that must be considered in the corresponding sulphur plant.

This presentation highlights the unique elements that must be taken into consideration when designing or operating a sulphur plant located in a refinery, gas plant or gasification facility. The following parameters are compared for each type of facility:

Acid Gas Quality

- Strength (H_2S concentration)
- Impurities
  - Ammonia
  - Hydrocarbons
  - Organic sulphur compounds
- Stability of flow

Utility Availability

- Hydrogen
- Natural gas
- Oxygen

When acid gas feed quality is poor, robust and reliable sulphur plant operation can be difficult to achieve. However, this presentation Illustrates that when appropriate utilities are available in the facility, acid gas feed quality obstacles can be overcome. On this basis, the features of each type of sulphur plant are compared to provide a perspective on relative ease of design, operation and reliability.

Figures 1-4 provide an overview illustration of each type of facility, with comparative parameters highlighted.

Figure 1. Refinery Sulphur Recovery Unit

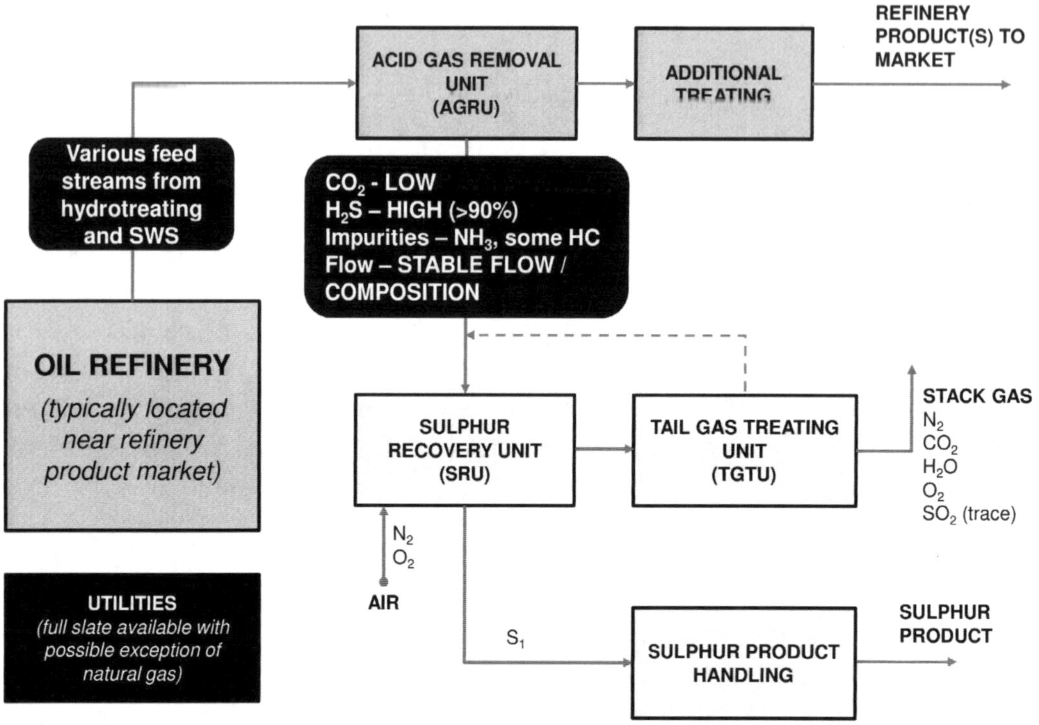

Figure 2. Gas Plant Sulphur Recovery Unit

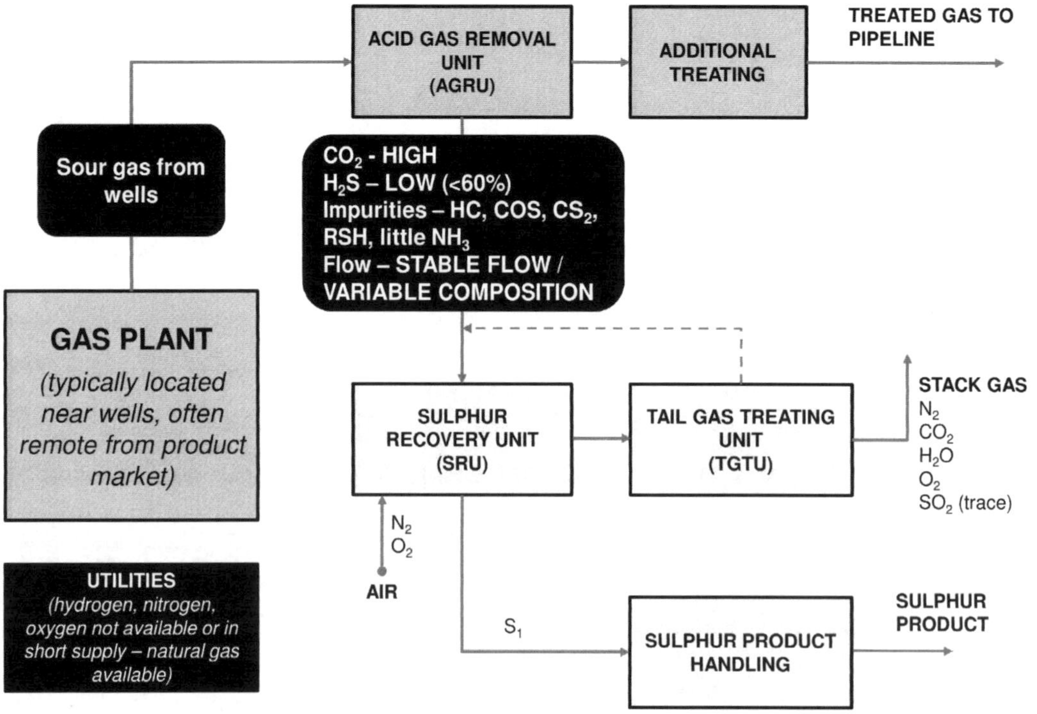

Figure 3. Gasification Facility Sulphur Recovery Unit

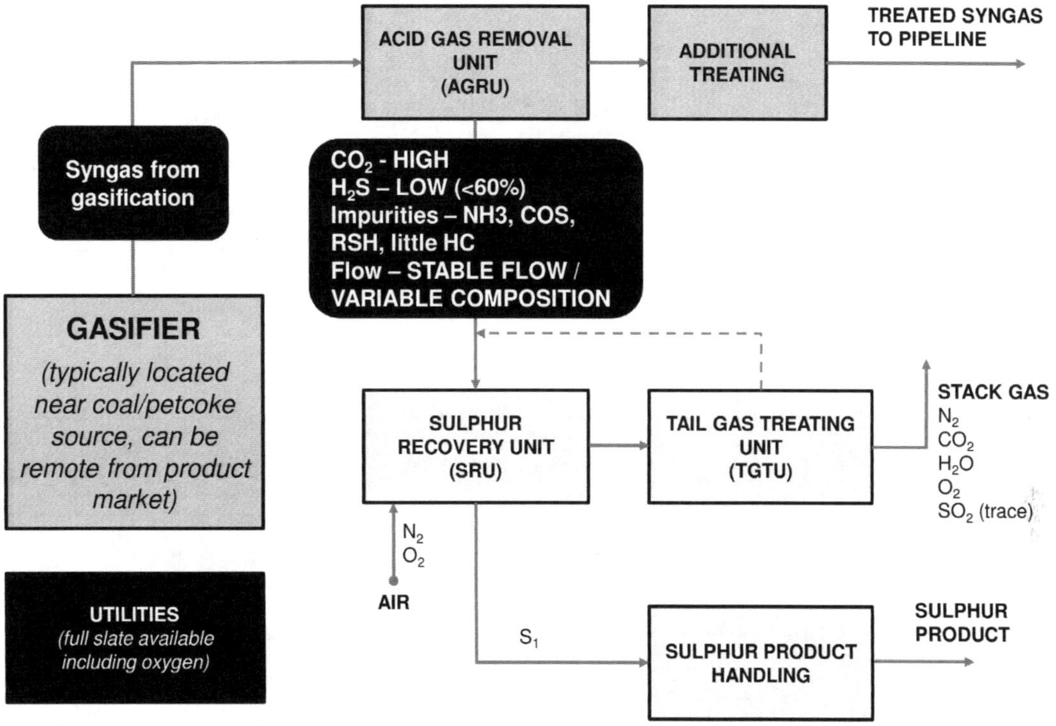

The ease of operation ranking reveals that refinery sulphur plants typically present the fewest obstacles to achieving robust design and reliable operation, due to their high acid gas feed quality, as compared to other types of facilities. Gas plants typically produce the lowest quality acid gas feedstock and often do not have the appropriate utilities available to balance this, thus ranking the lowest in terms of ease of design/operation. Gasification facilities typically produce a somewhat difficult acid gas feedstock but have a full suite of utilities (including oxygen) available to counteract this. Thus, gasification facilities rank in between refineries and gas. The ranking approach employed to achieve these results is illustrated in Figures 4 and 5. Higher rankings indicate more favorable conditions and lower rankings indicate less favorable conditions.

Figure 4. Sulphur Plant Ease of Design/Operation Comparison

FACILITY TYPE	ACID GAS FEED QUALITY					UTILITY AVAILABILITY			
	Acid Gas H_2S vol%	Impurities			Acid Gas Stability	H_2	Nat'l Gas	N_2	O_2
		NH_3	HC	Organic Sulphur					
REFINERY	5	1	4	5	5	5	2	4	3
GAS PLANT	1	4	1	2	2	1	5	1	1
GASIFIER	2	2	3	2	4	5	5	5	5
Weighting	30%	15%	15%	10%	30%	20%	50%	---	30%

Figure 5. Sulphur Plant Ease of Design/Operation Ranking

	ACID GAS FEED QUALITY	UTILITY AVAILABILITY	OVERALL
REFINERY	4.25	2.90	7.15 (MOST FAVORABLE)
GAS PLANT	1.25	3.00	4.25 (LEAST FAVORABLE)
GASIFIER	1.55	5.00	6.55

Although refinery sulphur plants have their challenges (primarily related to ammonia destruction), the ease of design/operation ranking illustrates that refinery SRUs are actually less problematic than SRUs in other types of facilities.

In summary, although many SRUs may look alike, SRUs in some types of facilities create greater challenges than others. Because of this lack of understanding that "all sulphur plants are not created equal," there are some facilities in the world that are not performing well.

Lastly, despite the fact that this presentation provides a generic comparison of the general advantages/disadvantages for different types of sulphur recovery units, it is important to remember that every solution is unique and requires experience and "know-how" to select the most reliable, robust solution with lowest lifecycle cost.

REDUCING CO2 IN THE RECYCLE ACID GAS FROM MDEA-BASED TAIL GAS TREATING UNITS

Ben Spooner, P. Eng

Farsin Derakhshan, P. Eng

INTRODUCTION

The purpose of an amine based tail gas treating unit (TGTU) is to recycle any left-over sulphur components in the tail gas of a Claus sulphur recovery unit back to the front end of the plant, rather than incinerate it. The remaining sulphur components are converted to H_2S and removed from the gas using an amine solution.

CO_2 is also present in tail gas streams and will be partially co-absorbed together with H_2S. This CO_2 co-absorption should be minimised. Any CO_2 removed in the TGTU will be recycled to the SRU front end along with the H_2S, which has negative consequences for the reaction furnace and overall sulphur recovery.

This paper discusses the effect of CO_2 on sulphur plants and how to minimise CO_2 co-absorption or to maximize CO_2 slip through the amine absorber of the TGTU.

Amine Experts believes that a well designed MDEA TGTU should be able achieve a minimum of 85% CO_2 slip.

EFFECTS OF CO_2 ON THE SULPHUR PLANT

The motivation behind maximizing slip though the tail gas treater is to prevent CO_2 from re-entering the sulphur plant. There are several reasons why it is advantageous to minimize CO_2 in the SRU feed, the most important involving the reaction furnace temperature, COS and CS_2 formation and sulphur plant capacity:

1. <u>Reaction furnace temperature:</u> CO_2 does not burn in the reaction furnace, and therefore will lower the furnace temperature. This will have negative effects on ammonia and aromatic destruction, which will then have consequences for the converter beds and the overall sulphur recovery.

   Maintaining reaction furnace temperatures is critical in maintaining smooth and trouble free operation of a sulphur plant. If processing Sour Water Stripper gas, the destruction of ammonia is of primary concern and reaction furnace temperatures of 2250°F+ are required in order to avoid ammonia salt formation and plugging of downstream process equipment. If processing acid gases containing aromatic hydrocarbons, reaction furnace temperatures of 1920°F+ are required to properly combust the benzene, toluene and xylene components. Incomplete combustion of these compounds can lead to poisoning of the downstream catalyst bed. Finally, in order to maintain stable flame operation, a reaction furnace temperature of 1700°F is required.

Though some cracking can occur in the reaction furnace, CO_2 is generally considered an inert compound, and cools the reaction furnace flame. As seen in Figure 1, diluting an acid gas stream with increasing amounts of CO_2 results in a rapid reduction of the reaction furnace temperature.

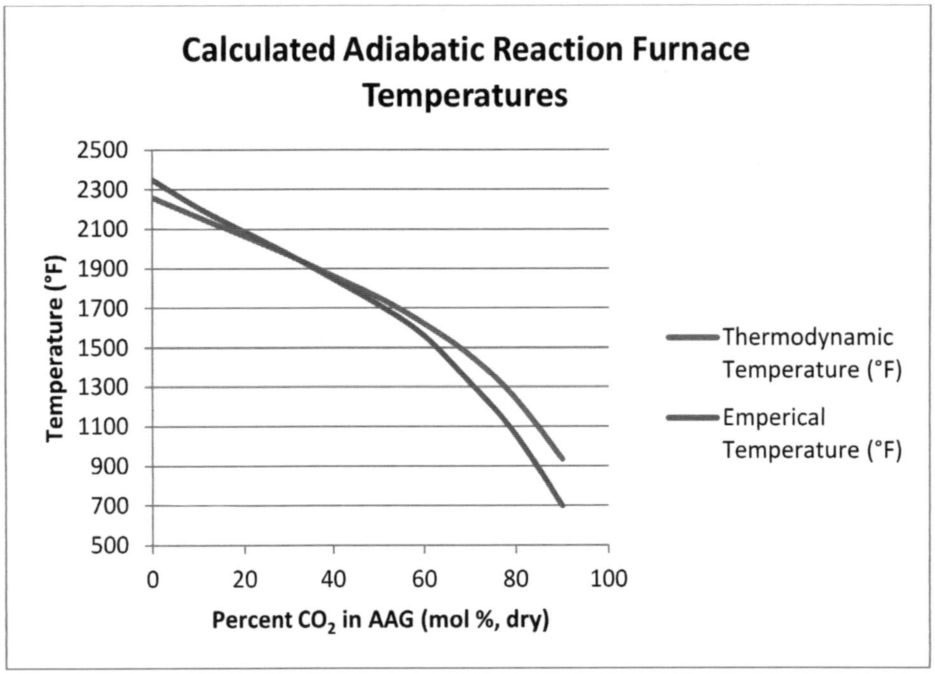

Figure 1 - Calculated Reaction Furnace Temperatures

Figure 1 shows that for a simple acid gas containing varying amounts of H_2S and CO_2 and 1% hydrocarbons, significant operating problems maintaining the required reaction furnace temperature start to occur at very low concentrations of CO_2 in the amine acid gas (AAG). With a CO_2 content of less than 5% (based on this example), it becomes difficult to maintain the required reaction furnace temperature for good ammonia destruction. At a CO_2 content of 30%, BTEX breakthrough could start to occur, and with a CO_2 content of over 50%, problems maintaining a reaction furnace flame become pronounced.

Although not necessarily detrimental to Claus plant operation, higher CO_2 content in the acid gas will require modifications to the operation of the reaction furnace in order to achieve and maintain the required operating temperatures, sometimes at a high capital investment cost or to the detriment of the overall efficiency. These items could include the addition of air and feed gas preheaters, oxygen enrichment, split flow reaction furnace designs, installation of high intensity or other specialty burners, co-firing with fuel gas or acid gas enrichment.

2. COS Formation: Although the chemistry inside a reaction furnace is quite complex and chaotic, one certainty is that CO_2 will partially "crack" in the furnace, resulting in the formation of CO, COS and (indirectly, though a drop in temperature) CS_2. Figure 2 shows the impact that CO_2 has on COS and CS_2 formation rates, using the same simplified acid gas composition discussed previously.

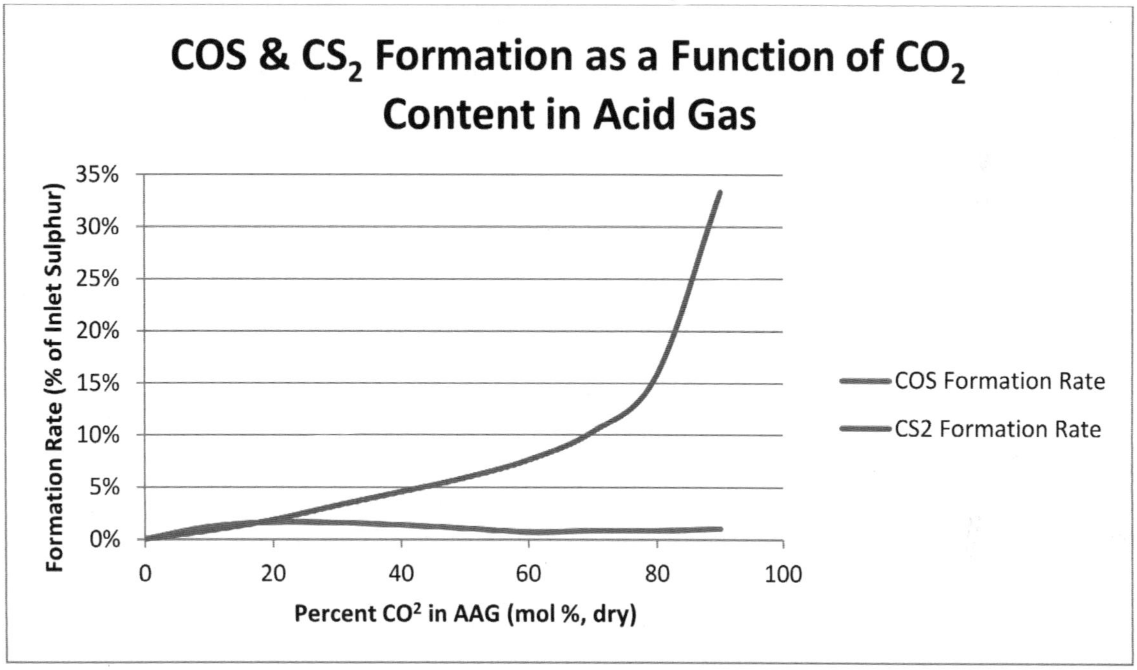

Figure 2 - COS & CS_2 Formation

As can be seen in Figure 2, the increasing CO_2 concentration of the acid gas results in ever increasing formation rates of CS_2. CS_2 formation has been shown to decrease with increasing temperatures, though it is not clear whether simply less CS_2 is formed at higher operating temperatures or whether CS_2 is formed then quickly hydrolyzed[(3)]. However, the addition of CO_2 cools the reaction furnace temperatures which results in increased CS_2 formation. CS_2 is especially unwanted since it binds up two sulphur molecules.

COS and CS_2 formation in the reaction furnace are important, because once formed, COS and CS_2 do not participate in the modified Claus reaction. These compounds must therefore be converted/hydrolyzed back to H_2S downstream of the furnace, either in the first Claus converter bed (operating at high temperatures, typically 600°F to 630°F) and/or utilizing special and often expensive catalyst, or they can be converted back to H_2S in the TGTU hydrogenation reactor. Since amine based TGTU's do not pick up COS and CS_2 in the absorber, any unconverted COS and CS_2 will result in increased sulphur emissions and reduced overall recovery efficiencies.

3. Overall gas capacity: In a world of low sulphur fuels, the demand for increasing the existing processing capacity of refinery sulphur plants is becoming increasingly important. Since CO_2 simply flows straight though the sulphur plant and does not participate in the Claus or modified Claus reactions, it simply takes up space and reduces the amount of sulphur-bearing gases which could otherwise be processed. A sulphur plant already operating at or slightly above its design capacity would quickly run into trouble with an increase in the CO_2 content of the feed gas.

   Apart from the issues related to furnace temperature and increased COS and CS_2 formation, operating a sulphur plant at higher-than-design throughput would see higher operating pressures that could result in:

   - problems with air blowers not being able to deliver sufficient air at the higher operating pressures,
   - heat exchangers not being able to sufficiently cool process gases resulting in more sulphur vapour losses,
   - problems with mass velocities through condenser tubes, resulting in liquid sulphur carryover (both of which would require more reducing gases in the TGTU hydrogenation bed)

   In order to address some of the reduced capacity issues, processing companies would have to consider technologies such as oxygen enrichment or the construction of additional/bigger sulphur plants, which would have a significant impact on the operating or capital budgets of refineries.

EFFECTS OF CO_2 ON THE QUENCH WATER OF THE TGTU

High CO_2 tail gas streams act to lower the pH of the quench water, which can decrease the strength of the protective iron sulphide film. The water will, therefore, be darker in colour and could cause fouling and plugging of the quench system. Ideally, the quench water has a pH of 7-8, however in plants with high CO_2 levels the pH is typically between 6 and 7. Some plants try to correct this by regular caustic addition to the (partly) circulating quench water. As a significant quantity of the quench water is not recycled, caustic injection is, at best, a temporary solution. Regular caustic injection can result in strong pH fluctuations which destabilize the protective FeS film.

CO_2 REMOVAL WITH MDEA

CO_2 does not react directly with the MDEA molecule; instead it dissolves and reacts in the water portion of the solution.

$$\mathbf{CO_2 + H_2O \rightarrow H_2CO_3 \quad (carbonic\ acid)}$$

$$\mathbf{H_2CO_3 \rightarrow H^+ + HCO_3^- \quad (bicarbonate)}$$

$$\mathbf{H^+ + R_1R_2R_3N \rightarrow R_1R_2R_3NH^+}$$

$$\mathbf{CO_2 + H_2O + R_1R_2\ R_3N \rightarrow R_1R_2R_3NH^+ + HCO_3^-}$$

The reaction between CO_2 and water (carbonic acid formation) is a "slow" step; it takes time to occur. Once the carbonic acid is formed, however, the MDEA reacts with it quite quickly and the bond will not be broken again until the amine is regenerated. The removal of CO_2 with a tertiary amine like MDEA is therefore kinetically limited by the reaction rate in the first step.

H_2S on the other hand, reacts directly with the MDEA molecule;

$$\mathbf{H_2S + R_1R_2R_3N \rightarrow R_1R_2R_3NH^+ + HS^-}$$

The reaction between H_2S and the amine is a very fast or instantaneous reaction. This means that H_2S removal is almost always equilibrium limited. Each contact stage or tray in an absorber reaches the H_2S equilibrium between gas and liquid.

The difference in chemistry between the H_2S and CO_2 removal is the key to understanding how H_2S can be removed with a minimum amount of CO_2.

The best strategy to minimize the CO_2 removal with MDEA is to prevent the CO_2 – water reaction from occurring by:

1. Optimizing the amine temperature
2. Optimizing the amine strength
3. Optimizing the amine circulation rate
4. Optimizing the amine feed point of the absorber

5. Choosing a more selective solvent (formulated MDEA or Flexsorb)

1. AMINE TEMPERATURE

As with most chemical reactions, the higher the temperature, the faster the CO_2 reaction takes place. In a TGTU, lower lean amine temperatures will minimize CO_2 pick up because it reduces the reaction between CO_2 and water. It should be noted that high amine temperatures (above 140°F) will also slip amounts of CO_2. Unfortunately, at high temperatures H_2S will also be slipped due to equilibrium limitations. Therefore, operating the absorber at high temperatures cannot be used as an operating strategy.

A low absorber temperature slows down the kinetics of the CO_2 reaction. H_2S removal occurs through a different mechanism which is much less affected by temperature. Therefore, the amine temperature should be kept as low as possible, though typically not lower than the inlet gas temperature. As the gas is coming from the TGTU quench tower, it is saturated with water. An amine temperature lower than the gas temperature would condense water and dilute the amine solvent. It should be noted that the normal temperature guideline for amine systems (maintain 5-10°F temperature difference between gas and amine) is not applicable for amine based TGTU's.

Normal operation is to maintain the lean amine temperature at or within 1-2 degrees of the inlet gas. 90-100°F for both the amine and the gas is ideal.

2. AMINE STRENGTH

Because water is one of the reactants, the more of it in the absorber, the more CO_2 will be absorbed into the amine solution. MDEA can be successfully operated at up to 50% strength (the other 50% being water). Above this strength the viscosity of the solution becomes too high and will negatively affect the mass transfer of H_2S into the amine.

Minimize the CO_2 hydrolysis by minimizing the water content of the MDEA solution. **Maintain solvent strength between 45-50%,** all the while utilizing a proper filtration program to ensure the viscosity of the amine does not get too high. Minimizing SO_2 breakthrough out of the SCOT (or SCOT-type) CoMo reactor will reduce the heat stable amine salt build-up in the system and reduce the chances of viscosity increase due to excess heat stable salt formation.

3. AMINE CIRCULATION RATE

The CO_2 reaction with water (to form bicarbonate) is a kinetically limited reaction. This means that CO_2 builds up at the gas-liquid interface and only reacts with the water and amine as it slowly diffuses to the bulk of the amine solution. In a trayed tower, higher circulation rates increase the height of liquid on each tray, as the weir creates a flow obstruction. The amine "stacks up" against it and provides a larger surface area for CO_2 absorption. In a packed tower, higher amine flows create higher hold-up of amine into each section of packing. The more liquid in the tower, the higher the CO_2 removal will be, as there is more gas/liquid interface for the CO_2 to be absorbed. Over-circulation of the amine is the single largest contributor to poor CO_2 slip in TGTU's.

Furthermore CO_2 increases the loading of the amine which, especially in low pressure TGTU applications, takes up very valuable acid gas holding capacity. Under certain conditions, it is possible that CO_2 will force the amine to release previously absorbed H_2S.

Optimizing (in most cases reducing) the amine circulation rate will always result in a decrease in H_2S and an increase in CO_2 in the treated gas which is exactly the goal.

In Figure 3 below, the effects of circulation rate are shown. As the amine rate is decreased from 200 gallons per minute down to 150, the CO_2 in the treated gas increases by 1,000 ppm, whereas the H_2S decreases from 45 to 40 ppm.

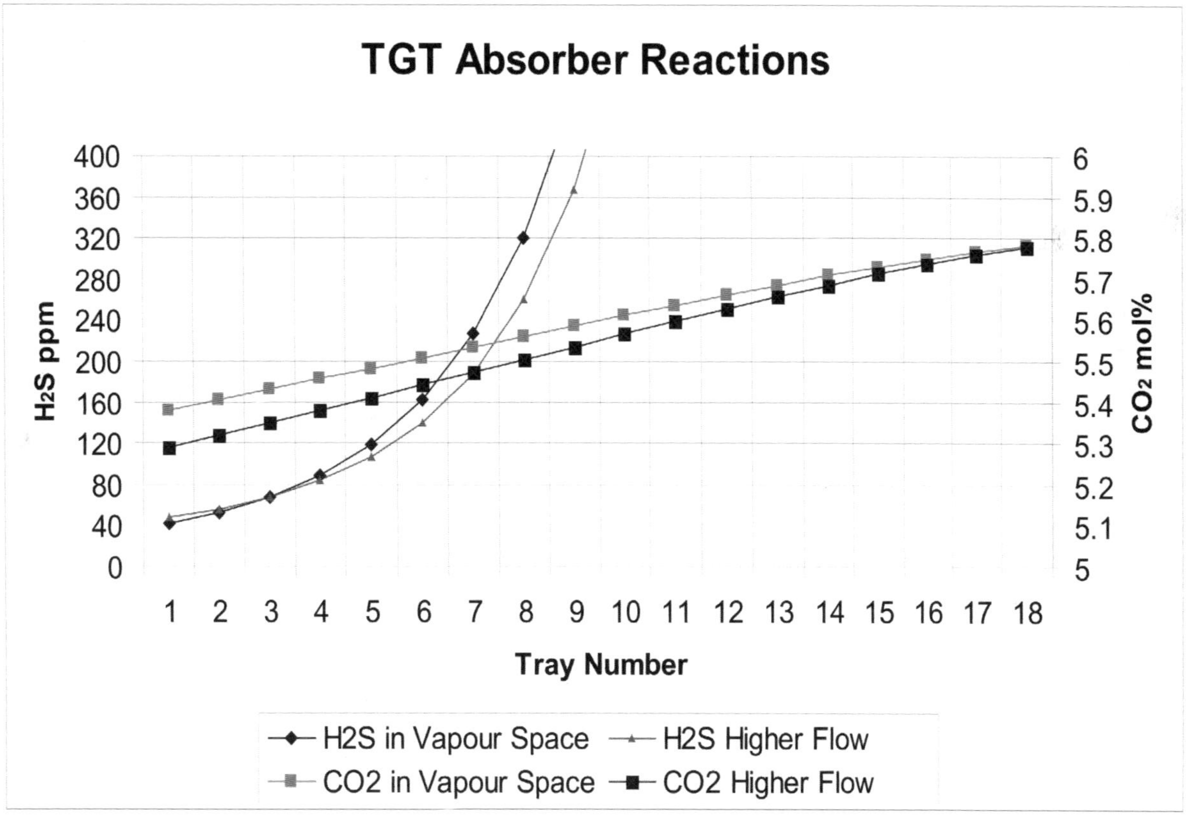

Figure 3: Effect of an increase in circulation rate on H_2S and CO_2 removal

The circulation rate should initially be targeted for a rich H_2S loading of 0.05 mol/mol (once the amine strength has been set, see point 2 above), and slowly decreased with a final target of 0.1 mol/mol. This should be re-enforced by the use of a reliable amine plant simulator.

4. AMINE FEED POINT INTO ABSORBER

TGTU absorbers often have multiple inlet points for the lean amine solution. Several inlet points are normal and can be used to increase or decrease the interfacial contact area between the gas and the amine solution. A shorter contact time (less interfacial contact area) will result in less CO_2 absorption.

The contact time between the CO_2 and amine depends on the internals of the absorber and the height at which the amine is injected. In a trayed absorber, contact time on each tray is determined by the weir height and number of trays. If packing is used, contact time depends on the height and size of the packing. In either case, lowering the injection point of the lean amine will reduce the contact time and interfacial contact area between the gas and amine, thus increasing the amount of CO_2 slip.
Neither the weir height of trays nor total height of packing is a parameter which can be adjusted during normal operation. Changing either of these requires shutting down the system and performing large scale maintenance. This is why most TGTU absorbers have multiple feedpoints built into the design.

It should be noted that there are differences between the selectivity of a packed tower compared to a trayed one. This is due to hydraulics of the internals and the corresponding relationship with mass transfer. A trayed tower has the liquid phase which is highly agitated; packed towers are opposite, whereby the liquid flows over the packing relatively smoothly. The gas flows are turbulent in both. Either type of internals can result in decent CO_2 slip, but the choice between the two must be carefully thought out and researched in the design stage. Because the net liquid hold-up in a packed bed (1-6% of tower volume) is lower than in a trayed tower (8-12% of tower volume) the lower liquid hold-up can result in less interfacial contact area and therefore lower CO_2 absorption. In general, Amine Experts has tested higher CO_2 slip rates on packed towers compared to trays, but this is by no means a universal rule.

A TGTU absorber has typically three lean amine injection points; one at the top tray or top height of packing, and two more at successively lower intervals. The balance is to inject at a high enough point to remove all of the H_2S necessary, but no higher. This procedure is best predicted first on a simulator, as shown in Figure 4 below. Using this type of chart, it is apparent where appropriate injection points into the absorber are, depending on the required H2S specification.

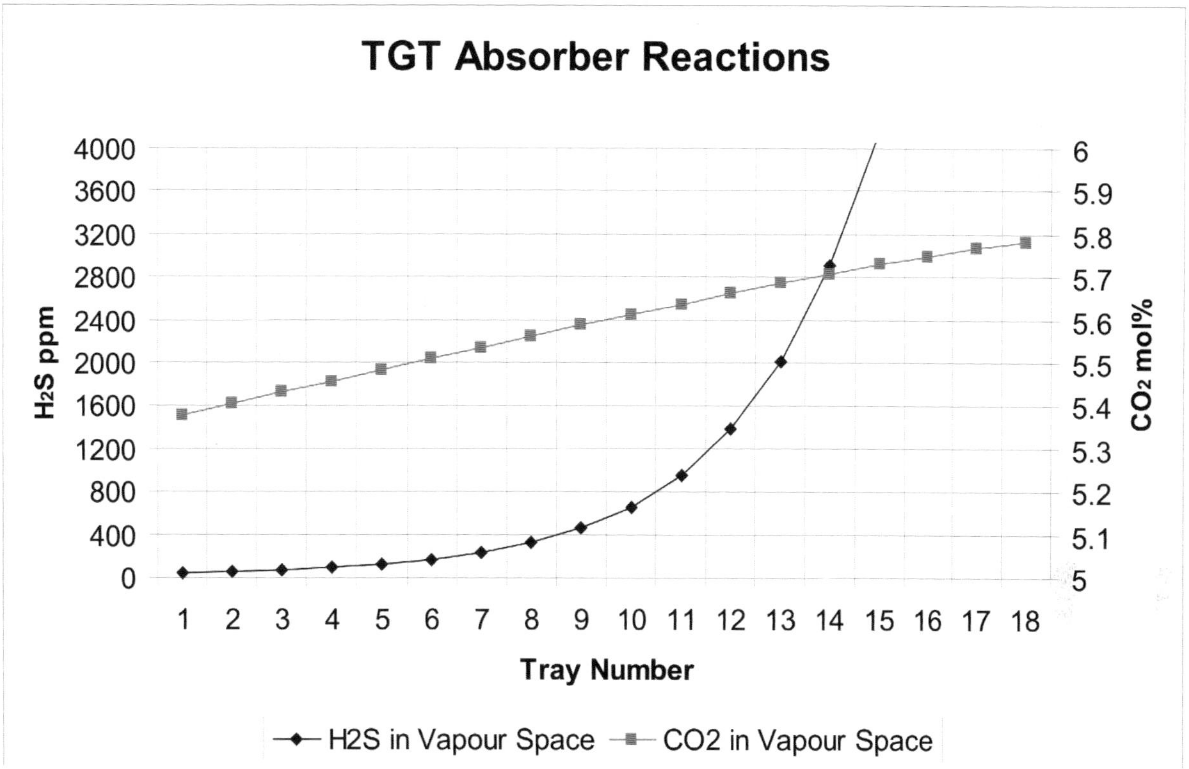

Figure 5: Simulated acid gas removal tray-by-tray in a TGT absorber

If possible, lower the amine injection by one feed point and measure the treated gas H_2S content. If it is acceptable, and simulations agree, continue lowering to another feed point. The main concern is the treated gas H_2S content.

5. CHOOSING A MORE SELECTIVE SOLVENT

Hindered Amine

The most selective solvent for TGTU applications is a hindered amine like the Flexsorb® SE/SE Plus from ExxonMobil Research and Engineering Company. A hindered amine is normally a secondary amine with a bulky group which hinders the direct reaction with CO_2. These molecules combine the low CO_2 reaction rate of a tertiary amine with the base strength of a secondary amine. Particularly the high base strength is very useful at the low pressure of a TGTU absorber because it makes it possible to have a significantly higher rich amine loading than with MDEA. The higher loading allows for a reduction in amine circulation rate, which improves again the selectivity. Flexsorb however is a much more expensive solvent than MDEA, nor is it as readily available. As well, it most likely requires a licensing fee.

Formulated MDEA

Enhanced MDEA formulations for TGTU's have been available for 25 years. These are normally the solvents that contain "pH suppressants" such as phosphoric acid. The reduction in pH allows for easier and deeper regeneration of the amine, especially for H_2S. These formulated solvents are known as: Ucarsol HS-103, GasSpec TG-10 or JeffTreat MS-300. The same effect of improved regeneration is achieved in MDEA solutions containing between 0.5-1 wt% heat stable salts.

By lowering the H_2S lean loading, the H_2S in the treated gas will also drop. This could enable the lean amine to be introduced into a lower feedpoint in the absorber, thus increasing the CO_2 slip as described above. Normally though, the aim of the improved regeneration is to obtain a lower H_2S level in the treated gas. TGTU's are, therefore, also often designed with more trays in the regenerator than normal amine units. This allows for deep H_2S stripping.

A second aspect of these formulated solvents is that they often contain anti-foam. The anti-foam reduces the liquid/gas interfacial contact area because it helps to break the gas bubbles in the solvent faster than normal. In this way it improves the selectivity even when the solvent is not "foaming" (there is always a few inches of foam on the trays of a normal working absorber). This also highlights the importance of minimising foaming problems to reduce the CO_2 pick-up of a solvent. Although TGTU amine systems do not often have foaming problems, it is important to eliminate the foaming as quickly as possible for good selectivity.

REFERENCES

B. Klint, “Hydrocarbon Destruction In the Claus SRU Reaction Furnace”; Sulphur Recovery textbook, Sulphur Experts, 2000

H.G. Paskall, “Reaction Furnace Chemistry and Operational Modes”; Sulphur Recovery textbook, Western Research, 1988

Sames, J.A., Dale, P.R. and Wong, B. (1987): “Evaluation of reaction furnace variables in modified-Claus plants”. Laurence Reid Gas Conditioning Conference, Norman, OK.

T.K. Khanamamedov, RH Weiland, “Upgrading Acid Gas Streams”; Sulphur magazine, Sept-Oct 2008

Mahin Rameshni P.E., “How to Improve the Efficiency of Tail Gas Treating in Sulphur Plants”; Worley Parsons

RH Weiland, John C Dingman, “Simulator Provides Guidance for Increasing CO_2 Slip in Gas Treating Applications”; OGT Inc

Selective H_2S removal: Case studies by using a new promoter system in amine units

Gerald Vorberg, Ralf Notz,
OASE® Gas Treatment Excellence
BASF SE, Ludwigshafen/Germany

Kumar Sugavanam,
OASE® Gas Treatment Excellence
BASF Corporation, Houston, TX

Abstract

Applications where selective sulfur removal is required have become an important topic over the last decades. In many applications, e.g. in acid gas enrichment (AGE), in tail gas treatment units (TGTU) as well as in natural gas applications (NG), the H_2S selective treatment including the production of an enriched high quality Claus gas has been one of the favored drivers so far. Meanwhile, the requirement for an adjustable CO_2 slip, the focus on tighter sulfur specifications along with a high degree of flexibility and savings in energy consumption have gained in importance as well.

The principles of selective removal of H_2S with amine-based solvents follow three major routes:

a) various design options and absorber internals, affecting the difference in CO_2 and H_2S mass transfer kinetics

b) severely hindered amines, controlling the selectivity for H_2S removal primarily in the absorber, and

c) generic and promoted tertiary amines, focusing more on the regeneration ability, and thus leading to tighter H_2S specifications.

In particular the combination of a) and c) is widespread and the most common option in industry.

Based on two case studies, this paper will describe the possibilities and leverage of selective designs by using an advanced promoter system for enhanced selective treatment with tertiary amines.

Principles

H_2S selective acid gas removal with amine based solvents is widespread in the industry. Particularly, the use of "generic" amines such as Methyl-di-ethanol-amine "MDEA" is well known in combination with straight forward designs. Although, being not a part of this paper, advanced solutions with sterically hindered amines incorporate useful properties: e.g. fast kinetics and high acid gas load at low acid gas partial pressure and a high H_2S selectivity due to the steric hindrance of the molecule itself. However, in some cases those advantages may be progressively reduced at increasing acid gas partial pressures e.g. most high pressure natural gas applications. Moreover, the relatively high price level of hindered amines is often part of economic considerations between CAPEX and OPEX.

As mentioned above, designs and current operations with generic solutions became very popular due to their "process simplicity" and the broad availability of the amine itself. However, the level of severity has changed making higher demands on existing or even grassroots plants and their subsequent designs. This raises the question, how safe are today's predictions and safety margins?

For instance, a reliable prediction of the "H_2S selectivity" or the CO_2 pick up respectively is not a trivial aspect in the simulation as its impact on sizing and performance can be substantial. In particular, plants being designed and operated at the edge need more attention, but what does this mean in reality? By nature, low partial pressure and/or high temperatures are "natural enemies" of H_2S selective operations and can create headaches for designers and engineers. The achievement of tighter H_2S specifications in low pressure tail gas service or a high degree in operational flexibility of natural gas plants are the new challenges, especially in warmer regions. While tail gas plants usually remain more or less at the same feed gas conditions over their lifetime, the complexity in natural gas plants may strongly increase for the following reasons:

Firstly, a natural gas plant should be suitable for all operational scenarios including summer and winter cases.

Secondly, the feed gas specification of those plants can vary over its lifetime. However, for plants using their treated gas as sales or pipeline gas, the treated gas specification is often fixed to a few ppmv of H_2S and 2-3 vol-% of CO_2. By means, a certain flexibility in selectivity is mandatory.

Thirdly and one should recognize that it is not only the feed gas specification, but also the feed gas rate that may change by the addition or closure of wells and last but not least contractual considerations, requiring different turndown rates. As the dimensions of absorbers and the type of internals are usually fixed, this can trigger substantial fluctuations in terms of H_2S selectivity itself. Usually, good engineering practice allows the addition of safety margins, but to what extent?

From a design perspective, high performance internals are able to cover a wide range of gas liquid ratios while multiple amine feed points provide flexibility in absorber height. However, for an existing plant, operators only have a limited degree of freedom as listed below to adapt the operation to the new situation:

1) Changes in amine feed point (if possible),
2) variations in circulation rate,
3) adaption of the reboiler steam feed rate and possibly a change in lean amine temperature.
4) Further measures, however, with minor impact are changes in the pressure of the flash vessel and the regenerator.

But what about using a promoter or an adaptive promoter technology in the amine systems?

For a long time, the simple acidification of amines was a very "handy" option to boost H_2S selective applications by lowering the H_2S treated gas specification as well as the energy demand based on the following effect.

H_2S is a stronger acid than CO_2, and has a stronger binding energy at the stripper temperature. Consequently, amine regeneration of H_2S down to trace levels becomes progressively more difficult. In particular, designs with very low H_2S specifications in the low ppm-range require a promoter in order to release H_2S more easily from the solution. In this regard, MDEA acidified with phosphoric-, sulfuric- or other acids, becomes regenerated more easily. Even the unintentional "acidification" by formation of heat stable salts (HSS) and their acids - acetates, formates, sulfates, etc. - will also lead to a very efficient H_2S regeneration "boost". However, the level of acidification has its constraints, as there is a strong impact on the equilibrium curve of the acid gas solubility in the solvent.

Today, an alternative for acidified amines is provided by an advanced promoter technology from BASF, which incorporates the advantage of an acidified amine blend, but without the disadvantage of a reduced acid gas capacity. Its mechanism is based on influencing and shifting the pH- and temperature dependent equilibrium load into the regime being required for the individual application.

Case Studies

Two cases studies have been selected to demonstrate the capability of such a promoter technology when working at the edge.

Case 1: Tail Gas Treatment

A tail gas treatment unit (TGTU) has been originally designed for a H_2S treated gas specification of 250 ppmv. Over the years, concepts have been developed to achieve tighter H_2S specifications below 100 ppmv due to upcoming new regulations and a new total sulfur management of the entire plant. Meanwhile, the CO_2 in the feed gas to the TGTU has slightly increased to 51 vol-% over the years. A stepwise acidification of the amine has been applied to reduce the residual amine lean load and the subsequent H_2S specification in the treated gas. However, the reduced acid gas loading capacity required a compensation by increasing the amine circulation rate. As a matter of fact, CO_2 coabsorption increased as well, moving the maximum acid gas capacity closer to the equilibrium. As a consequence, operability of this plant became highly sensitive as H_2S breakthrough could occur suddenly. Meanwhile, a tighter H_2S specification below 50 ppmv has also been considered. By taking current operational data into account, a combined optimization by acidification (H_3PO_4) and circulation rate adjustment at maximum reboiler duty have been considered to get as close as possible to the new requirements. Simulation data have shown a well delimited optimum, however the pinched vapour phase H_2S concentration profile in the absorber clearly demonstrates this as only an academic approach. By contrast, the application of an advanced promoter system from BASF with mentioned properties and its respective performance show a significant higher safety margin combined with less sensitivity on external fluctuations, such as lean amine temperature.

<u>Case 2: Natural Gas Plant (winter/summer scenario)</u>

A natural gas plant has been designed to meet sales gas specification, by means < 5 ppmv H_2S and < 2 vol% CO_2. Feed gas specification has changed over the years due to some new resources. CO_2 concentration increased above 10 vol-% while the H_2S specification remained at the same level. As the absorber height was limited to 25 trays, an increase in circulation rate was the only viable measure to match the CO_2 specification. During the winter season, this scenario worked properly at lean temperatures of around 113 °F (45 °C). However, during the summer season (lean amine temperature up to 132 °F, 55 °C), this kind of operation consequently did not meet specifications. This was mainly that due to limited reboiler and lean rich exchanger duty, a further increase in circulation rate only led to an increase in lean amine loading and a subsequent failure of the H_2S specification. The addition of an acid to the amine (as regeneration promoter) has been considered to lower the H_2S lean amine loading and to match the H_2S specification. But now the problem was shifted to a lowered CO_2 equilibrium load. In other words, the drop in pH value showed a significant impact at elevated temperatures and could not be compensated by the lower overall acid gas lean load. Consequently, plant capacity needed to be reduced by 8 percent.

As circulation rate and reboiler duty have been already at their limits, an alternative design has been provided with the advanced promoter system from BASF. The higher equilibrium load (compared to acidified amine) has provided sufficient capacity to meet the CO_2 specification of 2 vol-%. Furthermore, the improved H_2S regenerability at a slightly reduced reboiler duty gave enough room the meet the H_2S specification even with a certain safety margin.

Summary

Advanced H_2S selective acid gas removal technologies have become increasingly important for existing and new tail gas- and natural gas- treatment applications.

While tighter H_2S specifications are a particular target of tail gas units, the design and technology of world scale natural gas plants focus more on an adjustable CO_2 slip and a high degree in operational flexibility.

Today's plants require a very high degree in flexibility for turndown rates, changing feed gas specifications and conditions. In addition, very tight treated gas H_2S specifications in the low ppm range (e.g <5 ppmv) are becoming mandatory.

As low pressure (e.g. in tail gas units) and elevated amine temperature (e.g. in natural gas plants in summer operation) are "natural enemies" of H_2S selective designs, generic amines and straight forward designs might easily come to its limits and thereby not provide the optimum solution for meeting needed specifications.

MDEA-based technologies are attractive due to competitive solvent prices and generic designs. For OPEX, CAPEX and specification purposes, acidic promoted systems are in use to overcome the relatively high H_2S binding energy and to lower residual H_2S loading in the amine.

However, current acidified state-of-the-art systems also have their limits, especially when operating at the edge as acidification will also trigger an inverse effect on the entire acid gas capacity.

Alternatively, a new advanced promoter technology from BASF has been presented based on two case studies. It provides a substantial increase in regeneration ability while hardly affecting absorption capacity.

(34d) Sour Water Strippers Exposed

Ralph H. Weiland , Optimized Gas Treating, Inc., Clarita, OK and Nathan A. Hatcher, Optimized Gas Treating, Inc., Buda, TX

Sour Water Stripping: The Process Exposed

Ralph H. Weiland, Nathan A. Hatcher

Optimized Gas Treating, Inc.

Sugar Land, TX

ABSTRACT

The sour water system is often considered to be a refinery's sewer. It consists of water used for quenching and scrubbing HDS cold separators, sulphur plant vent gas, blow-down from amine regenerator reflux sections, and waste water from a variety of other operations. Its major components are ammonia and hydrogen sulphide with usually lesser amounts of carbon dioxide and heat stable salts (HSSs are mineral acids), and sometimes amine.

Sour water strippers (SWS) are moderately-large reboiled towers (30 to 60 trays) in which ammonia and other gases are removed from the sour water by steam stripping. Heretofore they have been designed using equilibrium stages. However, tray efficiencies have remained obscure with quoted values anywhere from 15% to 40%, a factor of two range. Consequently, designers have less than complete confidence in the reliability of their final design. The consequence of uncertainty is either overdesign and, therefore, excessive costs or sleepless nights because of an underperforming unit.

Recently a mass transfer rate-based simulation model has become available for designing and troubleshooting sour water strippers. In this paper, we use the model to determine tray efficiencies for ammonia and H_2S stripping, how they vary across the height of a tower, and what operating variables affect them, and how. We also predict quantitatively how the presence of heat stable salts and caustic injection for pH control affect the treat-ability of sour water to specified residual levels of ammonia and H_2S.

(34e) Fate of Ammonia in Refinery Amine Systems

Nathan A. Hatcher, Optimized Gas Treating, Inc., Buda, TX and Ralph H. Weiland, Optimized Gas Treating, Inc., Clarita, OK

Fate of Ammonia in Refinery Amine Systems

Ralph H. Weiland, Nathan A. Hatcher & R. Scott Alvis

Optimized Gas Treating, Inc.

Sugar Land, TX

ABSTRACT

Ammonia ingress and accumulation in refinery and bio-gas amine systems is not a new problem. However, increasing utilization of advantaged crudes with higher nitrogen content may present challenges in today's capital-constrained operating environment. Many refiners have instituted guidelines for purging amine regenerator reflux water for corrosion control. Historically, this has been done empirically based upon periodic lab analysis and adjustment of the purge water rate.

The true amount of ammonia ingress, and its material balance across refinery amine unit is not a topic that has been discussed in great detail, because until now, rate-based mass transfer models have not been available. Using the well-known ProTreat rate-based mass transfer process simulator, this paper addresses the following questions. Where available, comparisons to plant data measurements are provided.

How much ammonia can accumulate based upon choice of regenerator operating conditions?

How much ammonia rejection into the amine acid gas does this correspond to, and is this a significant concern to downstream (i.e., sulfur plant) operations?

Can ammonia build to levels that will cause additional H_2S to be trapped leading to higher lean loadings, reduced treating performance, or even regenerator flooding?

How does the choice of HDS cold separator operating conditions impact the ingress of ammonia to the amine system; just how effective is water washing?

How much ammonia skates through refinery amine treaters?

Ultimately, we attempt to answer whether the ammonia balance on a refinery amine system can be fully characterized based upon the knowledge of a few simple parameters.

ADVANCED CONTROL OF MULTIPLE SULFUR RECOVERY UNITS

Darwin E. Logerot, ProSys, Inc., Baton Rouge, LA

Abstract

Sulfur Recovery in a Refinery entails collecting sour vent gases (containing H_2S) from a number of process sources, including Hydrotreating Units, Hydrocracking Units, Crude Units, and others, then converting the H_2S to a more useable form such as sulfuric acid or elemental sulfur. The feed rates and compositions to the Sulfur Recovery Units (SRUs) are not controllable at the SRU. In addition, most Refineries employ multiple SRUs operating in parallel. This gives rise to a number of process control challenges, including:

Load balancing among the various units
Response to changing loads (total feed load cannot be controlled)
Response to variable compositions
Air/Oxygen load controls
Response to one SRU shutdown (rapid load increase for the remaining SRUs)

This paper will present the ramifications of these control problems along with the methodology used for successful implementation of the advanced control scheme for each.

The Process

Before discussing the advanced control aspects of SRUs, a basic process description is in order. Although the control concepts included in this paper can apply to any sulfur recovery process, the specific type discussed in detail is the Claus process. A Claus unit converts hydrogen sulfide to elemental sulfur.

Figure 1, in the Appendix, shows an overview schematic of the process described herein.

The feed gases to the SRU come from multiple sources in the refinery. Each source will be one of two basic types, either from an amine regeneration unit or from a sour water stripper. The amine regenerators strip H_2S out of an amine stream, after the amine had been used to absorb H_2S from a sour gas stream in one of the refinery units. The sour water strippers remove H_2S and ammonia from refinery waste water.

The feed gas to an SRU is very high in H_2S, with the primary impurities being water vapor, carbon dioxide and trace amounts of amine, hydrogen and light hydrocarbons. The gas from the sour water strippers also contains ammonia. The gases from the two types of sources are usually kept segregated until they arrive at the SRU.

The first step of the process is partial oxidation of the H_2S in the feed gas to form SO_2:

$$H_2S + 1.5O_2 \longrightarrow SO_2 + H_2O \qquad (1)$$

The oxygen fed to the combustion chamber (shown as the partial oxidation burner) is usually supplied from fresh air by an air blower. Many units also supplement the combustion air with pure oxygen to increase residence time and reduce pressure drop.

Two combustion zones are usually employed. The ammonia-rich (sour water) acid gas is fed to the first zone, and the amine acid gas is split between zones. Details of the split are beyond the scope of this paper, but if not done properly, salts tend to form in equipment downstream of the burner causing corrosion and plugging.

The resulting gas mixture is then run over a catalyst bed to accomplish the reaction to sulfur:

$$2H_2S + SO_2 \longrightarrow 3S + 2H_2O \qquad (2)$$

This reaction is usually conducted in several stages, with sulfur condensation and re-heating steps between stages.

The goal of the reaction section is complete, or nearly complete, conversion of the H_2S in the inlet gas. The stoichiometry of reaction (2) dictates that complete conversion is only possible if the oxygen / H_2S ratio at the inlet of the combustion chamber allows exactly 1/3 of the H_2S to be converted via reaction (1). The stoichiometric ratio must be controlled closely; otherwise there will be either excess H_2S or SO_2 in the tail gas.

After the reaction step, there should be only a small amount of sulfur compounds (SO_2 or H_2S) in the tail gas. These compounds must be removed before the tail gas is allowed to vent to atmosphere. Removal of trace sulfur is accomplished by a reduction step, usually using hydrogen to convert any remaining SO_2 to H_2S, followed by H_2S absorption into a circulating amine solution and finally, a tail gas incinerator.

Control Problems

The primary control issues in an SRU arise from two basic areas. Each introduces its own set of challenges for the control system.

1) Process factors:
   - Oxygen to sulfur ratio control at the front end is critical and difficult. The composition of H_2S or SO_2 in the feed gases is usually not known and varies. If the ratio is off in either direction, the result will be excess H_2S or SO_2 to be handled in the downstream equipment. This may overload the reduction, absorption and incineration sections, and may result in excess sulfur emissions.
   - Pure oxygen is expensive, but use of too much air instead of oxygen reduces residence time and increases pressure drop, affecting the oxidation rate and reducing throughput capacity.

- Each unit has a set of operating and safety constraints, including maximum equipment capacity, MAWP, max temperature, maximum oxygen concentration, and ultimate emissions limits.
- Proper split of amine acid gas between combustions zones is important to avoid salt formation and corrosion.

2) Most refineries use multiple sulfur recovery units in parallel to handle sour gas from multiple sources, and the source gas rates are dictated by the upstream units. As a result of this factor, the control system must:
   - Respond to uncontrolled load changes. These can be rapid and significant.
   - Ensure that the wild load is properly balanced so that one train is not overloaded while another runs well below its capacity.
   - React rapidly to a sulfur unit trip, raising the loads on the other sulfur units to maintain refinery stability.

In summary, the control problem is a balancing act that must consider all the above factors, keep the units onstream within acceptable operating limits, and maintain sulfur emissions below the permitted maximum.

Control Solutions

The control scheme discussed herein that solves the above problems is based on advanced regulatory control (ARC). ARC basically is an advanced control scheme that does not rely on process models, only on PID control blocks, calculations and selectors. The heart of the control solution is a pressure based load balancing ARC, diagrammed in Figure 2. This scheme uses a header pressure controller as the master. The master pressure controller cascades a setpoint to a total feed gas flow controller. In turn, the total flow controller cascades to a separate feed gas flow controller for each parallel train. As header pressure rises, feed rate targets to each train are increased to handle the extra gas in the system. A key feature of this scheme is the use of adaptive gain in the total flow controller. This means that the proportional gain is adjusted based on the number of Sulfur Recovery units available. A Sulfur Recovery unit is considered unavailable if the unit is down, if the SRU feed flow controller is at a constraint limit, or if the SRU feed flow controller is not in cascade mode. As a result, the overall scheme can react rapidly and appropriately to a sudden change in gas handling capacity, such as a unit trip or a new unit coming on line.

Discussion of the control scheme would not be complete without including the constraint features. Each individual flow controller can be constrained by any one of a number of limits. The constraints can reflect unit capacity, maximum air or oxygen availability, equipment pressure and temperature limits, etc. If any of the pre-defined limits are reached, the train flow controller will no longer accept additional load requests from the total flow controller. In turn, the adaptive gain on the total controller will adjust itself, since a train at a constraint limit is treated as an unavailable unit. So, if one train is limited by a constraint, the control scheme will adjust with the other available trains. Figure 3 is a diagram of the constraint scheme.

Downstream of the header pressure and unit feed flow control, each SRU has its own set of ARCs, designed to keep the unit operation on target. The most critical of these is the oxygen to sulfur ratio feeding the combustion chamber. Control of the oxygen to sulfur ratio is accomplished by an air/O_2 ARC. This ARC first calculates total oxygen requirement based on the feed gas rates. Total oxygen is trimmed by the output of an excess H_2S / SO_2 analyzer and controller on the tail gas. The total oxygen demand is supplied by air first, then by pure oxygen. The output of the oxygen requirement calculation is the setpoint of the air controller. The air controller can be constrained by combustion chamber pressure. If pressure is at its high limit or if the air controller is approaching saturation, any additional load is handled by pure oxygen. Pure oxygen rate is controlled as the difference between the total oxygen demand and the oxygen supplied by the air controller. This scheme assures that the unit will maintain the proper ratio, and that pure oxygen is only used if truly needed. Figure 4 shows the air / oxygen control layout.

Proper acid gas split between combustion zones is also an important consideration. The detailed methodology for the split varies with unit design, and is beyond the scope of this paper. Basically, the control system within each SRU uses its total acid gas set point along with a calculation algorithm to split the total gas into two zone flow controllers.

Benefits

There are a number of benefits resulting from these control schemes. Perhaps the greatest benefits result from greatly improved feed header pressure control. This results in fewer impacts to upstream units, fewer issues when a SRU trip occurs, no need to send sour gas to flare and fewer permit excursions.

Another benefit is better air & O_2 control, which facilitates oxygen savings (pure oxygen is only used when really necessary). Improved control over air to feed gas ratio also increases average sulfur conversion and allows the unit to reduce emissions.

These benefits are realized from a solution that utilizes advanced regulatory control (ARC) without using model-predictive control (MPC). This means that no external software is required, no process models are needed, and the process response to control action need not be linear.

Before vs After

Perhaps the best indication of the success of the control scheme is a reduction in sulfur emissions excursions. Before installation several years ago in a refinery operating three SRUs, the controls did not react appropriately to upstream issues and did not effectively maintain proper oxygen to gas ratios. As a result, the refinery experienced an unacceptable number of operating incidents and had several air permit violations over the course of a year. After installation of this scheme, no violations were experienced in a full year of operation. The cost of the control installation was recovered in a few months from better operation and the cessation of violations. This refinery can now withstand the sudden loss of an SRU without incident.

Acknowledgements

The author would like to acknowledge the contributions of the following individuals, without whom this paper and presentation would not have been possible:

Dustin Beebe, President, ProSys, Inc.

David Solberg, Associate Engineer, ProSys, Inc.

Appendix

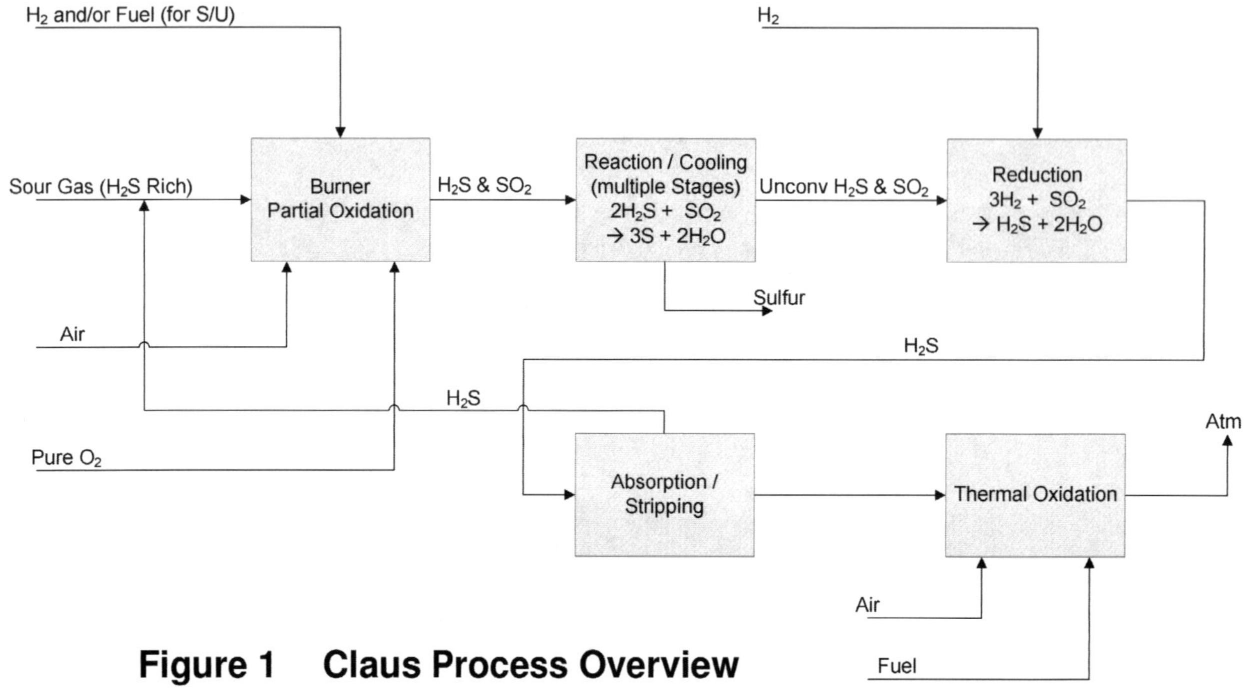

Figure 1 Claus Process Overview

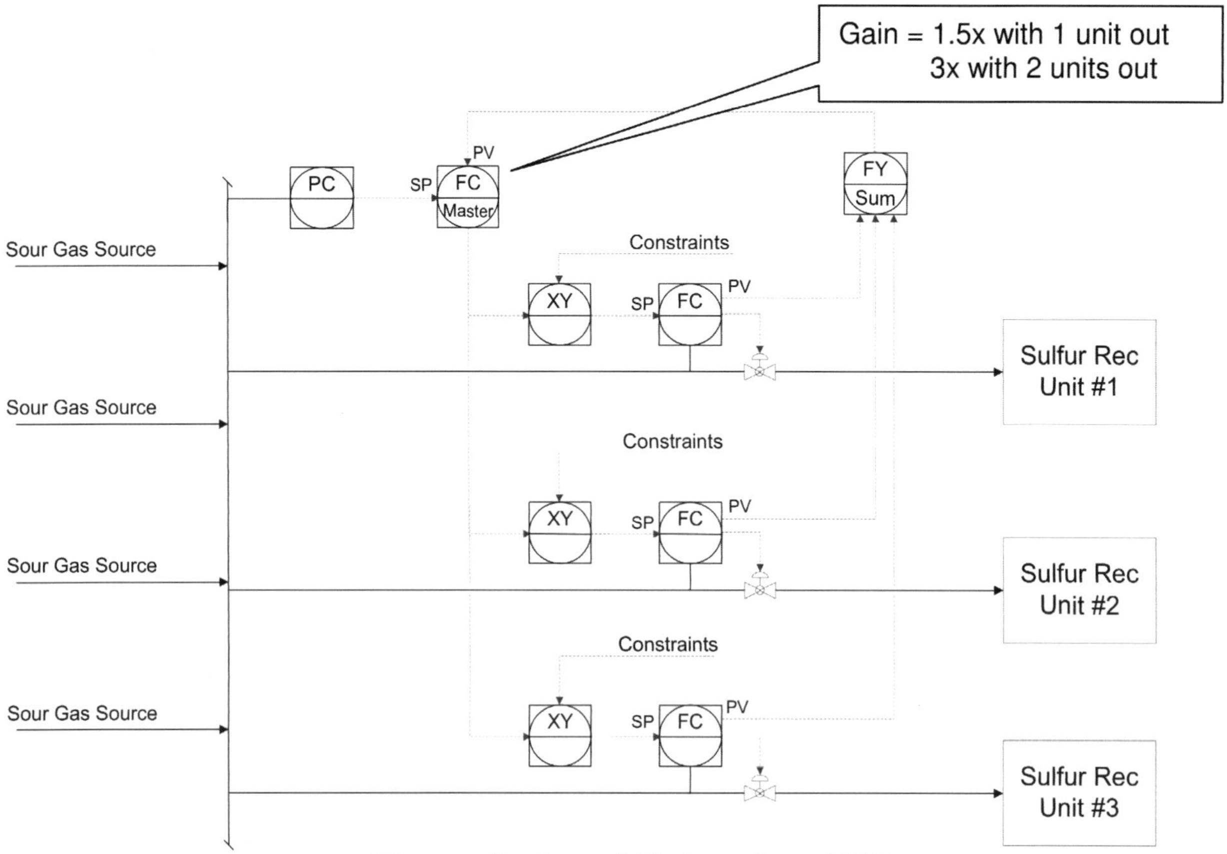

Figure 2 Load Balancing ARC

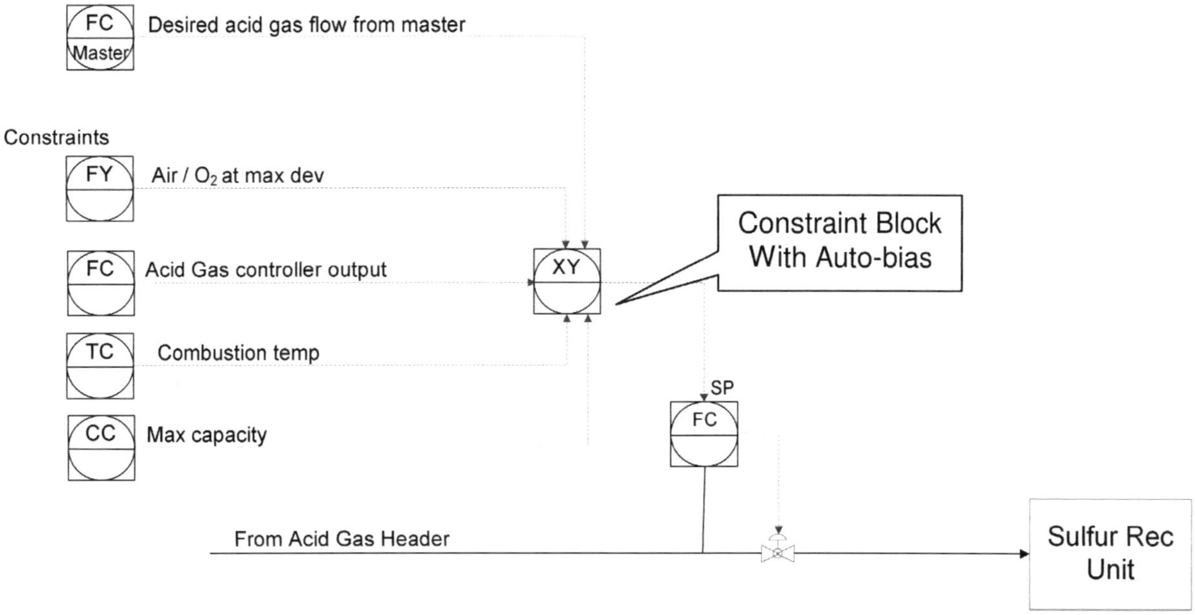

Figure 3 Constraint Configuration

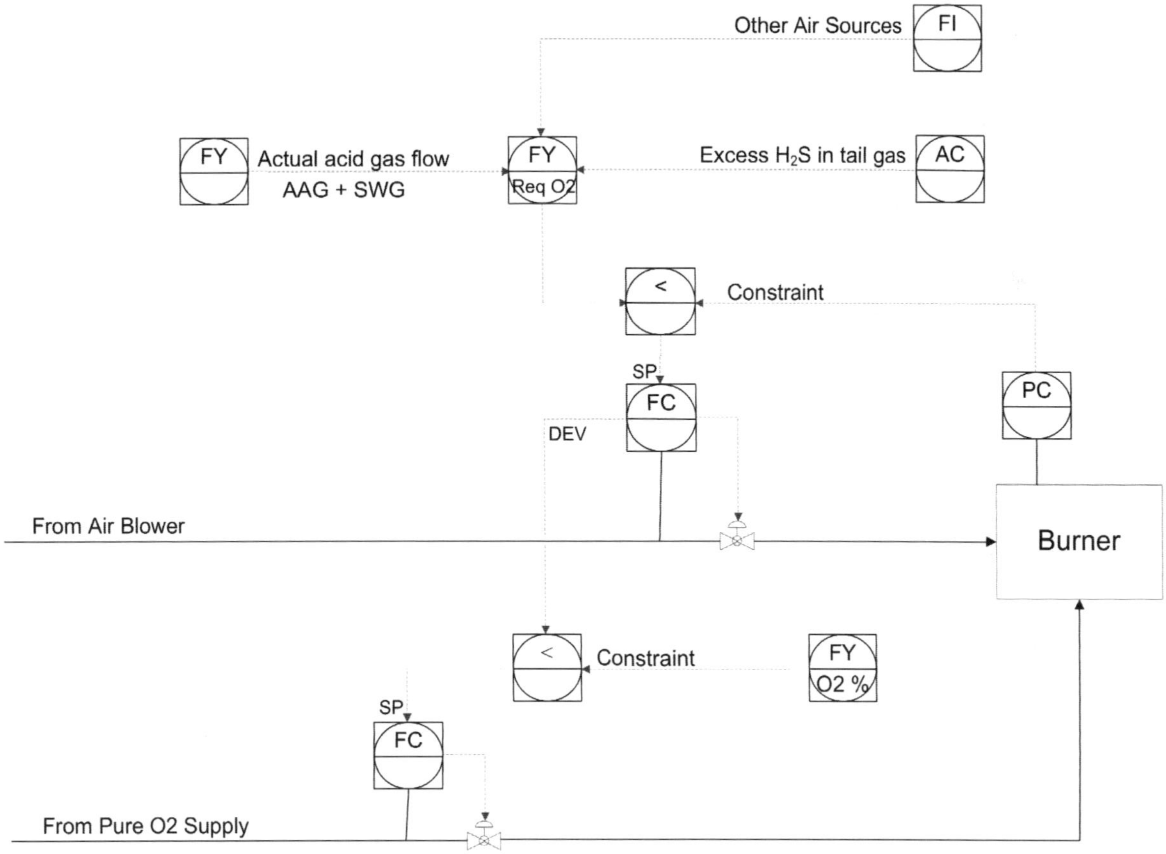

Figure 4 Air / Oxygen Control

Sulfur Recovery and Tail Gas Treating Process Gas Analyzers
Stigma, Sins and Salvation

Randy Hauer
AMETEK Process Instruments, USA

Zaheer Juddy
Analytical Instrumentation & Maintenance Systems (AIMS), UAE

John Sames
Sulphur Experts International, Barbados

Presented at the AIChE Spring Meeting
Houston, Texas
April 1-5, 2012

Abstract

Process Engineers have a strong inclination for qualitative and quantitative analysis of process streams. On-Line process gas analyzers are a fundamental requirement for control, optimization and energy efficiency but have a stigma with end users and maintenance managers, sometimes well earned. On-Line process gas analyzers comprise a relatively small share of the capital in a grass roots project but they require detailed attention if they are to be successfully implemented and fully exploited.

The paper discusses practical aspects of the suite of analyzers utilized on a SRU / TGTU complex as well as "best practice" recommendations based mostly on painful experience. The subject is one of the least understood facets of a project. Salvation can be found in the many small details and retaining process analyzer specialists at all phases of a project.

1) Introduction

It would be taken as an article of faith that a chemical engineer and an analyzer engineer would be linked together at every phase of a project. A chemical engineer by definition is most interested and has a large stake in qualitative / quantitative analytical data, the material balance is everything. Sadly this is not usually the case; we communicate through several layers, separated by a caste system.

The objective of the first part of this paper is to give an audience of primarily process design engineers a snapshot picture of process analytics. The authors are grateful for access to two technical papers from a major refining-chemical company who shared their "sins & stigma" experience, this part of the story should only come from an end user.

The second part is a review of analyzer best practices and process analytical innovation for SRUs and TGTUs covering five analyzer applications in some detail. Many of the innovations and enhanced features should be implemented at the FEED, EPC or systems integration (SI) stage but are often not. Better, earlier and direct communication between the process and analyzer disciplines is the solution.

2) Overview of the Process Analyzer Industry

It is worthwhile for a process engineer to have an idea of the breadth and scope of the process analyzer industry. The spend on an individual project is relatively small but the impact far outweighs the cost.

The Big Picture:

- The global process control enterprise is ~USD 136 billion/yr. Process Analytical Instrumentation (PAI) comprises only 6% of this amount, (~USD 8 billion/yr).
- Of this 8 billion the chemical process industries (CPI) accounts for ~70% of all process analyzer applications; further broken down into; USD 2.85 billion on maintenance, USD 1.75 billion on analyzers, USD 560 million on systems integration and USD 450 million on sample handling systems per annum.[1]

The Project Picture:

- According to a major international oil company, integration represents 55% and analyzers 45% of the total cost of an analyzer project. This somewhat contradicts the overall industry figures but can be explained by the fact the CPI spend more on shelters, no doubt influenced by hazardous area (explosion proof) design.[2]
- The 15-year cost of ownership of an analyzer system is equal to the total capital cost of the fully integrated system. Half of this cost is labor, the other half is parts. Of the half invested in parts, ~25% is consumables and ~75% replacement parts.[3]
- Shelter costs for chromatographs are on-par with the cost of the GC and this is when they are ganged eight to a shelter. This is the rationale for locating several analyzers in one shelter, which at times is a source of problem in itself.

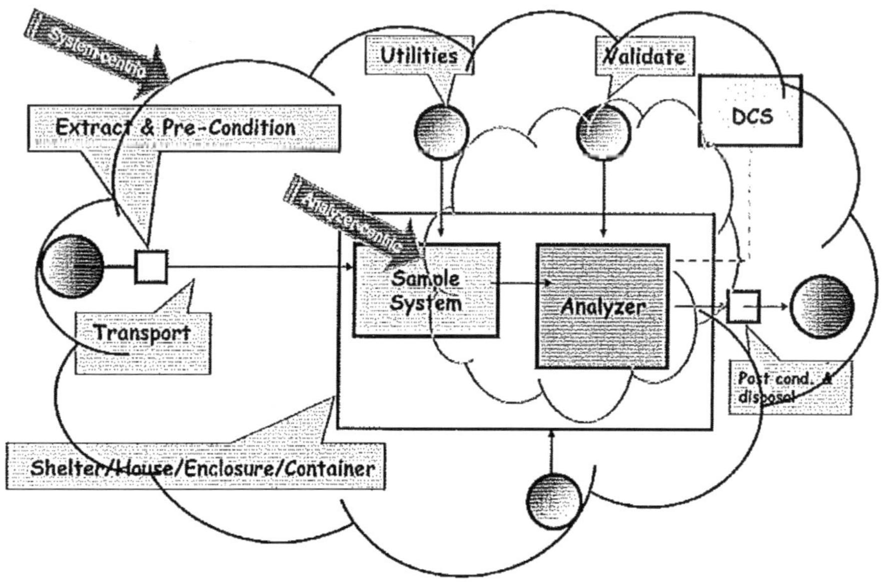

Fig. 1. Analyzer System Scope of Supply (Courtesy of Rob Dubois, "*by-line analytical*")

Trends and Generalizations in the Process Analyzer Industry:
The industry is characterized by widely diverging attributes. It is generally conservative about adopting new technologies, if it works, repeat, repeat, repeat.[1] On the other hand there are significant advances in analytical technology that are game changers in themselves. Here are some general trends;

- There is a revolution in spectroscopy with multi-component measurement capability competing with GCs.[2] Analyzers that are close-coupled to the process requiring very little integration are becoming common. Size and weight matter.
- The price of the shelter and HVAC dominate the price of an analyzer system. It is uneconomical to supply a shelter for 1 or 2 analyzers and large refining / chemical companies are moving away from GCs and analyzer houses for this reason
- One third of all analyzer systems are over-designed; one third are under-designed; perhaps one third are adequate.[4]
- The full capability and features of a process analyzer are rarely utilized. For example; over-range measurement of H_2S/SO_2 in tail gas, COS/CS_2 in tail gas, COS, (CS_2) in TGTU absorber off-gas and Ethernet web-enabled communications.

3) Stigma, Sins, Salvation

The Stigma
Process analyzers suffer a reputation of stigma even from the perspective of dedicated end user analyzer professionals. This is a list of the common complaints.[3]

- High cost (shelters), high maintenance
- Complicated installation, complicated operation
- Low reliability, questionable accuracy

The Sins

Part of the problem is one of experience at the design and EPC level of a project and the amount of time it takes to acquire adequate engineering skills to be addressing the variety of disciplines involved. Most qualified analyzer engineers are employed at the systems integration level and relatively few at the EPC level and practically none at the front end engineering design level. Some examples of how this impacts a project;

- If analyzer engineers had to pick one single problem area that is universal it would be piping design. Piping design done in the absence of a staff analyzer engineer is doomed for problems and rework.
- Instrument data sheets for analyzers that are out of date, 10 years or more, with only minor revisions in between.
- No provision for advances in process analytics, the temptation to define a measurement using a single analytical principle, Gas Chromatography being an example. The process GC is quite simply over applied as a default when a GC vendor is doing the systems integration.
- Critical evaluation of sample system design for specific applications is lacking. Most sample systems are designed based on duplicating previous projects with new features added haphazardly.[3]
- Relative to spending on DCS and discrete devices there are proportionally many more instrument and DCS engineers than analyzer engineers at the EPC level.
- Modern analyzer communications are sadly lacking in an SRU-TGTU project.[4]

The Salvation

The solution lies in retaining an analyzer professional at the EPC and FEED level during a project, any project of any size that has analyzer tags above the "1 to 5" level in terms of complexity (see below). The client must make intervention mandatory. This cannot be someone nominated on an ad hoc basis but rather a recognized full time specialist.

Why is this a problem? No one actually graduates as an analyzer engineer. It is a profession populated by chemical, electrical, instrumentation and mechanical engineers supplemented by physicists, chemists and spectroscopists who migrate to the profession. The skill set can take ten or more years to acquire and demand outstrips supply.

Following are the metrics used by a major oil company based on a three year statistical study of over 10 refinery, oil & gas and sulfur recovery plant complexes.[6] It serves to demonstrate complexity and the nominal maintenance requirements for process analyzers.

Complexity Factor	Type of Analyzer	Estimated Man-hours/month Maintenance
1~5 (Simple)	pH, conductivity, gas detection, O_2	2
6~8 (Physical Property)	Boiling point, flash point, freeze point, RVP, viscosity, etc	3
9 (Environmental)	CEMs SO_2, CO, H_2S, Opacity,	2.5
10~15 (Complex)	Tail gas, GC, Mass Spec, NIR, FTIR	4

Table 1. Grouping of Analyzer Categories for Maintenance Purposes

4) Sulfur Recovery and Tail Gas Treating Analyzer Applications

Following are the four principal on-line analytical measurements in a SRU-TGTU complex as well a new application being driven by safety. Two other applications (sulfur pit vent gas and SO_2 breakthrough to the TGTU) are not specifically covered here

4.1) SRU Acid Gas Feed (H_2S / HC / CO_2 / NH_3)

Feed (acid) gas analyzers are widely applied if not well-understood or fully implemented. Approximately 15% of existing SRUs and slightly more from new design have an H_2S analyzer and now more common, a combined analyzer for H_2S, THC (total hydrocarbon) NH_3, and CO_2. Some analyzer solutions have been over-designed and relatively few have been put into closed loop control.[7] Recently a major SRU engineering designer introduced a complete system including control algorithm and analyzer to achieve the desired result of continuous closed loop control realizing a 0.2% long term improvement of recovery efficiency (15% reduction in emissions) and mitigation of HC upsets to the process.[8]

Solutions have been developed that combine UV and IR detection to measure the H_2S and total hydrocarbon as well as the CO_2 content thereby providing an almost "full analysis". At first glance, a total hydrocarbon (THC) result would have limited value for feed forward control, however, infrared measurement can quantify the carbon count up to C_6. The utility lies in taking the THC result, as total carbon count (methane equivalent) and converting it to "air demand" using the same scale so it can be implemented into the control loop.[11]

Feed gas analysis has possibly the most benefit for the TGTU. It is somewhat counterintuitive, but it is not the sudden appearance of hydrocarbons that makes for problems, but the sudden disappearance. In the absence of feed composition information, the first indication of hydrocarbons in the acid gas is H_2S in the tail gas spikes up, SO_2 goes to zero; the control response is to "add air". Conversely, when the hydrocarbon episode suddenly ceases, SO_2 in the tail gas spikes up and this potentially causes damage to the TGTU amine since it is difficult to reduce the air before SO_2 breaks through the reduction reactor in the TGTU. The ability to anticipate the end of an HC episode at the front end before it transits the SRU to the TGTU is valuable in itself.

Best Practices

- The measurement has to be real time (not a GC). Speciation of the HC is neither required nor possible given the real time constraint.
- Sample system: Make it "hot-wet", it is not desirable to remove the water and in fact it adds an unmeasured bias by reporting the analysis on a dry basis.
- Sample system safety: Buy a solution that incorporates return of the spent gas to process, purging with nitrogen for maintenance purposes. If it frightens the analyzer techs it will fall into disrepair.
- Sample point: As far upstream from the reaction furnace as possible, preferably 5 seconds or more of transit time which approximates the analyzer response. For simplicity and response place the analyzer in close proximity to the acid gas line.
- Sample point safety: Do not place the analyzer in a cabinet and certainly not in a house. The analyzer should be IP-65 for outdoor installation and free ventilation.

- Indicating control is sufficient to realize most of the benefits. A method of simple binary control (e.g. add +5% air for HC upset of x magnitude) can be implemented. Closed loop control with safeguards and customization can be purchased as a package.[8] Operator training is essential so the analyzer is "understood".

4.2) SRU Tail Gas (H_2S / SO_2 / COS / CS_2)

With relatively stable feed compositions, flows and properly operating feedback control of the combustion air flow, control of the excess air to within +/- 0.5% is generally achievable resulting in typical losses of 0.2% for a three stage SRU. Unstable feeds and poor or non-existent control can result in significant losses of 10% or more in the worst cases. The tail gas analyzer, feedback control accounts for up 2% to 4% of recovery efficiency and can contribute more to overall SRU performance than a third converter.[10]

Unlike an acid gas (feed gas) analyzer a case does not need to be made for a tail gas analyzer in the FEED they are considered mandatory. There are two types of tail gas analyzer design; close coupled ("top of the pipe") and sample line type)

Best Practices

- Heat integrity is everything: All too often the sample point has excessive surface area or mass that defeat the electrical heating function of the analyzer (restricted to T-3 norms). Steam jacketing, or ControHeat™ is essential if the nozzle is greater than ~6" in length. MP steam is much preferable to wet LP steam which is better described as a heat sink rather than a heat medium.[11]
- Safety: Do not allow a piping mistake to restrict reasonable access to the sample point. Make this a Haz-Op check list item reviewed by analyzer maintenance.[12]
- Safety: It is unforgiveable not to connect the tail gas analyzer (all SRU analyzers) to the Ethernet analyzer communications system and yet this is very seldom done.
- Vertical tail gas pipe: When a SRU-TGTU complex is designed as a single unit, seldom is consideration given for a proper sample point on the process pipe connecting the final condenser to the inlet of the TGTU. Mounting the tail gas analyzer on a vertical section of pipe is possible but not desirable. Vertical piping sections of 18" or less are totally inadequate for even a sample line based analyzer
- If the SRU is a sub-dewpoint, selective oxidation (Superclaus) type or utilizes TiO_2 catalyst then the tail gas analyzer should be fitted to measure COS / CS_2
- The indicating outputs of [%H_2S] / [%SO_2] should be ranged at least 0-5% so they can be used during upset conditions by the operators and give them confidence the analyzer is operating. This simple feature is never implemented at the EPC stage; we remain slaves to the instrument data sheet first configured circa 1970.[13]
- Moving a tail gas analyzer upstream: It can be done for compelling reasons (turndown, selective oxidation) but be aware of vibration and inline burner issues. Be sure to include the vendor before relocation of the analyzer.

Trends

- The majority of tail gas analyzer installations are top of the pipe (70%) as they save money on shelter and infrastructure. That said, climate, analytical performance and structure constraints often require a sample line based analyzer, in these cases do not make the sample lines longer than 30' if at all possible.

- Hot condensate back flush option for ammonia salts should be selected for all SRUs burning SWS acid gas. It may never be needed but invaluable when it is.
- For safety purposes many some users require "double block" isolation for the tail gas analyzer. This is possible with both types but with some compromise. Do not consider "bleed" with the double block, S_V complicates this and is vented back to process in any case.
- Simple shelter or cabinet: the cost of a single house for one analyzer is prohibitive.

4.3) TGTU (H_2 / H_2S / COS / CS_2)

The most important measured parameter in an amine TGTU is the hydrogen content in the gas exiting the reduction reactor. Good control and measurement of excess H_2 is important to provide stable operation and protect the amine from SO_2 breakthrough. H_2S is a secondary measurement and H_2S can be measured at one of two sample points (or both) depending on the design and requirements. The typical two sample points are the top of the quench tower and top of the absorber column. In addition, the measurement of COS can be made at the top of the absorber.

Up until ~1995 the process GC was the most widely applied analytical method to monitor hydrogen content. Although a GC gives accurate results, it has several disadvantages not the least of which is the technique is non continuous. The TCD detector does require separation to isolate the H_2; in effect it can be treated as a binary mixture. The application was first addressed with discrete TCD and UV analyzers, and later combined the two devices into a single "hot-wet" measurement with an application specific sample system.[14]

Best Practices

- Select a common sample point for the H_2/H_2S, for this combination it does not matter if it is the outlet of the quench or the absorber. If the design has H_2S measurements at both locations consider a redundant H_2 sensor, the cost is minor.
- If there is a need to measure COS/CS_2 the sample point needs to be at the absorber outlet and in fact the additional measurements are only a marginal price addition.
- Locate the analyzer as close to the sample point as possible but also utilize the pipe run to minimize sample line length and locate the analyzer close to grade.
- The sample system has the same requirements as the feed gas analyzer in terms of safety. The H_2S is one (to three) orders of magnitude less vs. acid gas but "frightening" just the same.
- If (environmental) procedures allow for the quench and absorber to be bypassed at start up a temporary sample point at the quench inlet can be designed in for the purpose to stabilize the SRU (H_2) before bringing the quench / absorber on line.[14]
- Amine additives can interfere with the TCD, the solution is to add a Glysorb filter
- As for any SRU analyzer, run Ethernet connection for safety purposes.

Trends

- Of the ~130 analyzers installed by a vendor~110 are at the top of the absorber, the balance at the top of the quench with a few at the inlet to the quench.
- Of the ~110 units installed at the top of the absorber only ~10 have COS, 2 have CS_2 as well, at least one installation is using the H_2S+COS+CS_2 as a pseudo CEM.

- Replacement of process GCs with a combined UV/TCD analyzer is prevalent; it is a mistake to replace a legacy GC with GC.
- SO_2 can be monitored before the quench (in combination with H_2). It requires water removal and has not been widely applied.

4.4) Continuous Emission Monitors (SO_2 / H_2S / Mass)

Depending on the process configuration, SO_2 emission can vary from ~8,000 ppm for a 3-stage Claus SRU; ~1,500 ppm for enhanced Claus; ~100 ppm for standard design TGTU; to single digit ppm SO_2 in some local air quality districts. The measurement standards vary slightly with most SRU/TGTUs in the U.S. measuring SO_2 on a "cold-dry" basis according to the EPA part 60 guidelines. There are alternatives to measuring on a cold-dry basis that are well suited to SRUs and some environmental jurisdictions consider an SRU to be a special case in this regard.

Best Practices

- Consider the CEM to be a specialty item. Consider it to be included in the suite of SRU-TGTU analyzers, not a commodity to be lumped in with all other CEMs
- What sets it apart is incineration of sulfur compounds and how they are measured. Formation of SO_3 is not uncommon and the sample system needs to be designed for this service.
- If the environmental regulations call for SO_2 to be measured in mass terms then the analyzer must use a "hot-wet" technique and not a "dry basis". It is fundamentally wrong to measure flow on a wet basis, SO_2 on a dry basis and use the product of these two parameters to calculate mass flow of SO_2.
- Consider dual ranges (up to 100:1 is possible) if the bypass of the TGTU or enhanced Claus process requires this, it is possible in one analyzer

Measurement of Residual H_2S in the Incinerator Effluent

It is well-known and accepted that equilibrium values of reduced sulfurs exist after incineration with the assumption these are negligible if the incinerator is operated according to design. Proposed US-EPA legislation, [sub part J(a)] was going to require measurement of residual H_2S for refinery SRPs (sulfur recovery plants, defined as SRU +TGTU), but the law was not been promulgated and is currently on hold. In addition, some gas plant SRUs, most notably in Alberta, Canada, have been allowed to increase the amount of reduced sulfur compounds in the interest of saving fuel gas in the incinerator and are required to measure un-combusted H_2S.[15]

There are several ways to make the measurement. Where the H_2S is present in higher concentrations such as the energy optimized incinerators in Alberta, H_2S is (and should be) measured separately from the SO_2. Where the incinerator is optimized for destruction of reduced sulfur compounds after a TGTU, the consensus method has been to oxidize any un-combusted H_2S and other S compounds and residual sulfur and measure as total SO_2 thereby "accounting" for the H_2S. By oxidizing, the analyzer sample system essentially completes 100% of the oxidation job of the incinerator. Some users have reserved the option of segregating H_2S and measuring separately from SO_2 which can be done, but the challenge of resolving 7-8 ppm H_2S in a background of ~150 ppm SO_2 adds complexity. In-situ TDLAS laser techniques have been applied to this application with some success.

4.5) Process Combustion Stoichiometry (O_2 / CO) for Start Up

During start-up (shut-down), as an SRU transitions through fuel gas warm-up to the introduction of acid gas, measurement of O_2 is required. The requirement for most refineries is to start up the entire sulfur recovery plant (SRU + TGTU) in a single sequence with zero tolerance for emission exceedences. Historically, operators have manually taken samples using a portable electrochemical type O_2 analyzer. While giving more or less satisfactory results, the requirement of more stringent operating limits and hazard exposure are reasons to consider a permanent solution.[16]

For this application the most favorable detection principle is tunable diode laser spectroscopy (TDLAS) which measures "actual O_2" and TDLAS has become a favored method of "process" oxygen analysis. Procedures vary as to the amount of free oxygen, recommended during a fuel gas warm-up burn but the strategy should result in no more than 0.25% residual O_2. CO is a very useful parameter for combustion stoichiometry, an absorber in the infrared spectrum and can be integrated into the O_2 analyzer.

The sample handling requirements are similar to a tail gas "ratio" analyzer. While the measurement is not required after acid gas has been introduced and in steady state, there is a period to which this operating mode must be anticipated and the sample system is designed for this duty. A single analyzer can switch between SRU trains or sample points depending on requirements .[16]

5.0) Conclusions, Recommendations

1. As a process design engineer you need to be in direct communication with process analyzer specialists and vendors (with your analyzer engineer sitting next to you).

2. Retain a permanent and dedicated analyzer professional at the EPC / FEED for all projects. The resource for this is www.ADSymposium.org (ISA Analysis Division).

3. Utilize the expertise of specialized vendors at an early stage of FEED design as well as EPC detailed design before the award of a Systems Integration contract. Do not allow good design up front to be subordinated to bad decisions at the end.

4. Be an informed consumer; be wary of (over) applying process GCs on tags where they do not belong. Stay actively involved with SI decisions, SI process and avoid "change order" costs this way.

References:

1) Walton, Stephen., T. McMahon, J. Tatera, "PAI 2009-2012, Analysis & Forecast of Global Markets for Process Analytical Instruments Products & Services", 55th ISA Analysis Division Symposium, New Orleans, Louisiana, April 2010

2) Walton, Stephen., T. McMahon, "Survey of Process Analyzer Systems Integrators", 2010.

3) Podkulski, D. and J. Gunnell." Stop Buying Analyzers". International Forum for Process Analytical Chemistry, January 2011, Baltimore, Maryland

4) Novak, D., "Process Analytics: Are There Dinosaurs Among Us? - The Stigma of Process Analytics". Panel Discussion, ISA Automation Week Houston TX, October, 2010

5) Dubois, R., "Process Analytics: Are There Dinosaurs Among Us? - Myths and Mistakes That May Contribute to Our Extinction". Panel Discussion, ISA Automation Week Houston TX, October, 2010

6) Prakash Rajagopal, Chevron USA. Conversation(s), January-February 2009.

7) Harris K. and R. Hauer. "Measuring Total Hydrocarbon and H_2S in Amine Acid Gas Streams." *Hydrocarbon Processing Magazine*, February 2009.

8) Henning A. and G. Molenaar. "Next Generation SRU Control with ABC+" *Sulphur Magazine*, no.326, (January/February 2010).

9) Myles T, S. Langridge, N Truong "The Advantages of Combined Infrared and Ultraviolet Spectroscopy for Process Measurements." 57th ISA Analysis Division Symposium, Anaheim, California, April 2012

10) Bohme, Gerald and John Sames. "The Seven Deadly Sins of Sulphur Recovery." Paper presented at the Sulphur Conference, Calgary, Alberta, Canada, October 1999.

11) Al-Misfer, A., S.Vedula, R. Hauer, Z. Juddy, "Process Analyzer Best Practices for Sulfur Recovery, Enhanced Claus and Tail Gas Treating Applications". SOGAT Conference, Abu Dhabi UAE, March, 2009

12) Harris, P., "Sample Handling Systems". O'Brien Analytical Users Conference, San Diego CA, February, 2010.

13) Hunt, C., J. Sames, R. Hauer. "Understanding Claus Process Upsets Using Your Tail Gas Analyzer," *Sulphur Magazine*, no.256, (May/June 1998).

14) Harris K., R. Hauer, D. Potter, P. Harris, and B. Lewis. "Field Experience with a Single NDUV + TCD Analyzer on Amine Based Tail Gas Treating Units." Presented at ISA Analysis Division Spring Symposium, April 2007.

15) Klint, B. "Incinerator Optimization/Stack Top Temperature Reduction."

16) AMETEK Process Instruments. 2008. *Continuous Measurement of Process Oxygen for Claus Sulfur Recovery Unit (SRU) Start-ups and Shutdowns.* Pittsburgh, PA.

MISAPPROPRIATION OF TRADE SECRETS BY DEPARTING EMPLOYEES OR OTHER AGENTS

Teresa J. Lechner-Fish
Chris P. Perque
Gardere Wynne Sewell, LLP
Houston, TX
www.gardere.com
Telephone 713.276.5500
Email tlechnerfish@gardere.com
Email cperque@gardere.com

I. INTRODUCTION

This paper discusses a common business tort: misappropriation of trade secrets by departing employees or other agents. A trade secret may be any information that can be used to economic advantage. However, to be entitled to trade secret protection, such information must be maintained as a secret. Valuable proprietary information can be lost when employees or other agents depart to a competitor. Accordingly, protection of such proprietary information must begin at the onset of the employee's or agent's employment and continue throughout the employment. Protection and enforcement of trade secrets is discussed for various jurisdictions with an emphasis on recent cases in California and Texas.

II. MISAPPROPRIATION OF TRADE SECRETS BY DEPARTING EMPLOYEES OR OTHER AGENTS

Most states have adopted some form of the Uniform Trade Secret Act (UTSA), however, Massachusetts, New Jersey, New York and Texas have not adopted the Act. A few states including Texas and New York refer to the Restatement of Torts.

Preliminary injunctions may be obtained in special circumstances to prevent irreparable harm to the trade secret owner. Examples of recent preliminary junction cases in California and Texas are discussed below.

A. California

Like most states, California has adopted the Uniform Trade Secrets Act. *See* California Uniform Trade Secrets Act, CAL. CIV. CODE §§ 3426 *et seq.* Under the California Uniform Trade Secrets Act,

> (b) "Misappropriation" means: (1) Acquisition of a trade secret of another by a person who knows or has reason to know that the trade secret was acquired by improper means; or (2) Disclosure or use of a trade secret of another without express or implied consent by a person who: (A) Used improper means to acquire knowledge of the trade secret; or (B) At the time of disclosure or use, knew or had reason to know that his or her knowledge of the trade secret was: (i) Derived from or through a person who had utilized improper means to acquire it; (ii) Acquired under circumstances giving rise to a duty to maintain its secrecy or limit its use; or (iii) Derived from or through a person who owed a

duty to the person seeking relief to maintain its secrecy or limit its use; or (C) Before a material change of his or her position, knew or had reason to know that it was a trade secret and that knowledge of it had been acquired by accident or mistake.

'Trade secret' means information, including a formula, pattern, compilation, program, device, method, technique, or process, that:

(1) Derives independent economic value, actual or potential, from not being generally known to the public or to other persons who can obtain economic value from its disclosure or use; and

(2) Is the subject of efforts that are reasonable under the circumstances to maintain its secrecy.

See § 3426.1(b) and (d); *see also ReadyLink Healthcare v. Cotton*, 24 Cal. Rptr.3d 720, 727-28 (Cal.Ct.App. 2005).

In *ReadyLink*, the plaintiff was a licensed healthcare service provider with a nurse staffing division, which provided healthcare facilities and professionals with nurses. *See ReadyLink*, 24 Cal.Rptr.3d at 723. The plaintiff recruited, interviewed and hired nurses as its own employees and assigned them to its customer healthcare providers who requested nurses. *Id.* The plaintiff invested time, money and effort in developing its database of nurses and healthcare provider customers. *Id.* at 723-24. The database was used to recruit nurses and to service its clientele. *Id.* at 723.

The plaintiff also employed a sales force to recruit nurses, secure new business and service existing customers. *See ReadyLink*, 24 Cal.Rptr.3d at 724. Plaintiff hired the defendant, and, about 6 months later, it moved him into a nurse recruiting agent position. *Id.* The defendant signed 1) upon his employment, an employment agreement and an employee confidentiality, non-disclosure and non-circumvention agreement; and 2) upon his move into the nurse recruiting agent position, an agent agreement and another proprietary and confidential information, non-disclosure and non-circumvention agreement. *Id.*

About 14 months later, the plaintiff fired defendant for stealing company records containing proprietary and confidential information. *See ReadyLink*, 24 Cal.Rptr.3d at 724. While employed with the plaintiff, the defendant was caught on videotape entering company private offices at night and copying payroll practices, confidential documents and notes. *Id.* The defendant authorized a search of his residence and storage locker. *Id.* During the search, the police recovered 1-2 boxes of plaintiff's stolen documents. *Id.* Although the police investigated the theft, the defendant was not criminally prosecuted. *Id.*

Upon termination of his employment, the defendant signed a declaration acknowledging 1) his misappropriation and improper use of plaintiff's proprietary and confidential information, 2) his attempt to form his own company in competition with plaintiff, and 3) his improper solicitation of plaintiff's employees to staff his company and investors to finance his company. *See ReadyLink*, 24 Cal.Rptr.3d at 724. A mere two weeks after his termination, the defendant began working at Registry Network, a competitor, despite agreeing in his declaration to not be employed at any other healthcare staffing company. *Id.* The plaintiff brought suit against defendant and Registry Network for, *inter alia*, violating the California Uniform Trade Secrets Act, alleging defendant stole confidential information regarding the company's finances, clients, employee contacts, payroll practices and business methods.[1] *Id.* at 724-25. The plaintiff requested injunctive relief, as well as damages, prohibiting the defendant from disclosing the plaintiff's proprietary information to Registry Network and from soliciting plaintiff's employees and customers. *Id.* at 725. The trial court granted a preliminary injunction. *Id.*

Upon interlocutory appeal, the defendant contended that the trial court abused its discretion in granting the preliminary injunction. *See ReadyLink*, 24 Cal.Rptr.3d at 722 & 726. The Court of Appeals affirmed the preliminary injunction order, as modified. *Id.* at 723 & 730. The Court stated:

> ...***in deciding whether to issue a preliminary injunction, a trial court weighs two interrelated factors: the likelihood the moving party ultimately will prevail on the merits, and the relative interim harm to the parties from the issuance or nonissuance of the injunction***.

Id. at 726-27 (quoting *Whyte v. Schlage Lock Co.*, 101 Cal.App.4th 1443, 1449-50, 125 Cal.Rptr.2d 277 (Cal.Ct.App. 2002)) (emphasis added). The Court found ample evidence supporting the trial court's finding that plaintiff would likely prevail on its misappropriation of trade secrets claim:

(1) in defendant's declaration upon termination, the defendant admitted that he misappropriated proprietary and confidential information from plaintiff, and that a disclosure of such information violated local, state, and federal law, and he also admitted that, as a consequence of misappropriating such information, it would not be possible for him to engage in the healthcare staffing business without indirectly or inadvertently using or disclosing such information;

(2) in plaintiff's declarations supporting the injunction, plaintiff presented overwhelming evidence that the defendant intended to start his own nurse staffing business in competition with plaintiff using plaintiff's trade secret information, and that he solicited plaintiff's employees and nurses using plaintiff's proprietary and confidential information such as plaintiff's databases containing lists of nurses, employees, and healthcare facility customers, compilations of compensation, employment preferences, contact information, nurse applications and tests, and plaintiff's per diem program;

(3) the supporting declarations sufficiently establish that plaintiff's trade secret information has potential economic value because the plaintiff went to great expense to compile the data, and that the information would enable a competitor to recruit nurses and employees away from the plaintiff; and

(4) the defendant entered into two employment agreements in which he agreed not to use or disclose plaintiff's proprietary and confidential information.

Id. at 728. Therefore, "it is reasonably likely that the plaintiff will prevail on its misappropriation claims." *Id.* at 730.

After finding that the plaintiff would likely prevail on its trade secrets misappropriation claims, the Court reviewed the relative interim harm to the parties from the issuance or nonissuance of the

---

[1] After plaintiff notified Registry Network of the defendant's theft of its confidential and proprietary information, Registry Network terminated the defendant's employment. *Id.* at 725.

injunction, and found ample evidence that defendant posed a threat to plaintiff's business operations at the time of the injunction hearing, including:

(1) the defendant intended to create his own competing business, and he had taken significant steps to carry out his business plan, even though he had agreed not to commit such illicit acts;

(2) shortly before his termination, the defendant was caught on videotape stealing plaintiff's documents, and a search of his home and computer revealed he had misappropriated plaintiff's documentation and information;

(3) the defendant acknowledged he misappropriated proprietary and confidential information, and he solicited plaintiff's employees and nurses using the information even though he had agreed not to use or disclose the information; and

(4) right after leaving plaintiff, the defendant secured a job at a competitor and continued soliciting plaintiff's employees and customers using plaintiff's proprietary information, while simultaneously pursuing his plan of creating his own healthcare staffing business.

Id. at 732. Because the plaintiff also satisfied the irreparable harm requirement, the Court affirmed the preliminary injunction order, as modified. *Id.* at 734.

California companies should guard the secrecy of their proprietary and confidential information by marking confidential documents as "confidential to company" and storing original documents in a vault, limiting access to information on its internal computer system to persons with appropriate system identification and passwords, requiring persons entering the company premises to identify themselves and to wear a badge, checking materials in and out of the premises, and using confidentiality agreements with employees upon employment and, again, upon reassignment to another position.

For particularly sensitive information, companies should use video cameras to record activity in areas where confidential information is kept, or where copiers are located. If an employee is caught stealing such information, the employer should require the employee to sign a declaration admitting to wrongdoing and agreeing to non-compete provisions upon termination.

California employees should avoid accessing their employer's proprietary and confidential information without its express or implied consent to do so.

B. Texas

Texas—like New York—has not adopted the UTSA—it applies common law referring to the Restatement (First) of Torts. To establish misappropriation of trade secrets, a plaintiff must show:

(1) existence of trade secret;

(2) trade secret was acquired through a breach of a confidential relationship or was discovered by improper means; and

(3) defendant used the trade secret without the plaintiff's authorization.

See IAC, Ltd. v. Bell Helicopter Textron, Inc., 160 S.W.3d 191, 197 (Tex.App. –Fort Worth 2005). Texas courts have adopted the definition of a trade secret from the commentary to Section 757 of the Restatement of Torts:

> Any formula, pattern, device or compilation of information which is used in one's business and presents an opportunity to obtain an advantage over competitors who do not know or use it.

See IAC, 160 S.W.3d at 197 (quoting *In re Bass*, 113 S.W.3d 735, 739 (Tex. 2003)); *see also* RESTATEMENT (FIRST) OF TORTS § 757 cmt. b (1939). To determine whether information is entitled to trade secret protection, courts consider the *Bass* factors:

(1) the extent to which the information is known outside the holder's business;

(2) the extent to which it is known by employees and others involved in the holder's business;

(3) the extent of the measures taken by the holder to guard the secrecy of the information;

(4) the value of the information to the holder and its competitors;

(5) the amount of effort or money expended by the holder in developing the information; and

(6) the ease or difficulty with which the information could be properly acquired or duplicated by others.

See IAC, 160 S.W.3d at 197-98 (citing *In re Bass*, 113 S.W.3d at 739); *see also* RESTATEMENT (FIRST) OF TORTS § 757, cmt. b.

In *IAC*, plaintiff was in the helicopter design and manufacture business. *See IAC*, 160 S.W.3d at 196. The defendants were in the helicopter maintenance and service business, and were trying to be in the helicopter replacement blade business. *Id.* In 2001, Jim Van Horn formed Van Horn Aviation, LLC to develop a 206B replacement blade. *Id.* Randy Stevens, principal of IAC, Ltd., agreed to manufacture and market the replacement blade. *Id.* At the time, IAC, Ltd. was a customer service facility (CSF) for the plaintiff. *Id.* As a CSF, IAC, Ltd. received proprietary data relating to the plaintiff's 206B rotor blade. *Id.* Stevens agreed to use the proprietary data disclosed by plaintiff solely for the purpose of repairing plaintiff's helicopters. *Id.*

Knowing that plaintiff's OH-58 blade is similar to its 206B blade, Van Horn purchased a Depot Maintenance Work Report (DMWR) for plaintiff's OH-58 helicopter from Newport Aeronautical Sales. *Id.* Van Horn used the OH-58/206B contour and twist data and 206B sealants and adhesives for the replacement blade. *Id.* at 197. When the plaintiff learned that defendants were making a replacement blade, it sued for, *inter alia*, misappropriation of trade secrets, and theft of trade secrets. *Id.* The trial court granted a temporary restraining order, and, about two months later, a temporary injunction. *Id.*

Upon interlocutory appeal, the defends contended that, *inter alia*, the trial court erred in granting the temporary injunction. *See IAC*, 160 S.W.3d at 197. The Texas Court of Appeals affirmed the temporary injunction. *Id.* at 203. The Court of Appeals stated:

To be entitled to a temporary injunction, the applicant must plead a cause of action and show a probable right to recover on that cause of action and a probable, imminent, and irreparable injury in the interim…

Id. at 197 (internal citations omitted) (emphasis added). The Court found that the plaintiff's evidence, when viewed in light of the *Bass* factors, supported the trial court's finding that the 206B data that plaintiff disclosed to IAC, Ltd. and the DMWR obtained by Van Horn are entitled to trade secret protection at least until a trial on the merits. *Id.* at 198. With respect to the 206B data, the plaintiff produced evidence showing the measures it takes to guard the secrecy of the data, the amount of effort plaintiff expended in developing the data, the data's value, and the difficulty with which the data could be duplicated by others:

(1) plaintiff guards the secrecy of its data by storing the originals of its drawings and specifications in a vault, posting security guards at its plants, requiring persons entering the plant to identify themselves and wear identification badges, checking material going in and out of the plant, limiting access to data on its internal computer system to persons with appropriate system identification and passwords, and using confidentiality agreements with its vendors to whom it supplies proprietary data;

(2) one of plaintiff's engineers testified that thirty to forty thousand hours were required to develop the design of the blade, and another thirty to forty thousand hours were needed to develop techniques for its production;

(3) one of plaintiff's sales and marketing employees estimated that the replacement blade market was worth approximately twelve million dollars a year; and

(4) expert testimony showed that it would not be possible to duplicate the blade specifications with the necessary degree of precision through reverse engineering.

Id. at 198. With respect to the DMWR, the plaintiff produced evidence of the measures it takes to protect the secrecy of the data contained in the DMWR and the extent to which the information is known outside of plaintiff's business:

(1) plaintiff showed that the DMWR was created using plaintiff's drawings that would have been marked "confidential to Bell" and that the DMWR stated on its face that it was not available through Attorney General publication centers; and

(2) plaintiff's evidence also revealed that the drawings used to create the DMWR were not available through a Freedom of Information Act (FOIA) request, and that Newport Aeronautical, Van Horn's alleged source of the DMWR, provided the DMWR only on a "need to know basis" and did not provide electronic access to it.

Id.

The Court also found that the plaintiff's evidence supported the trial court's finding that the defendant acquired the 206B data and DMWR through a breach of a confidential relationship or discovered it by improper means. *See IAC*, 160 S.W.3d at 199. With respect to acquiring the information through a breach of a confidential relationship, the plaintiff produced evidence of IAC, Ltd.'s probable breach of the confidential relationship with the plaintiff to obtain the 206B data:

(1) several witnesses testified, and plaintiff's Customer Service Facility Agreement showed, that plaintiff disclosed 206B data to IAC, Ltd. so that IAC, Ltd. could repair plaintiff's helicopters;

(2) IAC, Ltd. had an obligation to keep the plaintiff's data confidential pursuant to a confidentiality agreement;

(3) Stevens, the president of IAC, Ltd., was a partner with Van Horn in the defendant's 206B replacement blade project;

(4) IAC, Ltd. kept the 206B data in its library, and Van Horn had access to the data; and

(5) the contour and twist data for the 206B blade is identical to that for the replacement blade.

Id. With respect to acquiring the information through improper means, the plaintiff also produced evidence of Van Horn's likely improper means to obtain the DMWR:

(1) the DMWR was not available through Attorney General publication centers;

(2) the drawings upon which the DMWR was based were not available through a FOIA request; and

(3) Newport Aeronautics provided the DMWR only on a "need to know" basis.

Id.

In addition, the plaintiff produced some evidence that defendant used the 206B data and DMWR without the plaintiff's authorization:

(1) the defendants used contour and twist data contained in both the 206B data disclosed to IAC, Ltd. and the DMWR and that they used primer and adhesive specifications present in the 206B data;

(2) Van Horn created his certification plan to submit to the Federal Aviation Administration for the 206B project on the same day he purchased the DMWR;

(3) IAC, Ltd.'s was expressly prohibited from using the 206B data pursuant to a confidentiality agreement, and Van Horn was not authorized to use the data in the DMWR.

Id. at 199-200. "Because [plaintiff] presented some evidence to support each element of its trade secret misappropriation claim, it has shown a probable right to recover under this cause of action." *Id.* at 200.

After finding that the plaintiff would probably recover on its trade secrets misappropriation claim, the Court reviewed whether the plaintiff met its burden of showing that a probable, imminent, and irreparable injury would result before trial in the absence of a temporary injunction. *See IAC*, 160 S.W.3d at 200. Plaintiff's evidence showed that the defendants had possession of plaintiff's data entitled to trade secret protection and they were actively using that information to compete with plaintiff in the replacement blade market. *Id.* Further, plaintiff's evidence also showed that defendant's replacement blades were going to be available for shipment around June 15, 2004 and would be priced lower than plaintiff's blade:

> [Plaintiff's] Executive Director of Integrated Product Solutions testified that the creation of a lower-priced 206B replacement blade market would undermine the goodwill that Bell has with its clients, leading to an incalculable loss of business goodwill.

Id. at 200. Because the plaintiff also satisfied the imminent and irreparable injury requirement, the Court held that the trial court did not abuse its discretion by granting the temporary injunction. *Id.*

Texas companies should guard the secrecy of their proprietary and confidential information by marking confidential documents as "confidential to company" and storing original documents in a vault, limiting access to information on its internal computer system to persons with appropriate system identification and passwords, requiring persons entering the company premises to identify themselves and to wear a badge, checking materials in and out of the premises, and using confidentiality agreements with employees and vendors. For particularly sensitive information, companies should use video cameras to record activity in areas where confidential information is kept, or where copiers are located.

Texas vendors, that are subject to a confidential agreement, should guard the secrecy of the other party's confidential information as discussed above.

Chris Perque is a partner with Gardere Wynne Sewell LLP in Houston, Texas specializing in intellectual property and commercial litigation.

Teresa Lechner-Fish is an associate with Gardere Wynne Sewell LLP in Houston, Texas specializing in intellectual property and commercial litigation.

The views expressed herein are solely those of the authors and do not necessarily reflect the views of Gardere Wynne Sewell LLP or its clients.

(36b) Environmental Liability: Regulation, Compliance and Litigation Risk

Mary Ellen Ternes, McAfee & Taft, Oklahoma City, OK

Compliance with environmental regulations, while necessary and helpful in minimizing litigation risk, does not completely insulate industry from litigation. This presentation will review potential pathways towards unwanted administrative and judicial entanglements regarding which even the most compliant companies should be aware, including enforcement disputes, citizen suit actions and common law claims.

Title: Propellant Systems for Household Spray Products
Submitter's E-mail Address: cjabraham1@aol.com
Preferred Presentation Format: Oral
Keywords: Hazards, Standards and Total Release Foggers
Session Selection: Litigation Perspectives for Chemical Engineers
AICHE – Houston, TX April 1-5- 2012

PROPELLANT SYSTEMS FOR HOUSEHOLD SPRAY PRODUCTS

C. J. Abraham, Scientific Advisory Services, Ltd, Great Neck, NY

Abstract

The problem delineated in this paper involves consumers throughout the United States who use and are exposed to bug bombs or bug foggers (also known as total release foggers orTRF's). They are the most dangerous aerosol products on the market. We will focus on the defective design of the formulation, the state of technology available to eliminate the risks, the defects in the testing of the product, the false information that is submitted to the EPA and the lies stated in the application process and registration of pesticide products.

The most popular propellant system for total release foggers for indoor use consists of propane, butane and isobutene. Even though the propellent system consists of up to 35% of the ingredients, applications to the EPA for approval of insecticides designate the propellant system as inert. This designation is incorrect and fraudulent. In addition to the products being defectively designed, they contain a hidden hazard resulting in fires, explosions and traumatic burn injuries.

There are several resolutions which can make the product either completely safe or minimize the risk of consumers exposing themselves to explosions and toxic vapors. This involves alternative safer propellant systems and significant changes in the testing of the product, proper reporting of information in the MSDS's, proper testing and significant changes in the warnings.

Introduction

Over the last forty years this author has been involved with dozens of fires and explosions involving a popular propellant used in a number of products including total release foggers also known as TRF's. The propellant is composed of propane, butane and isobutene. The insecticide is purchase by the lower income segment of our population.

The last case that I was involved in was typical of what could happen with the TRF's. Two women, aged 25, and 26, were severely burned when the explosive and extremely flammable propellant from a popular insecticide product came in contact with an ignition source in the home of the mother of one of the women about seven hours after her mother released a number of canisters in her home. The women were burned over 70% to 90% of their bodies. Both women sued the manufacturer, retailer and distributor of the TRF insecticide.

Both women lived in a rural area of, Nebraska and the mother lived in Iowa. Prior to arriving at the mother's home, they all met at a store in between. The women were not informed of the canisters,. The women proceeded to the mother's residence to pick up two dogs.

Once the women arrived at the residence, they went in through the front door. The house needed cleaning up so both women decided to surprise the mother with a clean house. when she arrived sometime later.

Prior to the explosion, neither women detected any odor in the house other than the odor from the animals. After some time, both of them became exhausted and lay down on the couch. They woke up to the barking of dogs. They rose up to see what the dogs were barking about. When they were approximately three to four feet from the rear of the house, they heard the furnace "click" on. An explosion and flashover followed and both of them experienced the results and extreme heat of the flashover. As soon as they heard the explosion, one of them noticed that she was on fire. They went out the back door dropped and rolled on the ground.

It is important to note that neither of the women detected any odor of a liquid, chemical or vapor. That includes the gasoline which was located in a plastic container placed on the dryer in the washroom. There was also a very small amount of the gasoline remaining on the floor in the bathroom from a prior spill. Neither women used the hot water or adjust the thermostat. In addition, the lights in the living room, kitchen and bathroom were on when they arrived.

The mother never told her daughter or her freind that she set-off eight (8)TRF's throughout the house prior to her leaving. The mother's grandchildren were expecting to visit her the following week and they were allergic to spiders. The mother had purchased the TRF's at a "Dollar" store. The instructions suggest that the consumer can use one fogger per room. The label states that the four-can package is able to treat 2,000 cubic feet of unobstructed area. The subject house is approximately 1,200 square feet. There are no statements indicating what would happen if extra TRF's were used throughout the house.

The explosion and fire that occurred destroyed the home. Initially, the fire department did not perform a complete analysis with reference to the cause and origin. For example, if gasoline vapor was the cause of the fire, why wasn't the container of gasoline on top of the dryer involved with the fire? The fact is the gasoline container was still intact after the house was destroyed. The Fire Department concluded that the cause of the fire was the propellant from the TRF coming in contact with the ignition source of the furnace.

Background- The Product

Total release foggers can be purchased in many grocery, hardware stores and chain stores They are relatively inexpensive and have been used for several decades in lieu of using a professional exterminator. In order to properly use those products, one must disconnect all the electrical appliances. This includes refrigerators, freezers, lights and anything else that uses electricity. In addition, the furnace, gas stoves, hot-water heaters and all other sources of ignition must be turned off. It should be noted that all of the above is the responsibility of laypeople with limited or no education in this area. The total release foggers are turned on throughout the apartment or house and the area is left completely closed. The owners return a number of hours later with the expectation that the cockroaches, spiders and other insects are killed. The manufacturer's instructions state and assume that after a number of hours the previous explosive environment is safe to enter and that everything that was turned off previously can now be turned on again. Not one manufacturer informs the consumer that an explosion-proof ventilation system should be used to make sure that most, if not all, of the concentration of the explosive propellant is dissipated. The manufacturers also know that the average layperson does not have any type of gas-detection equipment, so he would not be equipped to even measure or detect explosive gases. The gases are invisible and are not odorized. Therefore, the consumer is exposed to a hidden danger that is known to the manufactures but not known to the consumers.

The Consumer Product Safety Commission (CPSC) has reported accidents involving similar propellant systems from the early 70s to the present time. One of the many cases that the undersigned has consulted on was reported in the United States District Court Decision: Allison Kowak, et al Plaintiffs vs. Fabrege U.S.A. and Precision Valve Corp., Defendants. Civ. No. CV 90-1919. Nov. 13, 1992. In addition, a demonstration was performed by the undersigned on CBS TV shortly after the appellate decision was published (812F.Supp. .942).

For many decades, thousands of explosions have occurred where total release foggers (TRF's) were used in apartments and homes throughout the United States. These explosions resulted in fires, destruction of property, burn injuries and death. They occurred mostly in apartments of the segment of our population that is made up of lower-income families that cannot afford professional exterminators.

In most instances, the TRF is composed of approximately 9% mineral spirits and 0.515% cypermethrin and a 35% combination of propane, butane and isobutane. The sole purpose of the propellant is to spread out and distribute the active ingredient, cypermethrin. In essence, the inert ingredients of the product account for 99.485% and the active ingredient accounts for 0.515%.

In addition to the above, major manufacturers, in every one of their applications to the EPA for approval purposes, indicated that their products incorporated only the active ingredients and not the propellants as the ingredients to be evaluated. In fact, the applications submitted stated that the inert ingredients, mainly propellants, amounted to 99.3% or more.

From a formulation standpoint, everything that is in the container for any specific use is the total amount of each ingredient as retailed to the public. For example, in all the applications of a TRF by more than one major manufacturer, propane, butane and isobutane should have been incorporated as part of the application process to the EPA and not designated as an inert ingredient. If it was an inert ingredient the thousands of explosions, fires and resulting injuries over the last 30+ years would never have occurred.

The Propellant

Manufacturers, in most instances, fail to place the EPA and the consumer on notice that the highly flammable and explosive gaseous propellants are invisible and much heavier than air. They also travel along the lower part of the floor unless disturbed. This creates a hidden danger to the consumers but is well known to the manufacturers. More importantly, those facts have been concealed from the EPA in the application and approval process.

What makes it more difficult to detect is the fact that the explosive formulation is invisible and not odorized. In addition, the vapor density of the propellant and active ingredients are heavier than air and tend to fall to the lowest part of a room or floor. In many instances, the vapor can actually flow down a stairway and wind up at the lowest part of the floor below which is usually the furnace. Contrary to the manufacturer's inferences, the TRF is not distributed evenly throughout a room from floor to ceiling. In addition, the system does not follow the ideal gas laws.

The flashpoint for the subject propellant, as stated on the Material Safety Data Sheet, is (minus) - 140ºF. The flashpoint of a volatile liquid is the lowest temperature at which it can vaporize to form an ignitable mixture in air. Thus, the propellant can ignite at temperatures much lower than normal room temperatures.

Propane has a molecular weight of 44.10 grams per mol as opposed to 28.97 for air so it is about 1.5 times heavier than air. Iso-butane and butane have a molecular weight of 58.12 grams per mole so it is about 2 times heavier than air. When released in air, the propane / isobutene propellant has a negative buoyancy and thus falls under the effect of gravity.

In his textbook "Evaluation of the Effects and Consequences of Major Accidents in Industrial Plants" (Publisher: Elsevier Science, 2008, Chapt. 6, pp. 195-248), Joaquim Casal discusses the flow of heavier than air gasses including propane. The difference in the molecular weight and the negative buoyancy and gravity cause the propellant to fall to the ground level and displace the air. The significant difference in the molecular weight does not allow for the homogeneous mixing of the propellant gas with air.

Aerosols released from a pressurized can will first move in the direction of release due to the momentum of the gas. (Momentum is a quantity in mechanical engineering expressing the motion of a body or system, equal to the product of the mass of a body and its velocity. If the direction of release is pointing upward, the gas will first move up until it loses its momentum under the effects of gravity. At that point, "if the material released is heavier than air, it will first descend to the ground and then spread radially" [Casal page 219].

If the heavier-than-air gas is released in an indoor space, the edges of the cloud will eventually reach the perimeter of the space and then conform to the shape of the bottom section of the space. As Casal states in his book, the mathematical models which predict the actual shape of the cloud as a function of time are complex and require a computer to solve. Most of these models do not account for obstructions in the room. Complications are added if there is any air flow in the room (from a fan, air conditioner, or open window) and if there is a temperature variation in the room from floor to ceiling (because hot air rises). What is known however, is that regardless of the exact shape of the edges of the cloud as time progresses, after several seconds of dispersion, the heavier-than-air gas will displace the air at the bottom of the room and fill the space.

The TRF industry relies upon J. D. Commerford's article entitled: "The Measurement of Propellant Obtained from the Discharge of Total Release Aerosols" published in the July, 1992 issue of the trade magazine "Spray Technology & Marketing." This article could not have been peer-reviewed because the scientific method used was incorrect and inaccurate. The author assumes that the TRFs follow the ideal gas law and the product is distributed throughout each room evenly from floor to ceiling. He also assumes that normal ventilation within all of the environments will dissipate the concentration explosive gases to render them harmless.

Dry air is treated as an ideal gas for scientific analysis even though it is made up of nitrogen, oxygen, carbon dioxide and several other trace gasses. Much of this is done to simplify the analysis and there are many chemical processes that rely on extracting the different gasses from air. Thus, in no case can mixtures of dry air and other gasses be considered to be a uniform ideal gas. As Casal states in his book and illustrates using measurement data from several sources, heavier-than-air gasses have a negative buoyancy in air. When released into the air, heavier-than-air gasses tend to fall under the influence of gravity. The uniform dispersion of the heavier-than-air gasses in an enclosed room cannot occur even after the long periods of time reported.

The heavier-than-air gasses will conform to the floor of the space, displace the air and the volume occupied will be equal to the volume of the heavier-than-air gasses. If the room has a large cross sectional area, the heavier-than-air gasses will occupy a height extremely close to the floor of the room. My expectation when I review the test data reported in the article is that the measurement equipment cannot measure the gas concentrations this close to the floor.

While *not completely incorrect*, the article is certainly misleading. For example, the author of the article says that, "the volatile propellants rapidly flash off leaving finely divided particles of concentrate" [page 29] and "the propellant flashes off into the enclosed air space" [page 28]. The use of the non scientific term "flash off" seems to indicate that the propellant somehow disappears from the room. This is misleading in that the propellant disperses but in no way disappears from the room until a window or door is opened or it seeps through some other opening.

The author of the article also states that there are three schools of thought about where the propellant goes, either falling to the floor, layering itself or being distributed uniformly distributed [page 28]. Again, this is very misleading since it implies that there is disagreement about what happens when a gas is released into an enclosed space. The real situation is that any of these occurrences are possible depending on the properties of the gas.

The author of the article also states, "Instantaneously the volume concentration of propellant in the air are not uniform, but in time the propellant is distributed to all parts of the room." [page 29] This is certainly true but he does not state the obvious implication of this, namely ***that the concentration of the propellant at a certain instant of time and at a particular location of the room can easily be between the lower explosive limit (LEL) and the upper explosive limit (UEL). At this point in space and at this instant of time, if the propellant comes in contact with an ignition source, IGNITION WILL OCCUR.***

There is also a problem with Cummerford's measurement technique in the distance from the propellant source. The sample points were positioned several feet from the source of the propellant. You can model what the exact shape of the plume is as the propellant leaves the can but for the most part, it will not be as you expect. The propellant will go upward because of its momentum as it leaves the nozzle of the pressurized can. Since the propellant is heavier than air, at some height gravity will overcome the momentum and the propellant will begin to fall to the floor. To measure the concentration of propellant where it may be between the lower explosive limit (LEL) and the upper explosive limit (UEL), you would want to put the sample locations within this plume. By placing the sample points several feet from the discharge site, the plume has a large cross sectional area and the concentrations of propellant will clearly be low.

There is a review an article entitled, "Propellant Distribution from Total Release Aerosols Discharged into an Enclosed Chamber, by Richard G. LeVasseur and J. D. Commerford which appeared in the April, 1997 issue of the trade magazine, "Spray Technology & Marketing". This is an update of the previous article for a different propellant and using a different Foxboro instrument for measurement. This article has the same problems as the previous one, namely that the measurement starts too late (although earlier than the previous article) and at a distance that will obviously yield low concentrations.

Lastly, Cummerford did not attempt to test the area with an ignition source because it would have been obvious to the undersigned that there would have been an explosion resulting in injuries to him and others. If that occurred, he would not have been able to write the article for the trade magazine.

It should be noted that the management of the defendant manufacturer used that information for their own product and was testified to be their technical director and research and development manager employed by the defendants. As stated previously,, the assumption that their product follows the ideal gas law is totally incorrect. This can also be demonstrated by taking, for example, a GLADES air freshener and expel some it into a room. The observation of the flow of the ingredients and the vapor follows Casals analysis in the referenced publication.

Chemistry Review of the TRF

A chemistry review of the documents of the manufacturer's reports that the highly flammable explosive repellent under pressure is "Combustible." The manufacturers also state that "exposure to temperatures above 130ºF may cause bursting." That is a common statement used on all products containing the propellant. The inference is that the product is not extremely flammable and/or explosive. The technical review branch of the EPA in 2006, reported that the label does not need the signal word: combustible.

The EPA relies on the expertise of the manufacturer to present a complete, accurate and true description of the chemistry, formulation and summary of the testing by the manufacturer.

ANSI Z129.1 1982 Standard titled For Hazardous Industrial Chemicals – Precautionary Labeling states that, "Each hazardous chemical must be considered individually in the light of its own characteristics and customary or reasonably foreseeable handling or use. In such cases, suitable statements should be developed to fit the situation." This is the basic standard of care that the major manufacturers failed to follow.

Since there are millions of organic chemicals and formulations, there is no government agency or international agency that is capable of evaluating each and every formulation. They rely completely on the alleged integrity and honesty of the presenter of the application and/or the manufacturer. In every case, the EPA relies on the manufacturer's integrity of their application in the approval process. They do not have the wherewithal or the facilities nor the time to test the formulations of thousands of applications. In fact, there is no government agency or international agency that is capable of evaluating each and every formulation submitted for approval.

A major corporation's research and development laboratory performed a "Container Stability Determination, Real Time and Accelerated Stability" in their EPA application. The tests also used the "Product Properties Guidelines, Group B-Physical/Chemical Properties." Even though the corporation defined it as a pressurized liquid, not once did they identify the propellant as an extremely flammable/explosive combination of gases when used in a foreseeable protocol. The manufacturers did not properly test the subject product and presented false information in their report.

In one of the cases the undersigned was involved in, the manufacturer reported that the "flammability" is "0" under the section "Hazardous Materials Identification System." The propellant was defined as a "hydrocarbon propellant" with CAS Numbers 75-28-5 and 106-97-8. These gases are a combination of a highly explosive and extremely flammable gas. In section 4 titled "Fire and Explosive Hazards Data," the defendant manufacturer designates that the- "Flammable Limits" is" N/A". One can conclude that it is highly probable that these statements in the MSDS is both fraudulent and false.

More than one manufacturer states that an odor neutralizer is used. This ingredient must also be listed as a component as part of the application submitted to the EPA. The undersigned could not find one document listing the odor neutralizer ingredient.

In addition, manufacturers' MSDS for their TRF products which are described as a total release indoor aerosol insecticide, is formulated with more than 90% of isobutane, butane and propane and is housed within a pressurized container. Under Section III, the MSDS indicates that the vapor density is greater than 1, when in actuality, it varies from 2 to 4. The defendant manufacturer also stateed under "Flammable Limits," that the product is not flammable. That communication to anyone using their products is totally false, misleading and fraudulent.

The EPA's response to all of the misleading information resulted in a suggested hazard statement by them on the label reading, "Contents under pressure. Do not store near heat or open flame. Do not puncture or incinerate container. Exposure to temperatures above 130ºF may cause bursting." There was no mention of the propellants being extremely flammable and explosive because the manufacturer never placed the EPA on notice of that fact and the consumers, with that limited information, released the insecticide throughout the areas of their residence.

Other documents submitted to the EPA by the manufacturer stated that the active ingredients were tetramethrin and cypermethrin. The total of the inert ingredients as listed were 99.30%. The inert ingredients were not defined or identified. However, the manufacturer knew that when their product was released into a room, the "inert ingredients" combined with the active ingredients and distributed the active ingredients within a specified area. In order to obtain the true toxicity of the product, the pressurized liquid concentrate should have been identified. As such, and as elaborated in the "Executive Summary" submitted to the EPA, the information would be questionable. There were many other areas in the EPA application where the flammable characteristics of the repellent, once released, were not properly reported and listed by the manufacturer as not applicable (N/A).

The EPA

The EPA has continually ignored their own experts,, as well as fire officials who are concerned about the potential dangers of explosions and/or fires caused by routine usage or storage of consumer aerosol products. Instead, the EPA responds to pressure from the manufacturers of these products who are resistant to changing warning labels. Extrapolating from data the New York City Fire Department collected concerning the increase in fires and explosions caused by aerosols, an internal EPA memo estimated an excess of 500 fires and/or explosions annually in this country."

In the 1970's, scientists discovered that the chlorofluorocarbons (CFCs) used as propellants in aerosol cans (among other things) were a leading factor in the depletion of the ozone layer. As a result, hydrocarbon propellants (i.e. propane, butane, isobutane) and ether propellants replaced CFCs in aerosol products. The down side to this alternative is that hydrocarbons are extremely flammable substances, and it appears that the public has not been sufficiently warned and educated about their safe use. The

Environmental Protection Agency (EPA) has significant evidence that the propellants can pose significant hazards in normal household use,and yet they have not revealed this information to the public. It is up to the EPA to respond to consumer protection needs. Unfortunately it appears that it reacting to pressure from aerosol manufacturers. The EPA should resist this pressure and change warning labels on many aerosol products, such as total release foggers and space-spray pesticides to warn consumers of the potential dangers.

In response to this danger and an increase in the number of accidents caused by aerosol products, in 1987 the EPA proposed new regulations to improve warning labels on them. But their proposed changes were opposed by the aerosol industry. In an EPA memo one official states, "Industry, through the CSMA [Chemical Specialties Manufactures Association], indicated that they would initiate a lawsuit against PR 87-8 [the EPA's recommended labeling requirements] if it were not rescinded." After much debate, the government succumbed to the legal threats and withdrew their proposal.

The business lobby has been able to cut off all warnings that could aid in decreasing the number of fires and explosions caused by aerosol products. An internal EPA memo states:

"Incident data from the CPSC [Consumer Product Safety Commission] indicates that 60% of pesticide aerosol incidents resulted in explosions that caused injuries requiring medical attention. . . many incidents have resulted due to abuse of the aerosol container.However, if the aerosol cans were obviously marked as 'Flammable' or 'Extremely Flammable' some of the abuse might be prevented. It should be noted that the "abuse" described is foreseeable and known to the manufacturers.

Aerosols using the highly flammable/explosive propellant are used daily in bathrooms, kitchens, and garages, In most cases, the user has no knowledge concerning the propellants in the spray can. According to an article published in Aerosol Age magazine, "Well over half of all aerosols are marked 'FLAMMABLE' or 'EXTREMELY FLAMMABLE'. . .". The article goes on to state, "90% [of aerosol products] contain flammable propellants and over two out of three contain flammable or combustible concentrates." The increase in the number of fires and explosions is do to the increased use of the combustible hydrocarbon propellants.

Aerosol containers are regulated by three different federal agencies: the Environmental Protection Agency (EPA) which regulates pesticide products, the Consumer Product Safety Commission (CPSC) which regulates non-pesticide products, and the Department of Transportation (DOT) which manages bulk shipments of aerosol containers. The DOT has become very effective in regulating this industry. Test methods and labeling seem to be a high priority at the DOT. But the EPA and the CPSC have not had the same attitude toward this issue. While the EPA is the major focus of this report, the CPSC virtually has actually ignored the same consumer safety concerns (i.e. proper testing methods and improved labeling). The EPA in the past has made some effort to look into this issue.while the CPSC has done little to solve the problems aerosols have caused. Ironically, the CPSC is the one federal agency that collects all of the reports of fires and explosions resulting from aerosol containers and still does nothing to improve the protection of consumers. One must remember that it is the CPSC, as well as the EPA and the industry itself that are at fault for aerosol products accidents caused by the innocence of an uniformed consumer.

The EPA began looking into the problem created by chlorofluorocarbons (CFCs) in the mid-1980's. In 1987, the EPA released PR Notice 87-8, "SUBJECT: Label Improvement Program - Required Labeling for Products Containing Flammable Propellants".

PR Notice 87-8 proposed to change the labeling of aerosol products by adding the words, "PHYSICAL OR CHEMICAL HAZARDS" to products that contain propane, butane, isobutane, and dimethyl ether as one or more of the propellants.

This statement was followed by an extended warning label that gave information about use, storage, pilot lights, electrical equipment, incineration and high temperatures. Manufacturers were given until July 31, 1988 to change labels on products released for shipment and July 31, 1989 for all products sold to the consumer.

In response to this initiative, industry dug in their heels. In an EPA memo one official states, "Industry, through the CSMA, indicated that they would initiate a lawsuit against PR 87-8 if it were not rescinded." After much debate PR Notice 87-8 was withdrawn on December 10, 1987, less than ten months after it was published. The EPA released PR Notice 87-11 rescinding the former, "based on the lack of sufficient anecdotal accident information to define the scope of the problem with respect to flammability of aerosols containing the aforementioned propellants". The aerosol industry cast enough uncertainty and the issue to terminate the proposed changes. The termination of the proposed labeling was to be followed by an extended period of data collection and research by the EPA and the industry.

Why has the aerosol industry taken such a stance against the EPA and the consumer? A major reason for the aversion to the previously proposed changes revolves around profits. Changing testing methods and labeling will cost the industry due to the redesigning of aerosol products. Changes would alter the current way the production process occurs and the industry would have to brar the cost of the new labels. Prior labeling would have to be replaced in a specified period of time to be in compliance with regulations and to avoid any penalties.

An EPA memo described a conversation an agency official had with an official from S.C. Johnson & Son, the manufacturers of Raid. According to the EPA official, after discussing testing methods, the industry official stated, "his company does not need to worry about the effects of flammability products since its insurance company settles on any damages or injuries." This statement was also verified by an internal EPA Office of Pesticide Programs memo, "Industry does not appear to be concerned as their insurance companies pay all claims."

How many fires and/or explosions are enough to cause the industry to change its approach to the use of aerosol products? So far it appears that the industry relies on the excuse of "consumer misuse" as the primary reason for accidents caused by aerosol products.. Industry should ask itself why this abuse occurs and how the misuse can be prevented. Instead the manufacturers keep trying to find possible ways to get around the system. According to EPA officials, ways of getting around the system are known, "With current approved testing methodology, a 'nonflammable' rating of the hydrocarbons and ethers can be achieved by proper engineering of the nozzle and the delivery system." Montford A. Johnsen, Technical Editor for Aerosol Age magazine, agrees with this statement. "The nature of the flame projection test invites the formulator to make modifications of product or package to pass it and thus avoid labeling the product as 'Flammable'." Yet consumer misuse is industry's answer to the problem, even though they have been notified of numerous accidents that have resulted because of an uninformed consumer.

Why would the aerosol manufacturers contimued conduct business in this manner? The propellants in question (propane, butane, isobutane and ethers) are some of the least expensive propellants on the market. Alternative propellants would cost more which could eventually drive up the final cost to the consumer.

By providing a full explaination of the contents and the explosion/fire potential, the consumer might become more sensitive to the product and its inherent dangers.For the last eight years the EPA has tried to re-educate the public about the safety risks that could save lives and property. We believe changes are needed before the accident databases overflow with victims. This process may cost aerosol manufactures a small amount at first but in the long run it will save lives.

However, it should be industry's responsibility to educate the public if they cannot use existing technology to make their products safer. The problem is that technology is available that will make the product safe. Therefore, the exisitng products would be considered defectively designed.

Aerosol Propellant for TRF Insecticide and Health Hazards

In addition to the propellant system for TRF insecticides being extremely flammable and explosive, users are also exposed to other health hazards. Exposure to the propellant, once it is released, can cause drowsiness, have an anesthetic effect, cause eye burns and skin burns (frostbite), central nervous system depression, asphyxiation, headache, nausea, unconsciousness, rapid breathing, visual disturbances, muscular weakness, tremors, narcosis, eye damage, mild intoxication, loss of coordination, airways dysfunction syndrome, difficulty in breathing and death. The target organs are the central nervous system, lungs, eyes and skin.

Since the vapor density of the aerosol system is much heavier than air, the propellant tends to stay as close to the floor as possible. When that occurs, there is a lack of oxygen or air in the area adjacent to the floor and no insect is able to survive in that environment. Therefore, contrary to the claims that the defendants made to the EPA that the propellant is considered an inert ingredient, it is very effective in actually suffocating insects on contact.

The use of the propellant system in killing insects by suffocation was not considered by any of the defendants. This is a serious omission and should have been divulged in their EPA application. Furthermore, all of the testing on rats and other animals should have incorporated the use of the propellant rather than using a non-pressurized formulation. As stated in another section of this report, this is further evidence of why the testing performed by the defendants on animals was defective. Their application to the EPA for approval of the efficacy of the active ingredients.was not complete and, therefore, defective as to the results reported and/or not reported.

Asphyxiants, the TRFs and Risk Analysis

Propane, butane and isobutene, for example, are asphyxiants. There is no question that the lower flammability limits of propane and air was exceeded causing both an oxygen-deficient and explosive atmosphere. Butane is also heavier than air as well as isobutene. Inhalation of this combination of gases may cause rapid respiration, dizziness, fatigue and nausea. The manufacturers could have looked up these symptoms on the Internet. There is no question, in the opinion of this author, that they knew of the chemical, physical and health hazards associated with the propellant they used.

In addition to the above, the undersigned consulted on a litigated case in Florida about fifteen years ago where a professional applicator using a similar product wound up with neurological and respiratory injuries. The records for this case were unavailable to the undersigned.

Prior to creating the TRF insecticides, the manufacturers knew or should have known that the standard of care, and custom and practice in the chemical industry that involves hazardous products is that the manufacturer performs a risk analysis on the foreseeable uses and misuses of their product prior to placing it in the stream of commerce. If there is nothing available from a technical standpoint that would have made that product safer or reduce the risk to the user, then they had a duty to warn and place the user on notice of the inherent hazards and dangers associated with their product and methods of protecting themselves and others from foreseeable and possible harm. Based upon the documents reviewed by the undersigned, there is no evidence that the manufacturers performed a risk analysis and properly warned users against the foreseeable dangers and hazards.

If the manufacturers had performed a risk analysis, they would have known how to eliminate the fires and explosions that have regularly occurred over the last four decades. In almost all cases it was just a matter of time, from a statistical basis that the fires and explosions would occur. This is well exemplified by the thousands of explosions that have occurred over the last several decades up to the present time.

If they had researched methods of reducing the risk of the user being exposed to fires and explosions, they would have known that there were safer alternatives available.

Consumers

Consumers were expected to know that the propellant had gone after a certain period of time. This is not within their ability to do so based upon the physical and chemical nature of the product and the environment that the product was used in. The consumer would have had to use an explosion proof exhaust system placed at the doors with all the windows open in order to remove as much of the explosive propellant as possible. Because there are areas in their homes in which the propellant may be trapped, only a professional would know how to safely remove the explosive vapors. A professional would also be able to measure the concentration of the trapped or isolated propellant in any area of the home. The manufacturers are fully aware of thar fact but not the consumer.

The CPSC

The CPSC and litigated cases document the fact that these formulations explode and injure individuals. Through the use of the Freedom of Information Act, one can obtain the incidents without identifying the individuals or parties involved. However, the reports on the Internet estimate that there are over 500 fires and explosions in the United States. Further, the pesticides in bug bombs are also potentially toxic to humans and pets. Rid-a-Bug is fully aware of those reports. The EcoSMART website reports that there are at least 500 fires and explosions resulting from the use of bug bombs. That website was also available to every manufacturer and their management.

When an accident occurs involving a fire or an explosion, often the manufacturers of that product do not make a concerted effort to determine what occurred until litigation is initiated. It is at that time that they blame the plaintiff for misusing their product.. It should also be noted that when there is a fire or explosion where there is minimal damage or injury, those incidents are very seldom, if at all, reported.

In addition, there were no documents available whereby any of the defendants reported accidents and injuries as required to the CPSC. According to their own paperwork, they relied upon self-reporting by those injured. Since the late 1980's that has been unreliable and not a proper measure of the actual accidents occurring with the use of their products.

The defendants in this case have known for decades of reports of fires, destruction of entire apartment buildings (Salt Lake City, September 2009), explosions and burned victims resulting from exposure to propellant systems used in hair sprays, room deodorizers, repellent sprays for mosquitoes insects and other similar products. The only difference is that the other products use very little of the propellant for each application.

In addition to fires and explosions, there are also serious side effects affecting one's health. Consumers have experienced coughing, shortness of breath, airway dysfunction syndrome and upper airway irritation. Insect bombs usually contain pyrethroid and/or pyrethrins as active ingredients in addition to the extremely flammable and explosive aerosol propellants. Not one of the products is natural and in some cases of injury and death, the users followed instructions and vacated their premises for hours. In spite of this, there is one known death from insect bombs involving an infant found dead the day after the insect fogger was used in the infant's home.

The fact that the manufacturer did not list the whole formulation including the propellant system as part of the formulation in their application to the EPA and the public is, in my opinion, misleading and fraudulent. That EPA was never informed of the highly flammable and explosive nature of the propellant system in any of the applications. However, manufacturers, in their applications using the non-flammable propellant, made sure that the agency knew that it was "non-flammable." Their statements and applications speak for themselves.

It is a known fact that the courts do not recognize the self-reporting of injuries because most companies do not make an effort to know what occurred with each and every TRF that was sold. In addition, there is no evidence that manufacturers have reported any incident involving their products to the CPSC. In many instances, the lawyer representing the injured party is the one who makes inquiry.

Warnings and Instructions

The warnings and instructions placed on the subject products are defective. It is difficult to believe that the defendants claimed federal exemption when they misled the EPA with incorrect and fraudulent data regarding the extreme flammability and explosive characteristics of the Total Release Indoor Aerosol Insecticide.

The product does not include hang tags that incorporate the phrase **"DANGER DO NOT RE-ENTER UNTIL THE ROOM/AREA IS COMPLETELY VENTILATED WITH NEW AIR AND ALL FLAMMABLE/EXPLOSIVE GASES ARE ELIMINATED."** This tag should be placed prominently on the entrance to a room or house restricting entry until the area bombed is completely aired out. The hang tag statements would be in both English and Spanish.

In addition to the above, the warning on the subject product should also place the user and others on notice that the product is too dangerous to be used by anyone other than a professionally trained person. The warnings should include the following:

!DANGER!

THIS TRF FOGGER IS EXTREMELY FLAMMABLE AND EXPLOSIVE

THE PRODUCT MUST BE APPLIED BY PROFESSIONALLY TRAINED AND CERTIFIED TECHNICIANS

DISABLE ALL ELECTRICAL PRODUCTS, HOT-WATER HEATERS, FURNACES, REFRIGERATORS, STOVES, PILOT LIGHTS, SPACE HEATERS, OTHERWISE AN IGNITION SOURCE WILL IGNITE THE EXTREMELY FLAMMABLE AND EXPLOSIVE INVISIBLE VAPORS

IF THE PRODUCT IS APPLIED BY ANYONE BUT A CERTIFIED AND TRAINED TECHNICIAN, THERE IS A HIGH PROBABILITY THAT A FIRE AND/OR EXPLOSION WILL OCCUR RESULTING IN SEVERE BURNS AND POSSIBLE DEATH

AFTER 5 HOURS OPEN ALL WINDOWS AND DOORS AND APPLY EXPLOSION PROOF EXHAUST FANS FOR AT LEAST ONE HOUS TO MAKE SURE THAT THE EXPLOSIVE VAPORS ARE ELIMINATED

In addition to the above, a pictogram of a flame and explosion should also be placed on the label. The requirements for the pictogram can be found at 40 CFR 156.78(d).

Another condition that is not mentioned but is an important factor is the weather. In extremely cold climates, is difficult for the product to be used because one would have to open up the windows and doors of the residence, apartment or house, in order to completely eliminate all of the extremely flammable/explosive propellants. The problem is that the gaseous propellant is invisible and only a trained certified technician would be able to measure and determine whether all of the propellant has dissipated and is eliminated.

The Home Insect Control--The Manufacturer's Hype

Dr. Michael F. Potter of the Department of Entomology, College of Agriculture, University of Kentucky, reported in the paper titled, Limitation of Home Insect Foggers ("BUG BOMBS") that, "while foggers are convenient and require little effort or knowledge to dispense, they seldom are effective against indoor pest problems." He goes on to report that "when applied in this manner, very little insecticide actually penetrates into cracks, voids, and other hidden locations where cockroaches, ants, silver fish, and most other household pests congregate and spend most of their time. The ingredients within "Bug Bombs also tend to be repellent, causing insects to scatter and move deeper into wall voids and other hard-to-reach areas."

It is a known fact among entomologists that when insecticides are needed to eliminate pest infestation results are much better if a spray, bait, dust, etc., is placed directly into the areas where the pests are likely to be hiding. They have found that targeted insecticide applications are much more effective against all varieties of household pests, including fleas.

In addition, when using TRF insecticides, the resulting pesticide residues wind up on exposed surfaces and constitute a health hazard. Occupants with asthma and other respiratory ailments react severely when, for example, pyrethrins and other irritating volatile compounds are used indoors.

Safer Alternatives

The technology for safer alternatives has been known for decades. In fact, room deodorizers and hair spray accidents involving fires and explosions using similar propellant systems as the TRF insecticides have been reduced significantly by changing the propellants to non-flammable gases or with the use of manual sprays.

DuPont's Dymel, as previously is a non-flammable propellant system that the defendants knew about and actually applied for an EPA registration approval of the same formulation as the subject product in this case using the DuPont propellant in the year 2000. In spite of this, they continue to use the highly flammable and explosive propellants which ended up severely burning consumers and destroying property.

It should be noted that several or more of the major manufacturers made application to the EPA using DuPont's Dymel 134A. They identified the actual propellant as part of the total formulation. Dymel, although toxic like the combination of extremely flammable and explosive ingredients used in the subject product, is non-flammable and inert. Major manufacturers all knew that this product was available and could be used safely by the general public.

Spectracide, distributed by Spectrum Group, a Division of United Industries Corp., manufactures a non-flammable insecticide that kills ants, roaches, spiders, fleas and ticks. In the EPA registration # 9688-176-8845, the active ingredient is lambda-Cyhalophrin. This product is disbursed by manual spraying and it more than documents the fact that manufacturers using the propane/butane/isobutane propellant had safer products available that would have eliminated the risk of fires and explosions.

The ORTHO Group manufactures and distributes an insecticide called Home Defense Max. The product is non-flammable and will not explode when exposed to an ignition source. The product kills ants, roaches, spiders, fleas, ticks, earwigs, silver fish, crickets, carpet beetles, firebrats and moths. The EPA registration number is 239-2663. The active ingredient is bifenthrin.

Hot Shot Pest Control Concentrate is also a non-flammable, one-ounce product and can be diluted with water to create up to two gallons. The active ingredient is opernethrin and the EPA registration number is 73049-154-8845.

It might suggested that selling TRF insecticides in smaller sizes and as individual "bombs" would reduce the risk. Is a smaller "bomb" safer than a larger "bomb?" The answer is definitely "No." The consumer is still disbursing the explosive and extremely flammable propellant. The smaller size will not significantly reduce the risk.

Conclusions

FIFRA §3(d)(1)(C) provides for an insecticide to be classified for restricted use. The EPA is concerned about unreasonably adverse effects on the environment.

EPA regulations at 40 CFR 152.170 present a more detailed criteria for making a FIFRA §3(d)(1)(C) determination for an end-use product. When deciding whether to classify a product for restricted use, the Agency considers the toxicity of the product along with other pertinent evidence of hazard and, if any serious hazards could reasonably be mitigated by restricting use. They examine whether labeling could adequately address those hazards. Many manufacturers do not divulge all of the risks and hazards associated with the product which is, in this case, the extreme flammability and explosive nature of the propellants used. The fact that many manufacturers made false, misleading and fraudulent statements in their application to the EPA including all of their testing data, should eliminate any claim they would make as to being federally exempted in their product labeling. It is obvious in all cases involving their EPA applications that they intentionally failed to place the EPA on notice by identifying the "inert ingredients" as being non-flammable as described above.

In addition, safety should never be an option with reference to a product being federally exempted. Manufacturers intentionally compromised the safety and welfare of consumers when they could have easily used the non-flammable propellant systems available for their product. Manufacturers should never place TRF insecticides using highly flammable/explosive propellants in the stream of commerce where laypeople can purchase it. They are not trained in the safe use and application of a highly flammable and explosive product.

The EPA was never given the opportunity to weigh the seriousness or magnitude of the relevant risks in order to consider whether restricting the use would be an appropriate means of reducing those risks. The benefits do not outweigh the risks when there are safer alternatives available to the general public.

Author Information:

C. J. Abraham, PhD, DSc, JD, P.E., FRSC, FTI, DEE (IH), BCEE has spent the last 40 years consulting internationally on a variety of personal injury cases. He has also invented and commercialized a variety of products. Over the last 40 years he has evaluated many household and industrial products with reference to their foreseeable uses and foreseeable misuses. He has also tested a large number of products for their flammable and explosive characteristics and created a number of warnings and instructions as well. He is certified and has diplomat status in a number of areas. His background and experience can be found at www.scientificadvisory.com.

(36d) Tips From 25 Years of Environmental and Product Liability Litigation

Gayle S. Koch, Axlor Consulting LLC, Cambridge, MA

Chemical engineers, by choice or chance, often become involved in litigation proceedings. Experience with various environmental and product liability litigation leads to lessons on how to be aware of litigation issues and better prepare for the experience. This session will cover concepts such as economic benefit, allocation of costs, quantification of costs under uncertainty, cost causation, disclosure, and risk assessment issues. In addition, general tips such as standards, documentation and communications will be discussed.

(21a) Application of Agent-Based Modeling (ABM) of An Integrated System Modeling Framework for Designing a Sustainable Industrial Park

Najet Bichraoui, School of forest resources, University of Maine, Orono, ME and Anthony Halog, School of Forest Resources, University of Maine, Orono, ME; Forest Bioproducts Research Initiative/School of Forest Resources, University of Maine, Orono, ME; Forest Bioproducts Research Initiative, University of Maine, Orono, ME

Attaining industrial sustainability at the local/regional level requires collaborative efforts from participating agents toward common goals including resource conservation, low carbon emission, production efficiency, economic viability, and corporate social responsibility. Our existing socio-technical systems should transition or evolve towards achieving system's sustainability. This study aims to operationalize the idea of system's sustainability by developing an Agent-based model framework for designing an industrial ecology-modeled forest biorefinery. Novel system-based integrative modeling tools with consideration of triple pillars of sustainability and stakeholders' interests should be developed.

Forest and forest product plays an important role in addressing climate change; Indeed, trees have a unique ability to sequester and store carbon and therefore serve as a tool for carbon sequestration. From an industrial perspective, this industry plays a major role in sustainable development because of its chief raw material—wood fiber—is renewable (Douglas et Al, 1998). Industrial production in the forestry sector, including biomass, represents about 2 percent of world economic activity - annually more than US$400 billion - and 3 percent of world trade in merchandise; with paper and board alone accounting for almost half of these totals (Wergens, 1995).

Forest-based product industry has been undergoing regulatory reform worldwide for more than a decade, aiming to reduce carbon emissions, fossil fuel use and economic efficiency. Market performance depends largely on how each market actor responds to market design, including land exploitation rules, market observations, and information access. It is often the case that big pulpwood mills have gained more from these structured markets than retailers or local communities, partly because they have increasingly more opportunities to adapt their behavior to changing market conditions in profit maximizing ways. Unfortunately, some of this profit-driven behavior exacerbates greenhouse gas (GHG) emissions. At the same time, interest has increased in use of market-based policy instruments for environmental purposes, for example, to introduce climate change into the energy sector and reduce GHG emissions.

These challenges are just one aspect of the problem. A need to mitigate ecological impacts, especially greenhouse gas emissions, and energy use have highlighted the need to develop methods capable of addressing economic and ecological uncertainties consistently within an integrated framework. Forest product industry and its network are an evolving system of complex interactions between nature, physical structure, market rules, and participants. Participants face risk and volatility as they pursue their goals and make decisions based on limited information and their mental model of how they believe the system operates.

The industrial eco-system approach will provide comparative information in terms of industrial system response (e.g., how small changes in industrial process from one actor, or a specific event can lead to unexpected scenarios?) and develop ways to use comparative data to predict system responses. This approach will define the system characteristics that make a response either site-specific or more general, and will identify which policy solutions are applicable to similar environmental, economic or social problems.

Implementing an ABM tool within an industrial ecosystem will require a shift in the current culture of researchers working at distributed sites with individual outcomes to a culture that includes the pooling of capabilities, sharing of information, materials, technology, and knowledge. This represents a new paradigm for traditional engineering research. Even though the role of specialized researchers will remain important in their respective fields, their findings will increasingly be applied across disciplines, and the relevance of their results will be valued as part of the whole. Moreover, the use of simulation modeling such Agent-Based Modeling (ABM) for this type of project will provide broader insights on the potential of dynamic simulation modeling for large scale projects on complex system that would otherwise require infrastructure use to test the feasibility of the model.

Agent-based models, allow bottom-up simulations of organizations constituted by a large number of interacting parts. Thus, industrial ecosystems constitute an obvious field of application. This contribution explains what agent-based models are, reviews applications in the field of industrial ecosystems and focuses on a simulator of infra- and inter-firm communications.

From an industrial network perspective, ABM seems to be useful to model large scale system, by feeding the system with rules corresponding to the assumptions of what is most relevant regarding the situation within the industrial eco-park and then watch the emerging behavior from the agents' interactions.

The focus of this study is on showing how this concept of sustainable assessment can be achieved on a practical realization. The project will initially focus on pulp and paper industry in the state of Maine. The system boundary (industrial eco-park) sets the background for a promising application of regional sustainable evaluation. In the next chapter we will show how the forest-based product industry can be seen as a dynamic system, how an industrial eco-park can be built around this regional industry.

For this model the actors (turtles) will be a timber stock, a pulp plant, a stand-alone biomass facility a cement plant, each one of them either produce or consume pulp and lignin.

The goal is to see how they can work in symbiosis (see previous paragraph). What are the conditions for them to allow transfer of flow of material between their plants. The main forces will their sensitivity to regulation of Co2 emissions, and subsidies incentives. Also, there is a wide diversity of agents within the system, agents use different strategies, have different capacities, use different generation technologies, have different motivation, and could be physically located at different locations. In summary, the interactions between the industry partners have the intrinsic features of a complex system: a large number of participating agents, interacting on the basis of limited information and reacting to changes in demand (due to new trend such as newsprint drop and higher level of paper recycling for example). Key questions then arise: *What kinds of new material/energy synergies scenarios can emerge? At what conditions? And for how long these new scenarios can be sustained?*

The purpose of simulation modeling such as Agent-based modeling is to generate and explore alternative futures that may develop under different conditions. These simulators can explore various "what if" scenarios under different economic, social goals. They can show the possible evolutionary trajectories of given scenarios under different conditions and geographical context. An industrial eco-park organized around the forest-based product industry is a relevant area of application. Indeed, the industry due to its current economic challenges and inherent structure offer a variety of opportunities for synergies. Simulation modeling can be highly useful, especially in terms of policy making.

The ABM modeling approaches provide a coherent and robust approach in dealing with the complexity of industrial eco-systems, including the ability to address their inherent uncertainty. The combined modeling approach allows the analysis of policy-making tools and their ability to cooperate industrial ecosystem management, therefore providing solutions and clues to those organizations that have an interest in designing or promoting eco-industrial networks.

(AICHE 2012 Spring Meeting, Paper # 243102)

USING NANCLAYS AS BOTH COMPATIBILIZER AND REINFORCING FILLER FOR IMMISCIBLE POLYMER BLENDS: TOWORD MAKING THE PLASTIC WASTE RECYCLING INDUSTRY MORE COST EFFECTIVE

Abdulhadi A. Al-Juhani,
Department of Chemical Engineering, King Fahd University of Petroleum & Minerals (KFUPM), Dhahran, Saudi Arabia

Abstract

The focus of this research is to perform a comparative study for examining the effects of two types of hydrophobic nanoclays of different structures (Cloisite® 10A (C10A) and Cloisite® 15A (C15A)) on the solid and melt-state properties of an immiscible polymer blend composed of polypropylene and poly (ethylene oxide). Results obtained here were compared with our previous study (Al-Juhani, 2012, which focused on the use of hydrophilic Cloisite® 30B (C30B) nanoclay for the same polymer blend system), in an attempts to expose some light on the structure-property relationships that govern the bulk behavior of nanoclay-filled polymer blends. Results showed that, although clear compatibilization effect was observed by the addition of nanoclay irrespective of the type of nanoclays investigated, the bulk mechanical properties were strongly influenced by the hydrophobicity and structure of the organic modifier used.

Introduction

Plastic waste products accumulate annually in the environment in huge quantities worldwide and need proper waste management and cost-effective recycling approaches. One innovative polymer recycling approach is to fill a polymer blend (composed of more than homopolymer waste) with low loading of inorganic filler of high aspect ratio such as nanoclays, aiming to conceptually enhancing both the miscibility and the bulk properties of the recycled polymer blend at the same time. This recycling approach has two advantages. First, it eliminates the need of adding a costly conventional copolymer compatibilizer to the system, hence reducing the cost of the recycling process. Second, combining nanoclays with recycled polymer blends could produce advanced materials with broad range of properties by combination of the advantages of polymer blends (such as producing a new product with bulk properties better than the properties of the individual polymer components) with the superior merits of nanoclay itself (such as high stiffness and thermal stability, flame retardancy, and improved gas-barrier properties), and hence improving the profit of the recycling process. In other words, advanced materials could be generated by the combination of polymer blend technology and the nanomaterial technology, as well as this could help to save the environment from the accumulated plastic waste.

However, this added-valued products anticipated by the combination of nanoclays and polymer blends can not be obtained easily without a thorough and systematic experimental and theoretical study on the structure-property relationships that control the bulk properties of the nanoclays-filled polymer blends.

Therefore, we selected to investigate in details the effect of hydrophobicity and structure of nanoclays on the bulk properties of an immiscible blend consist of PP/PEO blend. In specific, the current study is a continuation of our previous study (Al-Juhani, 2012), where in that study the effect of hydrophilic Cloisite® 30B (C30B) on the bulk properties were investigated. However, in the current study, the effect of two types of hydrophobic nanoclays of different structure (Cloisite® 10A (C10A) and Cloisite® 15A (C15A) on the bulk properties of the same PP/PEO blend will be investigated. A comparative discussion will be made in this communication between the results of hydrophobic nanoclays (this study) and hydrophilic nanoclays (the previous study of Al-Juhani, 2012) whenever appropriate.

.

Experimental

Materials

The characteristics of the polymers used, polypropylene (PP) and poly (ethylene oxide) (PEO), were descried in details in our previous publication (Al-Juhani, 2012). Two types of nanoclays were used in the current study: Cloisite® 10A (C10A) and Cloisite® 15A (C15A), and both were purchased from Southern Clays Products Inc., USA. C10A had its surface organically modified with 2MBHT (dimethyl benzyl hydrogenated tallow quaternary ammonium), whereas C15A had its surface organically modified with 2M2HT (dimethyl di-hydrogenated tallow quaternary ammonium), and the tallow here was mainly composed of C18 (~65%). Both types of nanoclays had the same value of modifier concentration (125 meq/100 g clay). The d_{001} spacing of C10A and C15A were 19.2 Å and 31.5 Å, respectively.

Sample Preparation and Characterization

This study followed the same procedure for preparing and characterizing the samples as described in details in our previous publication (Al-Juhani, 2012). In brief, for preparing the samples, Haake PolyDrive melt blender was employed operated at a temperature of 200 °C and a rotational speed of 50 rpm for 10 minutes. The loading of C10A and C15A was varied at 0, 2 wt.%, 5 wt.%, and 10 wt.%, whereas a fixed 70/30 mass ratio of PP to PEO was maintained for all samples. After preparing the samples they were subjected to thermal testing by Differential Scanning Calorimetry (DSC), morphological testing by Scanning Electron Microscopy (SEM), dynamic melt rheological testing, and tensile mechanical testing. The characterization procedures were described in details elsewhere (Al-Juhani, 2012).

Results and Discussion

Thermal testing by DSC

The DSC crystallization curves for C10A-filled PP/PEO blends and C15A-filled PP/PEO blends were shown in Figure 1 (a) and Figure 1 (b), respectively. As shown in Figure 1, multiple crystallization peaks were observed for control unfilled samples, suggesting that fractionated crystallization did occurs for the immiscible PP/PEO blend without nanoclays. However, fractionated crystallization completely disappeared for the case of both C10A-filled samples and C15A-filled samples at the loading of 5 wt. % and above. The DSC crystallization curves for samples filled with hydrophobic clays of both types (C10A and C15A) very much resembled each others (Figure 1). However, at 2 wt.% loading, both C10A and C15A-filled samples behaved differently compared to what we reported previously for the hydrophilic C30B clay-filled samples (Al-Juhani, 2012). While two sharp PEO crystallization peak exist for the samples filled with hydrophobic C10A and C15A clays at 2 wt. loading (Figure 1), only one sharp peak exist for the samples filled with hydrophilic C30B clays at 2 wt. loading (Al-Juhani, 2012).

On the other hand, by analyzing the DSC melting endotherms, no significant effect of both C10A and C15A nanoclays could be observed on the melting temperature of PP and PEO (data not shown).

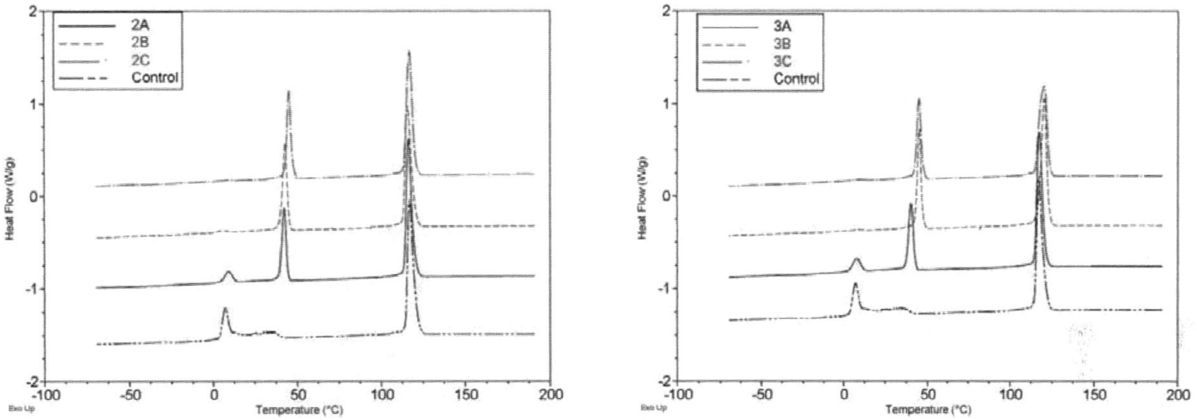

(a) C10A (b) C15A

Figure 1. DSC crystallization curves for PP/PEO blends filed with C10A and C15A nanoclay at various loading.

Morphological testing by SEM

The phase morphology of the PP/PEO blends with and without C10A and C15A nanoclays were examined by SEM, and representative images were shown in Figure 2. As shown in the figure, both types of nanoclays had sharp advantageous effect for fine-tuning and improving the morphology of the immiscible PP/PEO blends. The large domain size of the PEO droplets observed for unfilled samples (Figure 2a and 2b) became smaller and finer after the addition of nanoclays of either types at both 2 wt.% and 5 wt.% loading (Figure 2c-2f). This resembled the role of a block-copolymer compatibilizer for improving the compatibility of an immiscible polymer blends. However, at 10 wt. loading, better filler dispersion was observed for the case of C15A compared to C10A (data not shown). This observed improvement of the morphology of the PP/PEO blends by the addition of the hydrophobic C10A and C15A clays were in agreement with what we previously reported for the case of hydrophilic C30B nanoclay (Al-Juhani, 2012). However, we believed that the compatibilization mechanism was somewhat different, as to be discussed later.

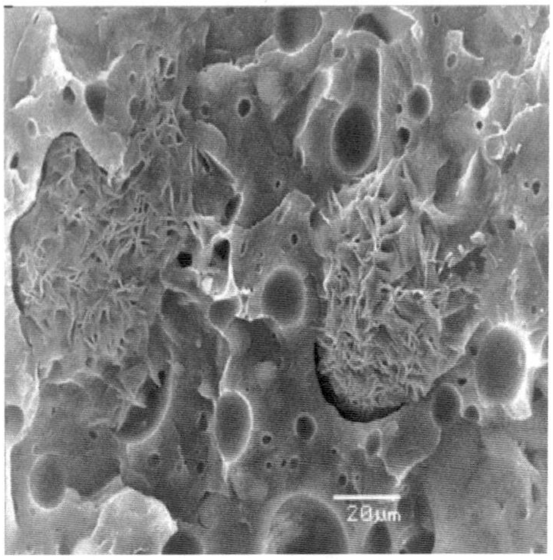

(a) Control sample (lower magnification)

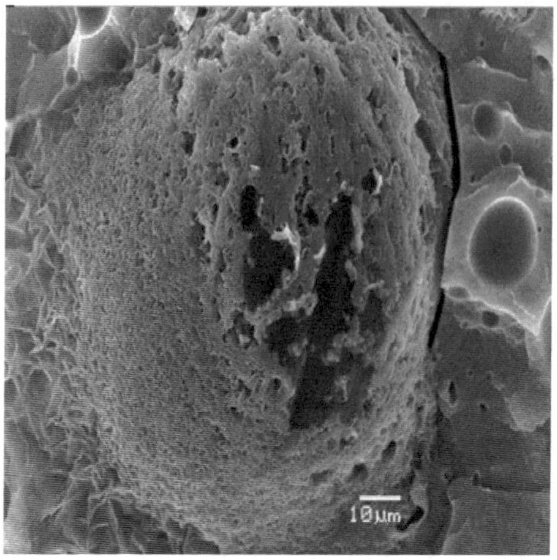

(b) Control sample (higher magnification)

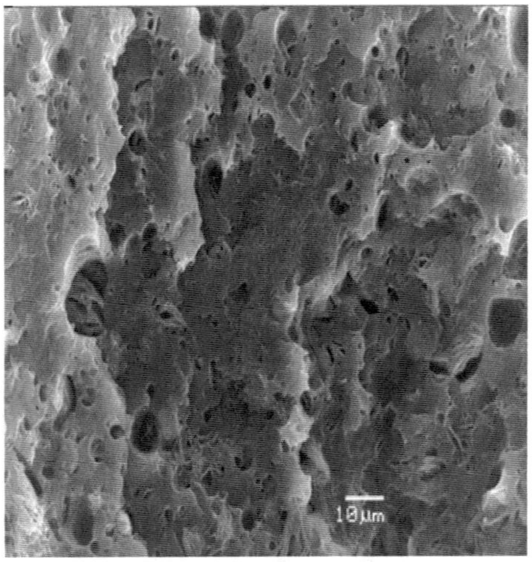

(c) with 2 wt% C15A

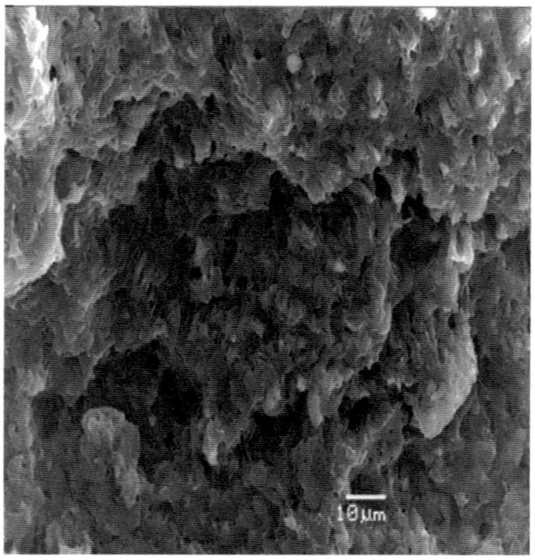

(d) with 5 wt% C15A

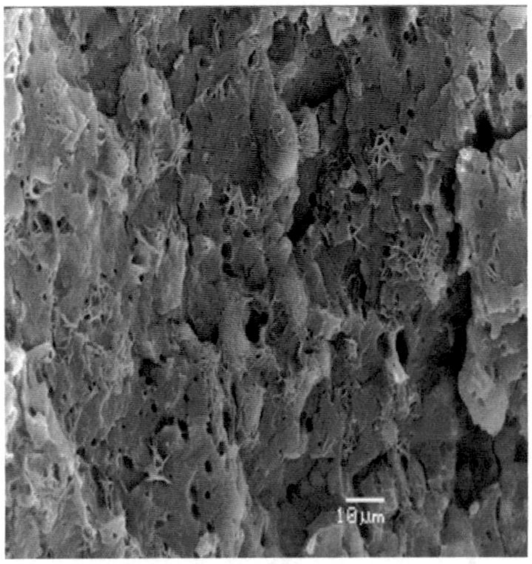

(e) with 2 wt% C10A

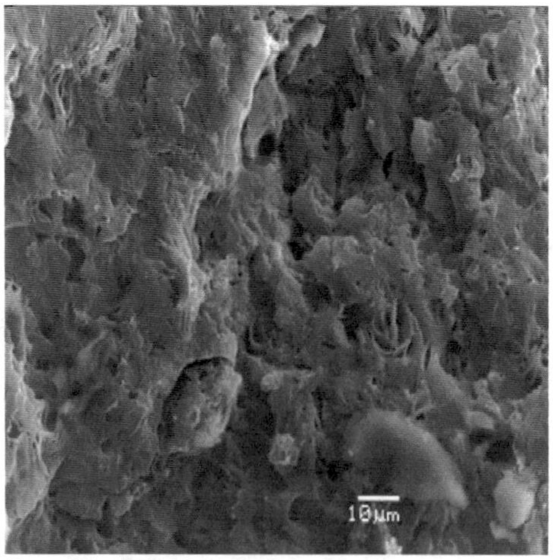

(f) with 5 wt% C10A

Figure 2. SEM images for PP/PEO blends filed with C10A and C15A nanoclay at various loading.

Melt rheological testing

As shown in Figure 3, both C10A and C15A nanoclays had sharp effect for increasing the viscosity and storage modulus of PP/PEO samples, and this increase was directly proportional to the nanoclay loading. However, an extra piece of information can be obtained by carefully inspecting the slopes of the storage modulus curves (Figure 3c and 3d) at low frequency range. Clear platue in G' values were observed for samples filled with C10A clays at all loading from 2 wt.% to 10 wt.%, while for the case of C15A clear platue in G' values only observed at loading of 5 wt.% and 10 wt.%. This indicate that physical interconnected network of C10A formed inside the polymer matrix at a loading of 2 wt.% and above, whereas network of C15A nanoclays formed at a loading of 5 wt.% and above. In other words, the network rheological thresholds for C10A and C15A were rather different here, most likely due to differences in the structure of the organic modifiers. Compared to our previous work of C30B/PP/PEO system (Al-Juhani, 2012), the network rheological threshold of C30B and C10A clays are similar (2 wt. %), whereas C15A showed a different behaviour.

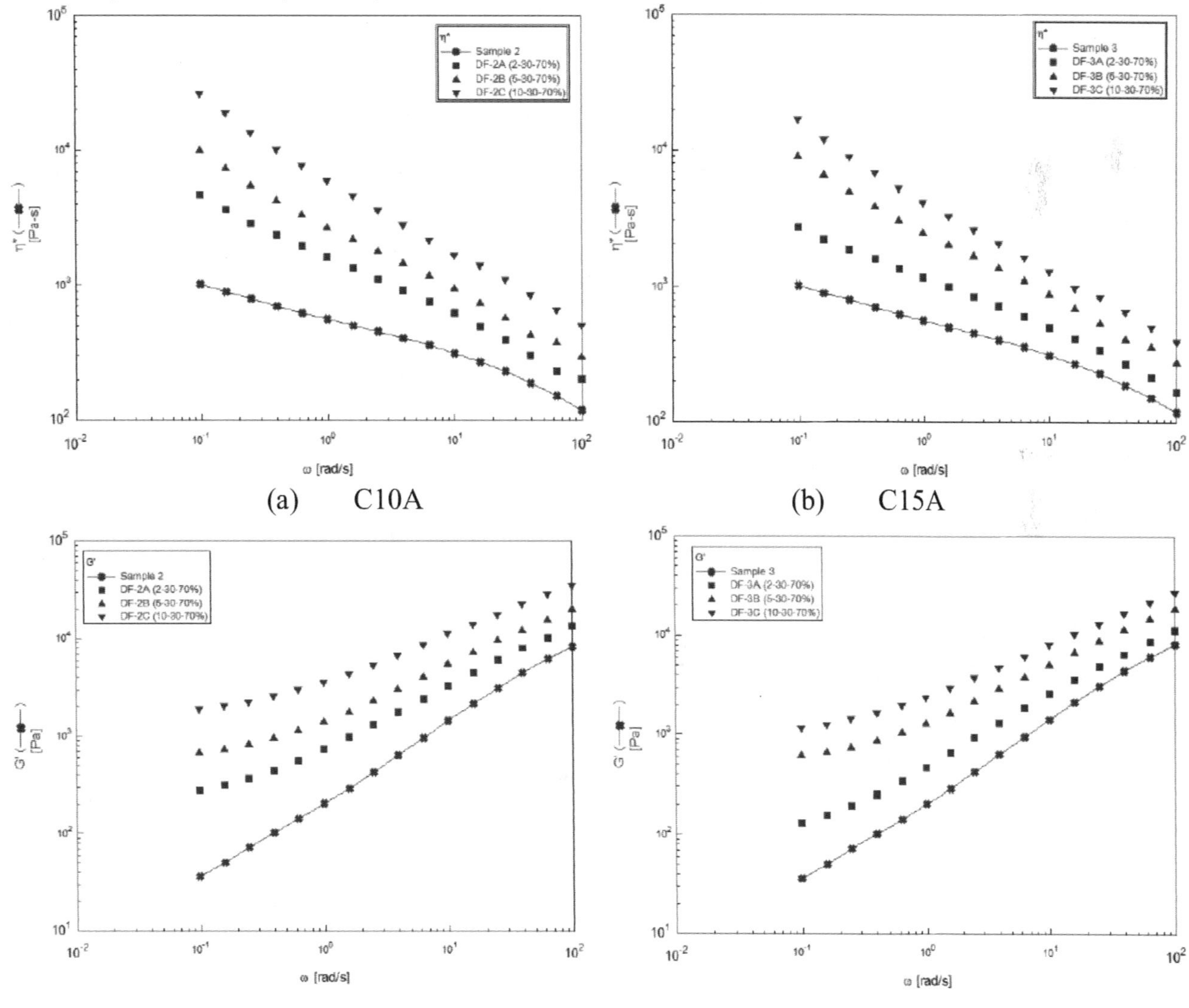

(a) C10A (b) C15A

(c) C10A (d) C15A

Figure 3. Complex Viscosity (a and b) and Storage modulus (c and d) as a function of frequency for PP/PEO blends filed with C10A and C15A nanoclay at various loading.

Tensile Mechanical Testing

The mechanical properties, obtained by analyzing the tensile stress-strain curves, were summarized in Table 1. The results for C10A-filled samples were expected and predictable, where the elastic modulus increased, the elongation at break decreased, and the ultimate strength decreased as the loading increased. However, the results for C15A, especially at the 2 wt.% loading, were unexpected. Compared to unfilled samples, the elongation at break at 2 wt.% loading of C15A considerably improved (increased), but the elastic modulus slightly decreased. At the higher loading of C15A nanoclays, the elastic modulus increased while the ultimate strength was kept similar to the corresponding values of unfilled samples.

Compared to our previous publication of the C30B/PP/PEO system (Al-Juhani, 2012), C30B at 5 wt% showed the highest improvement in the elastic modulus property amongst all the type and loading of nanoclays investigated, whereas C15A at 2 wt.% loading showed the highest improvement in the elongation at break property. The ultimate strength of C15A at 5 wt. % and 10 wt. % loading were better than that of C10A and C30B at the same loading. This showed that the bulk mechanical property of the nanoclay/PP/PEO blend was a strong function of the type of nanoclay used.

Table 1. Mechanical properties for PP/PEO blends filed with C10A and C15A nanoclay at various loading.

Sample #	Elastic Modulus (MPa)	% Elongation at Break	Yield Strength (MPa)	Ultimate Strength (MPa)
Control PP/PEO	485.29	9.80%	10.06	15.44
2 wt% C10A-PP/PEO	526.32	6.60%	9.47	12.99
5 wt% C10A - PP/PEO	549.09	7.40%	8.50	12.58
10 wt% C10A - PP/PEO	558.67	5.00%	7.61	11.95
2 wt% C15A-PP/PEO	461.69	16.00%	9.20	12.95
5 wt% C15A - PP/PEO	534.88	6.20%	10.07	15.34
10 wt% C15A - PP/PEO	548.46	4.20%	8.80	14.84

Further discussion on the structure-property relationship

The main finding in the current study for the C10A/PP/PEO system and the C15A/PP/PEO system, and comparing it with a previous study for the C30B/PP/PEO system (Al-Juhani, 2012), was the following. Although all of the three types of the nanoclays showed clear compatibilization effect by reducing the domain size of the PEO droplet phase, as confirmed by SME, the mechanism was different and the implications on the bulk mechanical properties were different. This had to do with the hydropbicity and structure of the organic modifier of the nanoclays, which could be the determining factors for the location and dispersion level of nanoclays inside the polymer blend. The hydrophobicity (in terms of solubility parameter) of all materials used in the current study was summarized in Table 2, whereas the structures of these materials were well-reported in literature. As shown in Table 2, the perfect match in solubility parameters of the PP phase and C15A nanoclay suggested that C15A most likely located in the PP phase and the level of dispersion would be high. On the other hand, for C10A nanoclay, it would be most likely located in the PP phase but the level of dispersion would be low due to the presence of the benzyl group in the structure of C10A. This was in agreement with SEM data, where SEM analysis confirmed that some aggregate exist at the highest loading of 10 wt% of C10A, whereas no aggregates was observed for the case of 10 wt. % of C15A (data not shown here).

Now, compared with our previous work for the C30B/PP/PEO system (Al-Juhani, 2012), the solubility parameter of C30B and its structure were much closer to the PEO phase, suggesting that C30B most likely located in the PEO phase and the level of dispersion would be high.

Consequently, the location of nanoclay had direct implication on the compatibilization mechanism. For the hydrophobic C10A and C15A nanoclays used in the current study, the compatibilization mechanism was mainly because C10A and C15A, which were located in the PP phase, acted as mechanical barriers to coalescence of the PEO droplets, therefore a reduction in the domain size of the PEO droplets was observed. This was different from the mechanism suggested in our previous work for the hydrophilic C30B (Al-Juhani, 2012), where it was suggested that, since C30B was located in the PEO phase, C30B caused the viscosity ratio of the droplet to matrix to increase considerably, therefore helped to suppress or delay the coalescence of the PEO droplets and hence a reduction in the domain size of the PEO droplets was also observed. Moreover, the location of nanoclay had strong effect on the bulk mechanical properties of the PP/PEO blend, as discussed in previous section.

Table 2. Solubility parameters (δ_i) of the nanoclays and polymers used in the current study.

Material	**δ_i ($J^{1/2}.cm^{-3/2}$)**	**Reference for δ_i**
C10A	16.2	Jang et. al, 2005
C15A	16.8	Sinha Ray and Bousmina, 2005
PEO	22.2	Chan et. al, 2006
PP	16.8	Michaels et. al, 1968

Conclusion

A comparative characterization study was performed for PP/PEO blends filled with two types of hydrophobic nanoclays of different structures (C10A and C15A). The results obtained here were compared with our previous work for filling the same PP/PEO blend with hydrophilic C30B nanoclay. Summing all results together, the following was worth to be mentioned. All the three types of nanoclays did show clear and similar compatibilization effect as confirmed by SEM which showed a significant reduction in the domain size of the PEO droplets after the addition of nanoclays of either type. However, the compatibilization mechanism and consequently the bulk mechanical properties were shown to be a strong function of the hydrophobicity and structure of the nanoclays used. Among all the types and loading of nanoclays investigated, 2 wt. % C15A showed the highest improvement in the elongation at break property. Moreover, the ultimate strength of C15A at 5 wt. % and 10 wt. % loading were better than that of C10A at the same loading. DSC crystallization testing showed that fractionated crystallization, which was observed for unfilled samples, only completely disappeared at loading of 5 wt. % and obove of either C10A or C15A. Melt rheological testing suggested that the network rheological threshold for C10A-filled samples and C15A-filled samples were around 2 wt. % and 5 wt. %, respectively. The results obtained could be a useful step for initiating a further detailed experimental and theoretical research in the area of structure-property relationships that control the bulk properties (especially the mechanical property) of nanoclay-filled polymer blends. The polymer recycling industry could gain high values of their recycled products if the optimum formulation could be predicted based on structure of nanofiller and the polymer components used.

Acknowledgment

The author would like to acknowledge the financial support of King Fahd University of Petroleum & Minerals (KFUPM) through the funded project no. SB100019.

References

1. Al-Juhani, A. A. (2012), "The effect of OH-modified nanoclays Fillers on the Compatibility and Properties of polypropylene/poly (ethylene oxide) blends" *Arabian Journal of Science & Engineering*, accepted.
2. Jang B N, Wang D, Charles A. Wilkie C A. (2005) *Macromolecules*, 38, 6533-6543.
3. Sinha Ray S, Bousmina M. (2005) *Macromol. Rapid Commun.*, 26, 1639-1646.
4. Chan A, Coppens K, Hall M, He V, Jog P, Larsen P, Koblinski B, Michael Read M, Rothe D, Somasi S, Uma Shrestha U (2006) Annual Meeting and Exposition of the American Association of Pharmaceutical Scientists, San Antonio, Texas, USA, Oct. 29–Nov. 2, 2006.
5. Michaels AS, Vieth WR, Alcalay HH (1968) *J Appl Polm Sci,* 12, 1621.

CHEMICAL DEGRADATION OF PCE RESIDUES: AN USEFUL TOOL FOR ENVIRONMENTAL PRESERVATION

A. Mondelo Rodríguez[1], Y. Mendoza Alvarez[2], J. M. Ameneiros Martínez[3] and E. Marquez Canosa[3]

[1] *CIMAB,* Carretera del Cristo No. 3 Esquina Tiscornia Casa Blanca, Regla, La Habana 11700, Cuba *e-mail: abelmr1@gmail.com*

[2] *GEOCUBA Oriente Norte, Carretera Central esq. Calle Holguín,Cuba*

[3]*Higher Politechnic Institute José A. Echeverría ISPJAE – CUJAE, Marianao Ciudad de La Habana, Cuba.*

Abstract

Chemical substances used in many of the industrial applications, are recalcitrant to conventional wastewater treatment process such as: aerobic, anaerobic treatment or chemical oxidation with chlorine or ozone. The development of Advanced Oxidation Process (AOPs) and Fenton like Process, become an alternative process for hazard and pollution reduction in wastewater from many industrial processes.

Fenton Chemistry has been proved as an effective method for the degradation of Halogen-Organic compounds such as TCE and PCE. This method is useful tool for the decontamination of many bacteriostatic or bactericide substances used in biotechnological process.

In the present work, PCE residues generated in a fractional distillation process at atmospheric pressure were degraded applying the Destructive Fenton Oxidation Process. Experiments at laboratory scale were conducted at 25 °C with hydrogen peroxide per iron II (ferrous) sulphate feed ratio of 5 and atmospheric pressure. It was observed a high reaction yield and an increase of concentration of chloride ions, as was expected; which it indicates the reaction proceeds as desired. After the process this substance was not detectable by TLC or HPLC; only for a reagent feed ratio of 2.0 it show a small quantity of the substance in the effluent, upper values of it offer a best destruction of the substance, adding periodically small volumes of hydrogen peroxide.

Resumen

Las sustancias químicas usadas en muchas aplicaciones industriales son recalcitrantes a procesos de tratamiento convencionales, tales como: tratamiento aeróbicos y anaeróbicos, oxidación química con ozono y cloro y sus derivados. El desarrollo de Procesos de Oxidación Avanzada (AOPs) y procesos tipo Fenton, se convierten en procesos alternativos para la reducción de la contaminación y el riesgo en aguas residuales de muchos procesos industriales.

La química de Fenton ha sido probada como un método efectivo para la degradación de compuestos orgánicos halogenados tales como tricloroetileno (TCE) y percloroetileno (PCE). Este método en una herramienta útil para la descontaminación de muchas sustancias bacteriostáticas y bactericidas utilizadas en procesos biotecnológicos.

En el presente trabajo, residuos de PCE generados durante su recuperación mediante destilación fraccionada a presión atmosférica fueron degradados aplicando el destructivo proceso de oxidación de Fenton Los experimentos a escala de laboratorio fueron conducidos a 25 °C con una razón de alimentación de peróxido de hidrógeno/sulfato de hierro II de 5 y presión atmosférica. En todos los casos se observe un elevado rendimiento de la reacción, y un aumento de la concentración de iones cloruros, como se esperaba; lo cual indica que la reacción procede como es deseado. Posterior al proceso de oxidación, la sustancia estudiada no fue detectable por cromatografía de capa fina (TLC) o por HPLC; solo para una razón de alimentación de reactivos de 2.0 mostró una pequeña cantidad de la sustancia en el efluente. Razones de alimentación de reactivos superiores a este ofrecieron una major destrucción de la sustancia, adicionando periódicamente pequeños volúmenes de peróxido de hidrógeno.

Key words

Fenton Chemistry, PCE residues, recalcitrant or toxic wastes and wastewater treatment

Palabras Claves

Química de Fenton, residuos de PCE, residuos tóxicos o recalcitrantes y tratamiento de residuales

Introduction

Some of products used in our modern society are too toxic to be disposed off without particular treatment. If not disposed off correctly, some cleaners, solvents, pesticides, and paints can contaminate a landfill, leak into the groundwater or contaminate the ocean resulting in tremendous risk for the safety and health of human being.

The release of contaminants such as solvents and other volatile organic compounds (VOCs) into surface impoundments or through other disposal practices has resulted in the contamination of the subsurface, and surfaces waters and groundwater[2].

Organics targeted for this category include[2,6,11]:

- Chlorinated Solvents (TCE, PCE, TCA, DCA, etc.)
- Munitions
- Pesticides
- Petroleum hydrocarbons (BTEX, PAH, TPH, diesel fuel)
- Wood preservative and
- PCB and Phenolics

Therefore, is convenient and important, analyze its origin and contaminant strength for reducing or eliminating it. PCE is a colorless, volatile aliphatic chlorinated compound, with an ether-like odor. It is slightly soluble in water and miscible with alcohol, ether, chloroform, hexane and benzene. It is reasonably anticipated to be a human carcinogen based on sufficient evidence of carcinogenicity in experimental animals.

In the groundwater PCE can persist for decades and travel in plumes with the groundwater flow. Microorganism in the soil and groundwater can slowly breakdown PCE over time. Because its lower solubility, immiscibility, relatively slower rates of dissolution and high density, it is extremely difficult to remedy PCE[12,13,14,15].

Since the Fenton reaction is a well-known potent oxidative reaction, many researchers have investigated the application of the Fenton reaction to degradation of various organic compounds[1-8,10,11]. Several studies have also reported that PCE was degraded by Fenton reaction.

Accordingly, the current study investigated the Fenton reaction for degrading PCE wastes from a PCE recovering through a distillation process.

Materials and Methods

Ferrous sulfate, hydrogen peroxide, and sulfuric acid were analytical grade and purchased from Merck.

Quantitative determination of PCE was performed through UV – Visible Spectrometry at 290 nm, using an ULTROSPEC LKB BIOCROM Spectrophotometer and 1 cm quartz cell, Reference cell: PCE free distilled water and A_{max} of 1.0.

The reaction proceed was followed by TLC analysis carried out on precoated Silica gel 60 sheerts Merck F254, using hexane – ethyl acetate 80/20. Detection was effected by charring with a 5 % hydroalcoholic solution after examination under UV light.

Chloride ions were quantified by the argentometric methods 407 A SMWW 19 edition 1995. Preliminary experiments were performed for screening the input variables and identify the most important subset of factors affecting the reaction yield. The temperatures and hydrogen peroxide per ferrous sulfate concentration ratio remains as the most important input variables to find the better yield values. This procedure conducted to a factorial 2^3 experimental design for finding the better reaction condition. It was observed that the pH of the reaction declines through the reaction proceeds. Thus, the starting pH was fixed at a value of 5.0 to reach a reaction pH of 3.5 and lower, after adding the Fenton Reagent. The iron catalyst was first added, and then the hydrogen peroxide was added slowly, in various small volumes. This procedure conduces to a moderated rise in temperature as the reactionproceeds.

The temperature were studied at 24 - 35 °C, at three levels (table 1), because of an increase in temperature will increase the reaction yield, but at values of 40 or upper, the decomposition of the hydrogen peroxide will become the main reaction. There also was experimented the reaction at 40 °C and the reagent consumption was upper, as was expected. In addition, the concentration ratios between the hydroxyl radical promoters (hydrogen peroxide / ferrous sulfate) were setting at 2,00 - 5,00 values at three levels. Iron catalyst was used as ferrous ions, and ferric hydroxide precipitated was re-used by acidifying it and send it back to another reaction; so, traces of iron can generate indefinitely amounts of hydroxyl radicals.

The Fenton reaction were conducted in an open vessel, at atmospheric pressure, with a water-ice bath to keep the temperature closed to the selected temperature value, and pH at starting reaction system of 5 - 5.5. A hard stirring was needed for a better oxygen mass transfer to the reaction mixture. The regeneration of the iron catalyst was carry out at a pH ranging of 8 - 8.5.

Results

In general form, an increase in the temperature was observed in all experiments, which it affords an increase in reaction rates. Nevertheless, higher reagent consumption was observed for temperatures of 30°C and upper. The most suitable conditions which afford a good reaction yield were 24 °C and a Hydrogen peroxide per ferrous sulphate concentration ratio of 5.0. There also was experimented the reaction at 40 °C and the reagent consumption was highest, as was expected. The reaction yield in all experiences was higher than 99 % based on the PCE reduction. Generally, the reaction takes place in 120 min. The pH in the reaction mixture decreasing as the reaction proceeds, as was expected, through the releasing of chlorides ions. After the process PCE was not detectable by TLC or UV - Visible spectrometry; only for a reagent ratio of 1.75 and 2.0 it show a small quantity of the substance in the effluent after 2 h reaction time, upper values of it offer a best degradation of the substance, adding periodically small volumes of hydrogen peroxide.

The process performed at 24 °C, atmospheric pressure and a ratio of 5,0 showed the best yield for PCE degradation. Average PCE concentration was 130 mg/I. After treatment, the PCE concentration was not detected in almost all experiences, except for a H_20_2/Fe^{2+} ratio of 1.75 and

2.0 yielding a PCE reduction of 90 and 93 %. Higher Temperatures values increase the reaction yield, but the reagent consumption was higher than when the reaction was conducted at 24 °C. In addition, the pH of the reaction mixture was decreasing as the reaction proceeds.

Table 1. Average Results of the Fenton Process Study.

Samples	2	5	8	6	1	4	A
H_2O_2 / $FeSO_4$	1.75	2.5	5.00	5.00	2.00	2.30	2.50
T (°C)	24	24	24	30	30	35	40
t (h)	2	2	2	2	2	2	2
H_2O_2 (mmol/L)	0.35	0.5	0.50	0.50	0.4	0.46	0.5
$FeSO_4$ (mmol/L)	0.20	0.20	0.10	0.10	0.20	0.20	0.20
% Rem.	90	> 99.9	> 99.9	> 99.9	93	> 99.9	> 99.9
PCEin	130	130	130	130	730	130	130
PCEout	13	ND	ND	ND	9.1	ND	ND
V_{sample} (mL)	500	500	500	500	500	500	500

Important: *It was performed several additions of H_2O_2 and one addition of $FeSO_4$* for controlling the temperature increasing.
Starting pH: 5 -6, reaction pH: 3 - 4, precipitation pH: 7,5 - 8,5.

The iron catalyst in the experiments was recovered by an increase of pH of the mixture and a filtration process. It also was re-used by acidifying the iron sludge to solubilizate it and sending it back to perform another experiment.

Conclusions

The Fenton process was studied for a wastewater containing PCE. All the experiences performed at 24 °C, atmospheric pressure and a ratio of 5.0 showed a good yield of the oxidation reaction. An increase of chlorides ions was observed, as the reaction proceeds, and a consequent pH declining, as was expected.

Acknowledgement

We want to thank to the Enterprise Group LABIOFAM for the financial support for the experiments. We also want to thank to the Enterprise of Quality Control of the Enterprise Group LABIOFAM for their help in all analysis techniques.

References

1) Balazs, G. B.; et al; Direct Chemical Oxidation: A non-thermal technology for the Destruction of Organic Wastes; FY 98 Final Report (Addendum to FY 97 Report) February 1998.

2) Nesheiwat, F. K. And Swanson, A. G.; Clean Contaminated sites using Fenton Reagent, Chemical Engineering Progress, Vol. 96, No. 4 (April 2000).

3) Bauer, R. And Fallmann, H.; The Photo-Fenton Oxidation – A Cheap and effective wastewater treatment method; Res. Cim. Intermed. 23, 341 – 354, (1997).

4) NAVFAC Y0187 Technology Demostration Program, NFESC Project Success, In-situ Oxidation of Chlorinated Organic Contaminants using Fenton´s Reagent; http://p2ashore.nfesc.navy.mil/success/FROP.pdf, revised February 21st, 2002.

5) Voelker, B. M.; Fenton Reaction and Oxidation of Aquatic Organics, Environmental Seminar, (April 27th 2001), http://web.mit.edu/civenv/parsonslab/faculty/voelker.html revised February 21st, 2002.

6) Innovative Technology, Summary Report, DOE/EM – 0484, Fenton's Reagent, October 1999.

7) Technical and Regulatory Guidance for in situ Chemical Oxidation of Contaminated Soils and Groundwater (ISCO-1) 2001; http://www.itreweb.org/ISCO.1.pdf; updated October 2002; revised Agoust 27th 2003

8) StrliC, M., Kolar, J. And Pihlar, B.; The effect of metal ion, pH, and temperature on the yield of oxidizing species in a Fenton-like system determined by aromatic hydroxylation; Acta Chim. Slov., 1999, 46 (4) pp 555 – 566.

9) American Public Health Association; Standard Methods for the Examination of Water and Wastewater; 19 Ed. Washintong APHA 1995.

10) Butterfield, I.M.; Christensen, P.A.; Curtis, P.P. and Gunlazuardi, J.; Water Disinfection Using an Inmobilized Titanium Dioxide Film in a Photochemical Reactor with Electric Field Enhancement; Water Res. 31 (1992) pp. 675 – 677

11) Nankung, K. C. and Sharratt, P.; Fenton and Photo-Fenton Process for the Treatment of Aqueous Wastes; Disertation in the Environmental Technology Center, Department of Chemical Engineering, University of Manchester, Science and Technology Institute (UMIST); Green Research Network; http://www1.york.ac.uk/res/gcrn/OXNamkung.ppt Revised March 27th, 2003

12) Tetrachloroethylene (Perchloroethylene) CAS No. 127-18-4, Substances profile; Report on Carcinogens, Eleventh Edition.

13) What is PCE? Navy Environmental Health Center, Environmental Program January 2003.

14) Serial Degradation of PCE by Delftia sp. N6 after dechlorination using Fenton's Reagent; Journal of Microbiology and Biotechnology 2006, 16 (11), 1734-1739.

15) Perchloroethylene (PCE) Degradation using Fine Metal Particles. S. Haredra and C. Vipulanandan, CIGMAT - 2005 Conference and Exhibition.

Environmental Performance of Products from Glycerol Conversion

Lady J. Gaitán Aguilera, Johanna A. García Ceron*, Luisa F. Martin Forero, Gabriel Camargo Vargas

Universidad Libre, Street 70 No. 53-40 Bogotá, Colombia
*johannaa.garciac@unilibrebog.edu.co

This study has a main objective it is made an evaluating impact process that will have the compounds formed from conversion of glycerol, with dates generated by EPIWEB software, will have life with the execution of this project, through an assessment of physical and chemical properties of different compounds such as Henry's law constant, rate reaction constant, water solubility, Octanol-water partition constant, amoung others. The results will be show in tabulate form (1) .

Table 1: Chemical Properties to Perform Environmental Screenings

ENVIRONMENTAL PROCESS	RELEVANT PROPIERTIES
Dispersion and fate	Volatility, density, melting point, water solubility, effectiveness of waste water treatment
Persistence in the environment	Atmospheric oxidation rate, aqueous hydrolysis rate, photolysis rate, rate of microbial degradation, and adsorption
Uptake by organisms	Volatility lipophilicity, molecular size, degradation rate in organism
Human uptake	Transport across dermal layers, transport rate across lung membrane, degradation rates within the human body
Toxicity and other health effects	Dose-response relationship

The methods described assume that a molecule is composed of a collection of functional groups or molecular fragments and that each fragment contributes in a well-defined manner to the properties of the molecule. These methods are generally described as group contribution methods, structure activity relationships called SARs or quantitative structure activity relationships called QSARs; these structures will be estimated by EPIWEB (2) . EPIWEB is the principal software in this investigation; this program estimates many variables like Octanol-water partition coefficient, Octanol-air partition coefficient, Henry's law constant, water solubility, and different aspects. Each variable is estimated from EPIWEB's sub-programs such as

1. KOWWIN (Octanol-Water Partition Coefficient)

Octanol-water partition coefficient is an equilibrium ratio of the concentration of a compound in octanol to the concentration of the compound in water; it characterizes the partitioning between hydrophilic and hydrophobic phases in the environmental and the human body frequently used as a correlating variable.

Table 2: Octanol -Water Partition Coefficient Results

COMPOUND	Log Octanol –Water partition coefficient
Acetaldehyde	-0.17

2- Propanal	0.19
Propanal	0.33
Benzene	1.99
Toluene	2.54

2. WSKOWWIN (water solubility of an organic compound using the compounds log octanol-water partition coefficient)

The WSKOWWIN Program (which is part of the EPI Suite) also estimates the water solubility of organic compounds at 25°C. However, the WSKOWWIN Program uses a different methodology (compared to WATERNT) to estimate solubility.

Table 3: Water Solubility Results

COMPOUND	WATER SOLUBILITY AT 25 °C (mg/L)
Acetaldehyde	1 $X10^6$
2- Propanal	2.2 $X10^5$
Propanal	3.06 $X10^5$
Benzene	1790
Toluene	526

3. HENRYWIN (Henry's Law Constant)

Henry's law constant is equilibrium ratio of the concentration of a compound in the gas phase to the concentration of the compound in the gas phase to the concentration of the compound in a diluted aqueous solution.

HENRYWIN estimates Henry's Law Constant (HLC) by two separate methods that yield two separate estimates. The Bond Method is able to estimate many more types of structures than the Group Method because it has a more extensive library of bond contribution values. HENRYWIN requires only a chemical structure to make these predictions.

Table 5: Henry's Law Constant Results

COMPOUND	BOND METHOD Atm –m3/ mole	GROUP METHOD Atm –m3/ mole	HENRY´S LAW CONSTANT Atm –m3/ mole
Acetaldehyde	6.78 X10-5	6 $X10\text{-}^5$	6.67 $X10^{-5}$
2- Propanal	3.58 X10-5	1.94 $X10\text{-}^5$	1.22 X10-5
Propanal	9.01 X10-5	9.48 $X10\text{-}^5$	7.34 X10-5
Benzene	5.39 X10-0.03	5.35 $X10^{-0.03}$	5.55 X10-03
Toluene	5.95 X10-0.03	5.73 $X10^{-0.03}$	6.64X10-3

4. KOAWIN (Octanol-Air Partition Coefficient)

The octanol-air partition coefficient (K_{OA}) is the ratio of a chemical's concentration in octanol to the concentration in air at equilibrium. It is useful for predicting the partitioning behavior between air and environmental matrices such as soil, vegetation, and aerosol particles.

Table 6: Octano l-Air Partition Coefficient Results

COMPOUND	Log Octanol –air partition coefficient

Acetaldehyde	2.224
2- Propanal	2.292
Propanal	3.113
Benzene	2.774
Toluene	3.296

5. BIOHCWIN (biodegradation half-life prediction)

A description of the BIOWIN Biodegradation Probability Program (which is part of the EPI Suite) is available in the on-line Help file for BIOWIN.

Table 8: Biodegradation Half-Time Prediction Results

COMPOUND	OVERALL OH RATE CONSTANT	HALF –LIFE (HOURS)
Acetaldehyde	16.9800 X10-12	7.559
2- Propanal	25.8200 X10-12	4.971
Propanal	21.9908 X10-12	5.837
Benzene	1.9498 X10-12	65.827
Toluene	5.2263 X10-12	24.559

6. KOCWIN (soil adsorption coefficient)

Koc can be defined as "the ratio of the amount of chemical adsorbed per unit weight of organic carbon (Koc) in the soil or sediment to the concentration of the chemical in solution at equilibrium" (Lyman, 1990); it is represented by the following equation (Lyman, 1990):

$$Koc = \frac{\left(\frac{\mu g\ adsorbed}{g\ organic\ carbon}\right)}{\left(\frac{\mu g}{mL\ solution}\right)} \quad (1)$$

The units of Koc are typically expressed as either L/kg or mL/g.

Table 9: Soil Adsorption Coefficient Results

	Koc	
	MCI Method	Kow Method
Acetaldehyde	1	3.219
2- Propanal	1	4.901
Propanal	1	10.52
Benzene	145,8	70.51

7. STPWIN (Chemical Fate in Sewage Treatment Plant)

The STPWIN program (in EPI Suite) is a version of the Sewage Treatment Plant (STP) Model originally developed by Mackay and coworkers at the University of Toronto (Clark et al. 1995), sometimes called the Toronto Model.

The STP Model is based on fugacity principles and like other such models, attempts to predict the fate of an organic chemical in an evaluative environment. In this case the evaluative environment is that of a conventional wastewater treatment plant that uses activated sludge secondary treatment. The model estimates the fate of a chemical present in the influent as it becomes subject to removal by evaporation, biodegradation or other degradation processes, sorption to sludge, and loss in the final effluent. The most critical and uncertain variable is the biodegradation rate and its dependence on biomass concentration.

Table 10: Chemical Fate in Sewage Treatment Plant Results

COMPOUND	Total biodegradation %	Total sludge adsorption %	Total to air %	Total removal %
Acetaldehyde	5.20	0.09	1.70	3.41
2- Propanal	7.57	0.09	1.67	5.81
Propanal	5.52	0.09	1.71	3.71
Benzene	68.94	0.04	1.11	67.78
Toluene	72.77	0.05	2.15	70.57

8. FUGACITY MODEL

In general, fugacity models predict the partitioning of an organic compound in an evaluative environment. A Level III model assumes steady-state but not equilibrium conditions

Half-Life Values

Half-lives are required for air, soil, sediment and water. The half-life for degradation of a chemical in water, soil, and sediment is determined using the ultimate biodegradation expert survey model of the BIOWIN estimation program. However, if a chemical is susceptible to abiotic hydrolysis under environmental conditions, the HYDROWIN program may be able to provide half-life values or an indication of a range of values that could be used. For chemical classes where HYDROWIN can calculate a half-life, it will be up to the user to decide on the pH and temperature that should be used for the half-life estimate and then the user must manually enter the value into the Fugacity, Half-life Selection option.

If the degradation of a chemical is likely to be a combination of chemical hydrolysis, photolysis and biodegradation, the half-lives should be converted to rate constants, the rate constants added together, and then an overall half-life from the summed rate constants should be entered into the Fugacity, Half-life Selection option.

Table 11: Fugacity Principles Results

COMPOUND	Air	Water	Soil	Sediment
ACETALDEHYDE	16,2	360	720	$3{,}24x10^3$
2 PROPANAL	12,7	360	720	$3{,}24x10^3$
PROPANAL	13,1	360	720	$3{,}24x10^3$
BENZENE	209	200	$1{,}8x10^3$	$8{,}1x10^3$
TOLUNE	43	360	720	$3{,}24x10^3$

9. WATER VOLATILIZATION MODEL (Water Volatilization half-lives)

The water depth, the wind velocity and current velocity of the river or lake are parameters required to estimate the volatilization (evaporation) rate of an organic chemical from the water body to air. The default values are:

Table 12: Default Values for Water Sources in EPIWEB

River Water Depth	1 m
River Wind Velocity	5 m/s
River Current Velocity	1 m/s
Lake Water Depth	1 m
Lake Wind Velocity	0.5 m/s
Lake Current Velocity	0.05 m/s

10. HYDROXYL RADICALS REACTION

This table shows the velocity constant and the half-life times by decomposition of the radicals OH create to these compounds. These radicals participate in multiples environmental reaction with the impact on environment and human health.

Table 13: Hydroxyl radical's reaction

	Over all OH Rate Constant (cm^3/ molecule – s)	**Half life (Hours)**
Acetaldehyde	16.9800 $X10^{-12}$	7.559
2- Propanal	25.8200 $X10^{-12}$	4.971
Propanal	21.9908 $X10^{-12}$	5.837
Benzene	1.9498 $X10^{-12}$	65.827
Toluene	5.2263 $X10^{-12}$	24.559

The conclusions are going to give with base of the numbers and the impact of the substances that are generated in different reactions of glycerol oxidation. Each compound in this case study was compared with a pattern or the main compound: GLYCEROL, so the characteristics like Henry´s law Constant, Octanol-Water Partition Coefficient, biodegradation, half-life time and soil adsorption coefficient are given to establish environmental impacts depending on the numbers thrown out by the software.
If the value is high in Henry´s law constant, the probability of going to the gas phase when is less the gas solubility in water will be high, therefore compound like Propanal, Acetaldehyde, 2 propanal have low volatility. Toluene and benzene has moderate volatility.

The results on Octanol-Water Partition Coefficient are the reason between the concentrations of a substance in a mixture of two immiscible solvents biphasic in equilibrium: n - octanol and water; the results obtained in the software EPIWED indicate the behaviour of each compound to study, concluding that toluene which has a coefficient of 2.54 is what makes lipid behaviour that can be absorbed by fatty tissues, while the Acetaldehyde has a coefficient of -0.17 which is closes to glycerol which has a coefficient -0.67 which makes having a hydrophilic behaviour.

Software EPIWEB is able to determined when a compound or substance is biodegradable, this tool is very important in an environmental impact study because it can help to determine how digestible in the environment and how long it taken in the decomposition to avoid the effects of contamination. Biodegradation results are attached to the half-life time; the persistence of compound at environment is directly proportional to the half-life time.

Half-life time is determined like the necessary time to carry out the biodegradation of the compound as well as benzene, the reaction is the slowest than the other compounds therefore it is the most persistence in the environment in relation with the other compounds. Also could establish that the main environmental impact in the soil is generated by the benzene as its level of adsorption determined by soil adsorption coefficient called KOCWIN in the software is higher almost 200 percent than the others compounds in this case study.

References

1. Allen D., Shonnard D., Green Engineering: Environmentally Conscious Design of Chemical Processes.

(2) EPA (United States Environmental Problems Agency) http://www.epa.gov/oppt/greenengineering/ accessed 10.2011

(3) EPA's Software, Estimation Program Interface (EPI) Suite <http://www.epa.gov/oppt/exposure/pubs/episuite.htm accessed between August to November 2011

(21f) Magnetic Carbon Nanocomposites: Synthesis and Environmental Remediation

Zhanhu Guo[1], Jiahua Zhu[1] and Suying Wei[2], (1)Dan F. Smith Department of Chemical Engineering, Lamar University, Beaumont, TX, (2)Chemistry and Biochemistry, Lamar University, Beaumont, TX

A synergistic process to recycle polystyrene (PS)/Ni@NiO polymer nanocomposites (PNCs) and to fabricate magnetic core@shell Ni@C nanoparticles is introduced. Different pyrolysis pathways of the PNCs have been suggested basing on the analysis of both gas and liquid pyrolysis products from gas chromatography/mass spectrometry (GC/MS). The pyrolysis pathway follows radical generation, recombination and hydrogenation of the PNCs, which generate more saturated structures. However, pure PS undergoes a radical generation and β-scission to form more unsaturated structures. The coke formed during the PS pyrolysis was used as a carbon resource to in-situ synthesize a graphite shell surrounding Ni nanoparticle surface. This Ni@C nanoparticle shows a significantly larger specific area (236.68 m^2/g) as compared to bare Ni@NiO nanoparticles (119.62 m^2/g) and can be stabilized in 1 M HCl for hours. A large saturated magnetization was also observed in the Ni@C nanoparticles. All these characters make it very useful in environmental remediation applications, with the advantages of higher adsorption capacity and easy magnetic separation process. The application of magnetic carbon nanocomposites in the environmental remediation is addressed as well with the detailed examples in Cr(VI) and As(III) removal from polluted water.

Reference:

Comprehensive and Sustainable Recycling of Polymer Nanocomposites, *Journal of Materials Chemistry*, 21, 16239-16246 (2011) Doi:10.1039/C1JM13044G

Determining the Rate of Biodegradation in Fuel Impacted Karst Aquifers

Roger Painter[1], Tom Byl[1,2], Lonnie Sharpe[1], Valetta Watson[1] and Tony Patterson[1]

[1] Civil and Environmental Engineering, Tennessee State University, Nashville, TN 37209, USA; rpainter@tnstate.edu, lsharpe@tnstate.edu, ahmadk19@gmail.com, justinvaughnharris@gmail.com

[2] United States Geological Survey, Nashville, TN 37211, USA; tdbyl@yahoo.com,

* Author to whom correspondence should be addressed; E-Mail: Rpainter@Tnstate.edu; 3500 John A. Merritt Blvd., Nashville, TN, 37209 Tel.: (615)963-5388; Fax :(615)963-5902 .

Abstract

Hydrologic and geologic characteristics of fractured rock aquifers have been described as not being suited for natural bioremediation because of small microbial populations. It is also widely perceived that karst groundwater often has insufficient residence time for biodegradation to occur. It is perhaps due to these perceptions that less research has been conducted for quantitative modeling of biodegradation in karst as compared to consolidated aquifers. Modeling biodegradation in karst is in the domain of non-ideal chemical reaction kinetics. For all non-ideal flow reaction zones, the effluent concentration response to an input of tracer must lie between the limiting cases of completely mixed flow and purely plug flow. The residence time distribution function (RTD) for tracer molecules in a single karst conduit or a complex system of conduits is a probability density function which can be interpreted to define the probability that contaminant molecules present at the influent at time equals zero will arrive at the effluent after a particular amount of time. To demonstrate this methodology the biodegradation rate of a contaminant (toluene) in raw karst groundwater from a BTEX impacted site in central Kentucky was quantitatively measured in batch microcosm studies and the extent of biodegradation of toluene in the same groundwater was measured for a complex flow system. The values of the pseudo first order rate constant (k') obtained ranged from 0.017 $(hr)^{-1}$ to 0.0210 $(hr)^{-1}$ compared to 0.0186 $(hr)^{-1}$ for the microcosm experiments. The close agreement between the values of k' obtained from the static microcosms and the ADE model indicate that the ADE does an adequate job of modeling the RTD for modeling biodegradation in karst aquifers. The values of k' obtained correspond to a half-life of less than two days for toluene and this has major implications to issues regarding natural attenuation of fuel impacted karst sites. The RTD approach presented here is applicable to biodegradation in any karst system that has been characterized with a tracer response curve from a conservative tracer study.

Key Words: Bioremediation, Karst, Resident Time Distribution, Advection Dispersion

---

1. Introduction

The high costs and technical difficulties associated with conventional remediation strategies make in situ biodegradation a promising approach for cleaning up contaminated aquifers [1]. The lack of studies examining biodegradation in karst aquifers may be due to the widespread perception that contaminants are rapidly flushed out of karst aquifers [2]. In highly developed and well-connected conduit systems, the rate of contaminant migration is expected to be much faster than the rate of biodegradation [3, 4]. However, the belief that contaminants are rapidly flushed out of karst aquifers is a popular misconception. Tracer studies suggest that large volumes of water may be trapped in fractures along bedding planes and other features isolated from active groundwater flow paths in karst aquifers [5]. In areas isolated from the major conduit flow paths, contaminant migration may be slow enough that biodegradation could significantly reduce contaminant mass if favorable microorganisms, food sources, and geochemical conditions are present [6, 7, and 8]. Researchers have also implied that natural bioremediation in karst or fractured rock is unlikely to occur because of the microbiological characteristics of karst aquifers; small microbial populations and low surface-area-to-volume ratio. Typical microbial numbers for material from unconsolidated aquifers have been reported to range from 1 x 104 to 1 x 107 cells per milliliter (cells/mL) [9].Studies have shown that water from bedrock aquifers also may contain microbial populations within this range. For example, total microbial populations of 9.7 x 105 to 8.5 x 106 cells/mL and heterotrophic bacteria populations of 3.5 x 103 to 5.0 x 105 cells/mL were detected in ground-water samples collected from a gasoline-contaminated karst aquifer in Missouri [10]. Greater than 70 percent of bacteria in consolidated aquifers are attached to solid surfaces. This fact may have led to the assumption that natural bioremediation in karst conduits is negligible because contact between attached bacteria and contaminants would be limited by the smaller surface area to volume ratio for karst.

The literature shows that efforts to mathematically model karst aquifers have followed two different and complementary approaches [11, 12, 13, and 14]:

1. A probabilistic approach that focuses on input events and output responses and takes into account local field observations of flow and transport processes.
2. A determinist approach that takes into account theoretical concepts of aquifer structure, the physics of each mechanism involved in the transfer and flow of water, and from that tries to simulate the hydraulic behavior of the aquifer.

There are four major conceptual approaches for deterministic models for karst groundwater:

1 The equivalent porous medium (MODFLOW);

2 The discrete fracture network;

3 The double porosity continuum;

4 The triple porosity approach – matrix, fracture and conduit.

MODFLOW should be used with caution because the degree of geologic heterogeneity exhibited by karst aquifers precludes the application of Darcy's Law and the use of average model inputs based on the assumed homogeneity of the subsurface [15]. Karst modeling is difficult due to the inability to adequately characterize the heterogeneity of the flow parameters, i.e. determine the geometry and hydraulic parameters of the high-conductivity conduit network [16, 17]. The second problem is related to the development of a computer code based on numerical methods. Strong heterogeneity of the flow medium and different flow regimes in high-conductivity channels and in the low-permeability matrix make the development of appropriate computer modeling codes difficult. The third problem is the transfer of the simulated results to a real system. Because of the higher degree of heterogeneity, uncertainties are much higher for karst systems than for most porous aquifers [18, 19]. Efforts are under way to develop more sophisticated deterministic models for karst and the triple porosity approach shows the most promise going forward since most karst aquifers have the common characteristics of storing most of the groundwater in the primary matrix and secondary fissure system and yet transmitting most of the groundwater through tertiary conduits.

When considering the adequacy and appropriateness of a probabilistic or deterministic modeling approach, it is important to be clear about the objective of the model. Deterministic models work well if the objective is to model hydraulic heads, groundwater fluxes and spring discharge. On the other hand, existing deterministic models for karst aquifers perform poorly for the prediction of flow direction, velocity and solute destination. Consequently a probabilistic modeling approach is appropriate for modeling biodegradation of contaminants in karst aquifers. The release of a contaminant into a karst aquifer and the eventual discharge of the contaminant at a down gradient spring is analogous to the operation of a non-ideal flow chemical reactor. Therefore a further justification for a probabilistic modeling approach is the fact that the state of the art chemical reaction engineering approach for modeling such a contaminant undergoing biodegradation is the application of a probabilistic biodegradation model based on an RTD derived from a field tracer study. Finally, another validation of a probabilistic approach for modeling biodegradation in karst aquifers is the wide use of quantitative field tracer studies to provide information regarding karst flow [20, 21].

Tracer studies are an integral step in determining the RTD of contaminants in a karst system. Much work has been done to develop models for complex groundwater and surface water systems which attempt to describe the RTD in terms of mathematical functions independent of tracer study information. The numerical transient model (OTIS) and the Solute Transport and Multi-rate Transport – Linear Coordinates (STAMMT-L) model are examples of such models [22]. However, a RTD based on tracer study data is often the single most important site specific information for modeling karst conduits. The RTD is not a complete description of a karst system but a given RTD is unique for a particular system. For large complex systems, representative tracer studies of particular conduits can be combined to provide a model of the entire complex karst system. The modeling of complex reaction systems as a number of simpler systems connected in series and parallel is an established method used for reactor design by chemical reaction engineers. In interpreting the results of a tracer study it is assumed that the contaminant experiences the same RTD as the conservative tracer except that the contaminant experiences biodegradation while in transient through the karst aquifer.

USEPA's QTRACER model is a probabilistic model that uses the one dimensional advection dispersion equation (ADE) to model the RTD for karst. In this context, it is a natural extension of chemical reaction engineering to use field tracer data to construct an RTD based on the ADE and couple the RTD to a chemical kinetic model to describe biodegradation in the karst aquifers. It is important to stress that the ADE is not being used here to model the physical characteristics of the aquifer flow. The physical characteristics of the aquifer are subsumed into the RTD and the ADE is modeling the RTD. The appropriateness of this approach is evident in the fit of the model RTD to the experimental RTD derived from the tracer study data.

2. Materials and Methods

Laboratory flow through karst microcosms shown in Figure 1 were constructed using a 20-liter glass reservoir and twelve 1-liter volumetric flasks arranged in three replicate systems of four flasks connected in series.

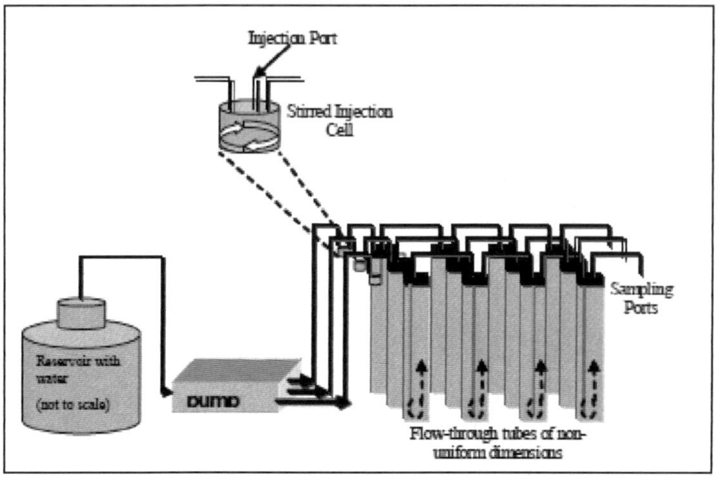

Figure 1: Laboratory flow through karst microcosms.

Water was pumped into all three systems by using a high-performance peristaltic pump. A stirred injection cell (10 mL volume) was placed at the entrance of each replicate system for the injection of tracer or toluene. During the conservative tracer studies Rhodamine dye was simultaneously injected into the stirred injection cells of the packed and unpacked systems. The Rhodamine concentration at the discharge port was monitored through time by collecting samples at 1-2 hour time intervals over a 4 day period. A fluorometer was used to quantify the Rhodamine in the water samples. The lower detection limit on the fluorometer was established at 100 parts per trillion.

Toluene was selected as the experimental contaminant because it is a component in most fuels and because previous work indicated pseudomonad bacteria, which are heterotrophic aerobic bacteria (HAB), from the Kentucky site could grow using toluene as a food source. The biodegradation experiments used water containing live bacteria collected from a 120-foot-deep well completed in a karst aquifer in south-central Kentucky. Bacteria counts ranged from approximately 700,000 bacteria per milliliter to 1.2 million depending on the well and sample collection time. These bacteria counts were derived using two methods, direct counts and BART growth tests, and the results of the two tests were within 20 percent of each other. Bacteria from the fuel contaminated part of the karst aquifer had a 5% lighter buoyant density and a wider range of sizes than the bacteria from the non-contaminated well. Additionally, bacteria isolated from fuel contaminated ground-water samples readily grew with dissolved gasoline as the only source of food. Static microcosms set up using aerated raw karst water and spiked with Toluene at 1 mg/L established a pseudo first order biodegradation rate constant of 0.0186 hr^{-1}. Sterile control microcosms had less than 10% toluene loss over the same time period.

Before the tracer study was initiated, the experimental systems were sterilized with bleach. The bleach was neutralized with sterile sodium thiosulfate. During the conservative tracer study, a constant flow rate of approximately 3 milliliters per minute (mL/min) was established for both systems. The water flowing through the abiotic system was sterile water that had a pH of 10. Previous work indicated that elevating the pH to 10 maintains an abiotic system. At the beginning of the tracer study, 300 micrograms (µg) of Rhodamine dye (1,500 µg of 20% wt/wt solution) was injected into each stirred injection cell. The Rhodamine concentration at the discharge port was monitored through time by collecting samples over a 4-day period.

Before the toluene biodegradation study was initiated toluene was injected into an abiotic system to investigate loss of toluene to vaporization and or adsorption, The experimental systems were sterilized with bleach. The bleach was neutralized with sterile sodium thiosulfate. Filter-sterilized toluene (87 µg) dissolved in 100 micro liters (µL) of methanol was delivered into the injection cell and pushed through the abiotic system with sterile water that had a pH of 10. Previous work indicated that elevating the pH to 10 maintains an abiotic system. Toluene concentration was monitored at the discharge port over the next 5 days. Water samples were collected in clean 40-mL volatile organic compound (VOC) vials every 1 to 4 hours. The water samples were immediately analyzed on a gas chromatograph (GC) equipped with a purge-and-trap system, silica-film capillary column, argon-carrying gas, and micro-argon ionization detector.

The karst groundwater containing live bacteria was then pumped through the system for four days to establish a bio-film on the glass surfaces. An 87-µg aliquot of toluene was then injected into the biotic system in the same manner as for the abiotic test. In order to document the presence of attached bacteria, glass slides were suspended in both the packed and unpacked systems. The suspended slides were removed prior to and at the end of the experiments and viewed using an epifluorescent microscope and the direct count method [23]. Bacteria counts using MPN and microscopic methods were used to confirm that bacteria covered the glass surfaces. Figures 1 and 2 shown below are representative of the photographs of the glass surfaces.

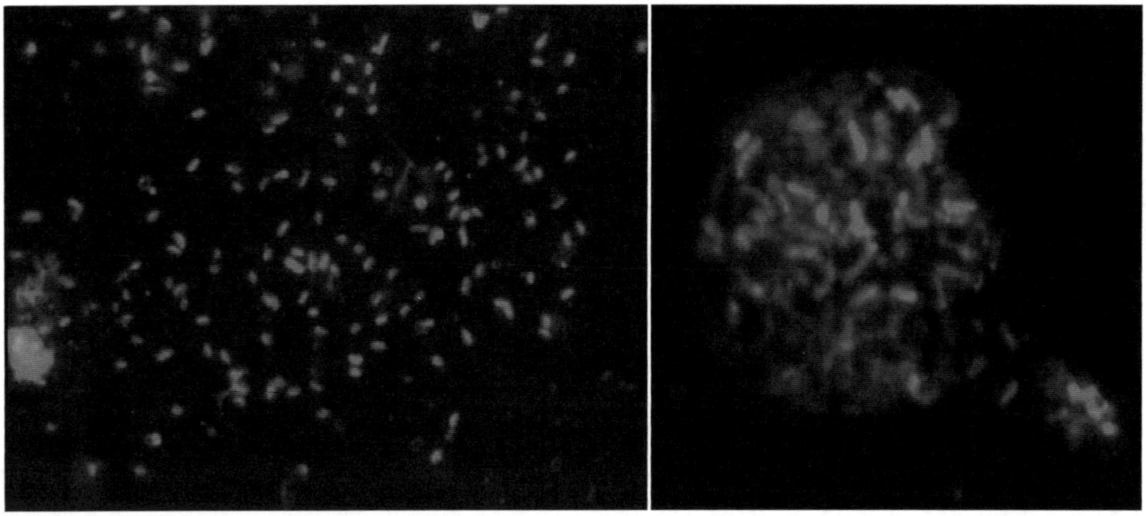

Figure 1. **(Left)** Bacteria (white objects) attached to the surface of the glass after 3 days of pumping water through the system (400x magnification, epifluorescent), and **(Right)** close up of a bacteria cluster on the surface of the glass (800x magnification, epifluorescent).

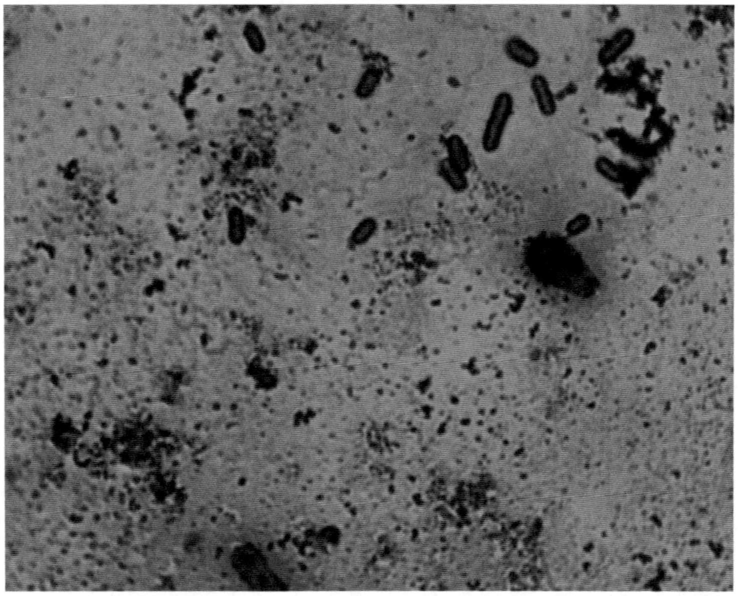

Figure 2. Free-living bacteria (dark objects) collected from the water column after 3 days. Flagella can be observed attached to the rod-shaped bacteria (1,000x magnification, bright field).

For a sparingly soluble contaminant (A) reacting with (B) the reaction may actually be second order or variable order:

$$\frac{dC_A}{dt} = kC_A C_B \qquad \text{Equation 1}$$

Where:
C_A = Concentration of Contaminant;
C_A = Concentration of Microbes;
k = Biodegradation Rate Constant;
and;
t = Time.

But in a case where the microbes are acclimated and at steady state and electron acceptors are not limiting the concentration C_B may not change appreciably while C_A does. Treating C_B like a constant:

$$\frac{dC_A}{dt} = kC_A C_B \Rightarrow k'C_A \qquad \text{Equation 2}$$

Where:

$k' = kC_B$ = pseudo first order rate constant and the biodegradation is modeled as pseudo-first order.

This does not however require that the mechanism of the microscopic chemical reaction be mono-molecular. The actual biodegradation rate is variable order and is a function of electron donor (carbon substrate) and microbial population. However, it is a common practice in chemical engineering to model complex reactions with pseudo first order kinetics. The following application of chemical reaction engineering theory to is strictly valid if the chemical biodegradation kinetics information is liner with respect to intensive fluid properties.

The residence time distribution function (E(t)) for contaminant molecules in a single karst conduit or a complex system of conduits is a probability density function (p.d.f.) which can be interpreted to define the probability that contaminant molecules present at the influent at time equals zero will arrive at the effluent after a particular amount of time. The RTD is depicted as a plot of E(t) versus time as time goes from zero to infinity.

The mathematical approach presented below involves coupling biodegradation kinetics to the RTD by application of the ADE. The transient form of the ADE was solved and fitted to the tracer study data to estimate the systems Peclet number. This Peclet number was incorporated into a second solution of the ADE with a chemical reaction term that was fitted to the toluene degradation data to estimate the pseudo first order rate constant.

E(t) was determined by injecting a pulse of a conservative tracer (rhodamine) into the system shown in Figure 1 at time (t) = 0 and then measuring the tracer concentration in the effluent as a function of time. At some time after injection of the tracer, consider an increment of time dt small enough that the tracer concentration exiting the reactor between t and t + dt is essentially constant. The amount of tracer leaving the reactor between t and t + dt is then in differential form:

$$dN = QC(t)\,dt \qquad \text{Equation 3}$$

Where:
Q = volumetric flow rate;
C(t) = concentration at time t;
and;
N = Mass

Where dN denotes the amount of material that has spent an amount of time between t and t + dt and Q is the volumetric flow rate which is taken to be constant. If N_0 is the amount of tracer injected at t = 0 then the fraction of material leaving the reactor between t and t + dt is

$$\frac{dN}{N_0} = \frac{QC(t)}{N_0}\,dt \qquad \text{Equation 4}$$

And for a pulse input the residence time distribution function is defined as

$$E(t) = \frac{QC(t)}{N_0} \qquad \text{Equation 5}$$

So that

$$\frac{dN}{N_0} = E(t)\,dt \qquad \text{Equation 6}$$

Noting that N_0 is the sum of all the tracer leaving the system

$$N_0 = \int_0^{\infty} QC(t)\,dt \qquad \text{Equation 7}$$

And combining Equations 3, 6 and 7 gives

$$E(t) = \frac{C(t)}{\int_0^{\infty} C(t)\,dt} \qquad \text{Equation 8}$$

E(t) is the pdf for the random variable C(t) and the first two moments of E(t), define the mean residence time and variance of the distribution. The mean residence time (t_m) is:

$$t_m = \int_0^{\infty} t\, E(t)\, dt \qquad \text{Equation 9}$$

And the variance (σ^2) is:

$$\sigma^2 = \int_0^{\infty} (t - t_m)^2\, E(t)\, dt \qquad \text{Equation 10}$$

For the non-ideal flow characteristics of karst E(t) and its moments are obtained by numerical integration of the tracer breakthrough data.

The advection, dispersion equation for the conservative tracer is:

$$\frac{\partial C(t)}{\partial t} = D_a \frac{\partial^2 C(t)}{\partial z^2} - U \frac{\partial C(t)}{\partial z} \qquad \text{Equation 11}$$

Where:
D_a = dispersion coefficient;
z = distance
and;
U = velocity.

D_a is a measure of the degree of mixing in the fluid due to molecular and eddy diffusion as well an indication of the nature of the fluids velocity profile due to viscous effects. This equation effectively superimposes mass transfer of the tracer/contaminant due to axial dispersion equal to $-D_aA_c(dc/dz)$ onto the convective mass transfer which is equal to $UA_{c.}$ It may appear that this model is capable of accounting only for axial mixing but it can be shown that the model accounts for radial mixing as well.

Equation 11 can be put in dimensionless form to facilitate its solution as:

$$\frac{1}{P_e}\frac{\partial^2 \psi}{d\lambda^2} - \frac{\partial \psi}{d\lambda} = \frac{\partial \psi}{\partial \lambda} \qquad \text{Equation 12}$$

Where,

$$\psi = \frac{C}{C_0};$$

$$C_0 = concentration\, at\, time = 0$$

$$\lambda = \frac{z}{L};$$

$$\theta = \frac{tU}{L};$$

$$and;$$

$$P_e \quad = \quad \frac{UL}{\lambda}$$

After a narrow pulse of tracer is injected, molecular diffusion and eddy diffusion cause the pulse to widen as the tracer molecules disperse randomly in all directions. This is depicted below in Figure 3. Consequently D_a and hence Pe are directly related to the resulting mean residence time and variance of the RTD obtained from the tracer concentration versus time data.

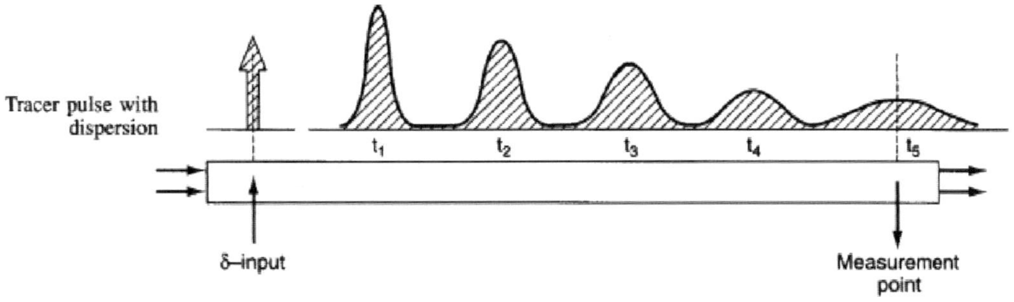

Consequently the solution of Equation 12 for Dankwerts boundary conditions for an open-open system gives the following relationship between the Peclet number (P_e) for the non-ideal flow system, the mean residence time (t_m), and the variance (σ^2) [24, 25]:

$$\psi = \frac{1}{2\sqrt{\frac{\pi\theta}{Pe}}} e^{\frac{-(1-\theta)^2}{\frac{4\pi}{Pe}}} \qquad \text{Equation 13}$$

$$t_m \quad = \quad \left(1 + \frac{2}{P_e}\right)\tau \qquad \text{Equation 14}$$

$$\frac{\sigma^2}{\tau^2} = \frac{2}{P_e} + \frac{8}{P_e^2}$$ Equation 15

Equations 14 and 15 often appear in the literature as equalities, in reality, the equations result from truncation of infinite series and are strictly valid only for large P_e. This is of practical importance since the whole concept of an apparent diffusivity in a dispersion-like equation breaks down for very low values of P_e where the flow approaches mixed flow. Mixed flow on the macroscopic scale is not achieved in typical karst. Fortunately P_e values for karst aquifers are relatively large due to their long effective path lengths. The Peclet number is a measure of the relative contribution of mechanical dispersion and diffusion to solute transport. It relates the effectiveness of mass transport by advection to the effectiveness of mass transport by either dispersion or diffusion. Peclet numbers below 0 4 indicate diffusion control; 0.4 – 6.0 suggest that diffusion and advection are in transition and thus approximately equal to each other; and greater than 6 0 indicates advection control . The Peclet numbers for karst are typically many times greater than 6.0. [26, 27]. In this study, P_e ranged from 12.01 to 14.80.

Values of σ^2 and t_m from the tracer study are used to estimate P_e and τ and these parameters are used to estimate the extent to which the toluene flowing through the system is biodegraded. The experimental value for the fraction of toluene biodegraded was obtained by numerically integrating the toluene concentration versus time data. The conversion as a function of time (X(t)) is:

$$X(t) = 1 - \frac{C_A(t)}{C_{A_0}(0)}$$ Equation 16

X(t) is the fraction of the initial contaminant biodegraded after some time (t) and since the RTD function (E(t)) is the probability density function (p.d.f.) for C(t), the mean or expected value of X(t) is:

$$X_m = \int_0^\infty X(t)\, E(t)\, dt$$ Equation 17

And noting that the rate of change of X(t) is independent of concentration and hence independent of the degree of mixing it follows that for first order kinetics:

$$X(t) \;=\; 1-e^{-k't} \qquad \text{Equation 18}$$

And the mean conversion is obtained by combining equations 17 and 18 [28]:

$$X_m \;=\; 1-\int_0^{\infty} e^{-k't}\, E(t)dt \qquad \text{Equation 19}$$

The steady state advection dispersion equation for the toluene undergoing biodegradation as it flows through the system is:

$$\frac{D_a}{U}\frac{dC}{dz^2}-\frac{dC}{dz}+\frac{r_A}{U} \;=\; 0 \qquad \text{Equation 20}$$

Where;
r_A = volumetric reaction rate.

This equation is linear for zero or first order kinetics and thus can be solved analytically. For higher order kinetics the equation must be solved numerically. As with the advection dispersion equation for the conservative tracer the equation for toluene undergoing degradation can be written in dimensionless terms as:

$$\frac{1}{P_e}\frac{d^2\psi}{d\lambda^2}-\frac{d\psi}{d\lambda}-D_A\psi \;=\; 0 \qquad \text{Equation 21}$$

Where D is the Damkohler number defined as:

$$D \;=\; C_{A0}^{n-1}\tau \;;$$

and for first order reactions (or assuming the reaction can be modeled by pseudo-first order kinetics):

$$D \;=\; k'\tau$$

Thus Equation 17 couples information from the RTD obtained from the tracer study to the chemical reaction kinetics term. In steady state mass transfer, the transient terms in the "Open-Open System" boundary conditions disappear and reduce to the Dankwerts "Closed-Closed" Boundary Condition. Furthermore, this equation is linear for first order kinetics and thus can be solved analytically. The solution for assumed pseudo first order biodegradation kinetics is [29]:

$$X_m = (1-\psi_m) = 1 - \frac{4qe^{(P_e/2)}}{(1+q)^2\, e^{(P_e q/2)} - (1-q)^2\, e^{(-P_e q/2)}} \qquad \text{Equation 22}$$

Where;
X_m = the fraction of toluene biodegraded;
and;
$q = \sqrt{1 + 4\, D/P_e}$

3. Results

The experimental values for the conversion (X_m) of toluene from the flow microcosms and the Peclet numbers from the conservative tracer studies allows for the calculation of the values for k'. The results from numerical integration of the concentration versus time data for the three tracer studies conducted are shown in Table 1.

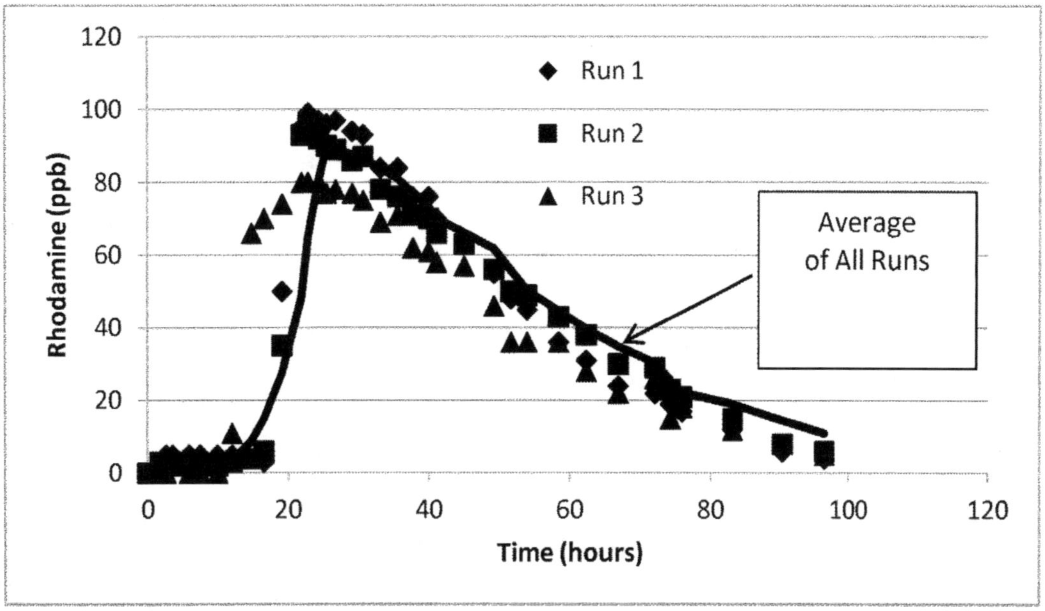

Figure 3. Concentration versus Time data for the conservative tracer test runs. This data was used to determine the mean, variance, and Peclet number for the laboratory karst system.

Table 1 Experimental Results for Tracer Studies			
	Run 1	Run 2	Run 3
σ^2 (hr)^2	297.7	319.4	348.7
t_m (hr)	41.13	43.13	39.65
Space Time (hr)	36.15	38.00	33.99
Peclet Number	14.50	14.80	12.01
Mass in (μg)	476.0	476.0	476.0
Mass out (μg)	461.2	428.8	441.9
Recovery (%)	96.0	90.0	93.0

The toluene data shown in Figure 3 was numerically integrated to determine the experimental values for X_m and the values of k' shown in Table 2 were calculated with Equation 22.

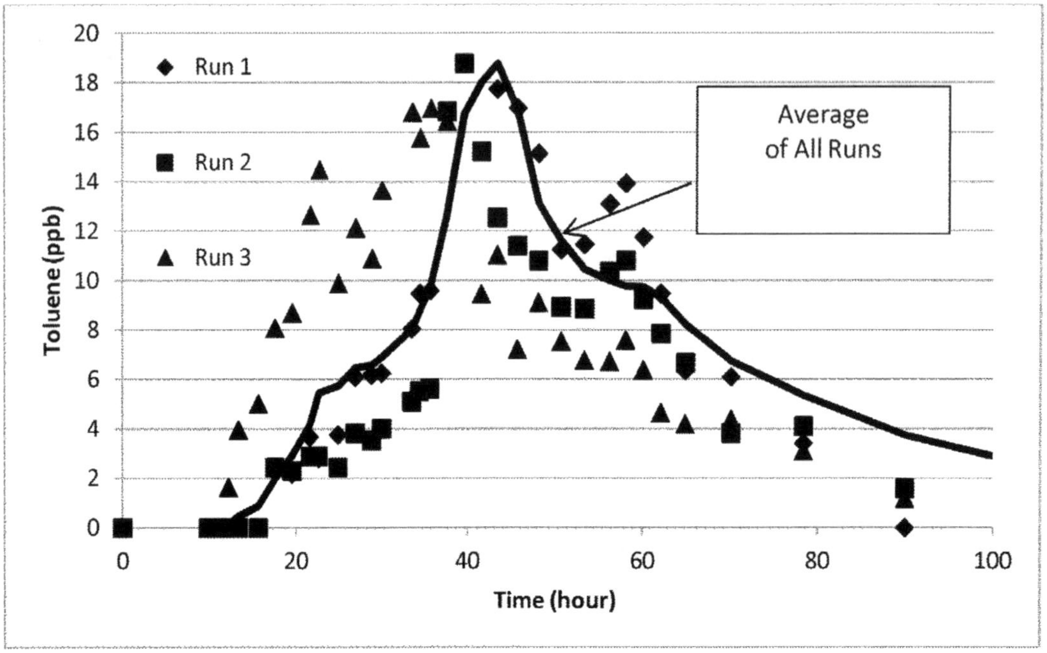

Figure 4. Concentration versus Time data for toluene undergoing biodegradation. This data was used to determine the pseudo first order reaction rate constant for the laboratory karst system.

Table 2 Experimental Results for Biodegradation Studies			
	Run 1	Run 2	Run 3
Toluene Injected (μg)	87.0	87.0	87.0
Recovery Abiotic (μg)	84.6	62.6	75.6
Recovery Biotic (μg)	41.5	36.5	41.6
Conversion (Exp.)	0.44	0.48	0.52
k' $(hr)^{-1}$	.0170	.0180	.0210

To facilitate the comparison of the experimental RTD to the model RTD a dimensionless RTD function was calculated as:

$$\boldsymbol{E}(\boldsymbol{\theta}) = \frac{\psi}{\int_0^\infty \psi \, d\theta} \qquad \text{Equation 23}$$

The experimental dimensionless RTDs are compared to the model RTD in Figure 5.

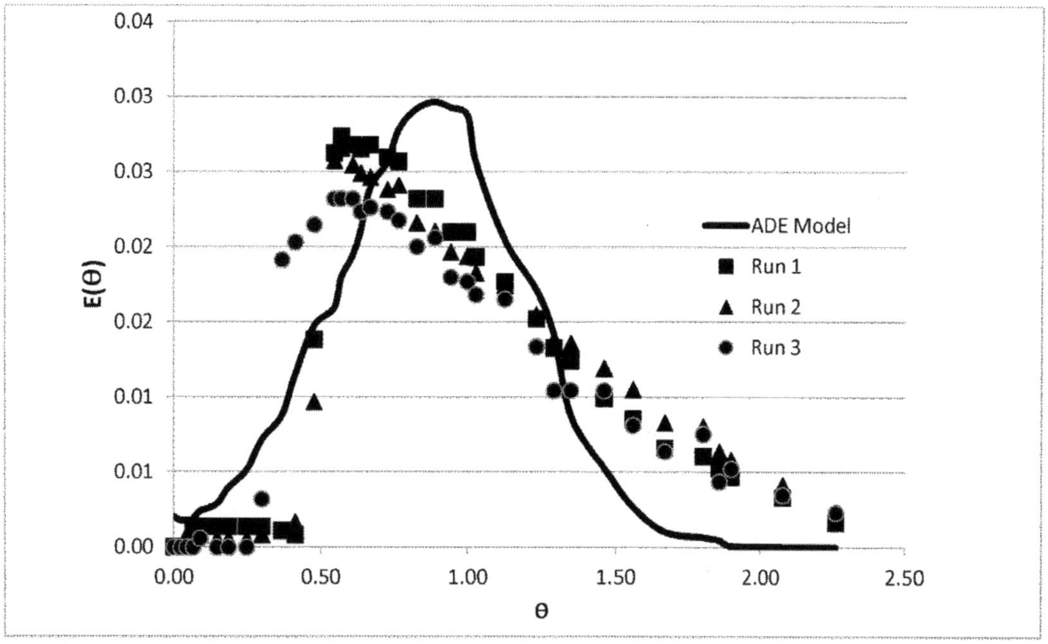

Figure 5. Dimensionless experimental RTDs obtained from numerical integration of the tracer data are compared to the RTD from the ADE.

4. Discussion and Conclusions

The scope of the results of this research is not limited to model development; it also addresses the fundamental issue of whether natural aerobic biodegradation occurs to a significant degree in karst. The perception that contaminants are quickly flushed out of karst aquifers is largely incorrect. Tracer studies for karst aquifers often indicate that these aquifers are characterized by diverse flow regimes and storage capabilities [30, 31]. The range of flow velocities between the least permeable and the most permeable portions of the aquifer is typically over 6 to 10 orders of magnitude [32].

An RTD approach for modeling biodegradation in karst aquifers is validated by the wide use of quantitative field tracer tests to provide information regarding karst flow. It is a natural extension of chemical reaction engineering to use field tracer data to construct an RTD and to model biodegradation in the system. The biodegradation of toluene in groundwater from a karst aquifer in central Kentucky was adequately described by pseudo first-order reaction kinetics. The values of k' obtained from the model ranged from 0.017 $(hr)^{-1}$ to 0.0210 $(hr)^{-1}$. The close agreement between the values of k' obtained from the static microcosms and the ADE model indicate that the ADE does an adequate job of modeling the RTD with regard to the area beneath the RTD curve. This area is proportional to the mass of the contaminant in the effluent. Comparison of the experimental RTDs to the RTD from the ADE in Figure 5 reveal that the solution of the ADE tends to be more symmetric than typical tracer RTD and fails to reflect the long upper tail that is typical of tracer study data. Methanol was used to promote dissolution of toluene to allow the delivery of a true concentration impulse input to the system. This poses a concern as the presence of methanol as a food source could augment biodegradation of toluene artificially. The close agreement of the reaction rate constants for the flow microcosm with the batch microcosm which did not incorporate methanol addresses this concern to some extent. However, this is a legitimate concern and future work may address this concern by conducting the experiment for a stepwise input instead of an impulse. This experimental approach would not require the injection of methanol.

The values obtained for k' correspond to a half-life of less than two days for toluene. This result has major implications for issues regarding the natural attenuation of fuel-impacted sites. This result indicates that the natural aerobic biodegradation of toluene will proceed to a significant degree in karst aquifers despite shorter residence times. This methodology is an adequate and appropriate approach for modeling biodegradation of contaminants traversing any karst system that is characterized by tracer study breakthrough data. The literature reports values of P_e ranging from 6.3 to 22 and in this study P_e ranged from 12.01 to 14.80. The appropriateness of Equation 12 for P_e in this range is directly verifiable by comparing the RTD obtained from the ADE to the experimental RTD from the tracer study data.

References

1. Bombach P, R. H., Kästner M, Fischer A. (2010). "Current approaches for the assessment of in situ biodegradation." *Applied Microbiology Biotechnology*, 86: 839-852.

2. Schwarz, K., T. Gocht, et al. (2011). "Transport of polycyclic aromatic hydrocarbons in highly vulnerable karst systems." *Environmental Pollution* **159**: (1): 133-139.

3. J.-C., Poty, F., Mahler, B. and Drogue, C. (2004), Colonization by Aerobic Bacteria in Karst: Laboratory and In Situ Experiments. *Ground Water*, 42: 526–533. doi: 10.1111/j.1745-6584.2004.tb02621.x

4. Ryan, M., and J. Meiman. 1996. An examination of short-term variations in water quality at a karst spring in Kentucky. *Ground Water* 34, no. 1: 23-30.

5. Field, M.S. Karst hydrology and chemical contamination. *Journal of Environmental Systems* **1993**, *22*, 1-26.

6. Byl, T.D.; Painter, R. Microbial Adaptations to Karst Aquifers with Contaminants. In Proceedings of American Water Resources Association (AWRA) Annual Water Resources Conference, Seattle, WA, USA, 9-12 November 2009.

7. Harvey, R.W.; Barber, L.B, II. Association of free-living bacteria and dissolved organic compounds in a plume of contaminated groundwater. *Journal of Contaminant Hydrology* **1992**, *9*, 91-103.

8. Harvey, R.W.; Smith, R.L.; George, L. Effect of organic contamination upon microbial distributions and heterotrophic uptake in a Cape Cod, Mass., aquifer. *Applied and Environmental Microbiology* **1984**, *48*, 1197-1202.

9. Ghiorse, W.C., and Wilson, J.T., 1988, Microbial ecology of the terrestrial subsurface: *Advances in Applied Microbiology*, v. 33, p. 107-172.

10. O'Connor, J.T.; Brazos, B.J. The Response of Natural Ground Water Bacteria to Ground Water Contamination by Gasoline in a Karst Region. In Proceedings of the Conference on Hazardous Waste Research, Kansas State University, Manhattan, Kansas, USA, 1991; Erickson, L.E. Ed.; Publisher: City, Country, Year; pp 81-293

11. Kiraly, L. 1998. Modeling karst aquifers by the combined discrete channel and continuum approach. *Bulletin Hydrogeology* 16, 77–98.

12. Teutsch, G. and Sauter, M., 1998.Distributed parameter modeling approaches in karst-hydrological investigations. Bulletin d'Hydrogéologie (Neuchâtel), 16, 99-109

13. Smart, C. and Worthington, S.R.H., 2004a. Groundwater in karst. In: J. Gunn (Editor), Encyclopedia of Cave and Karst Science. Fitzroy Dearborn.

14. Domenico, P.A. & Schwartz, W., 1998. *Physical and Chemical Hydrogeology* Second Edition, Wiley.

15. Quinlin, JF. 1989, Ground-water monitoring in karst terrains: recommended protocols and implicit assumptions: Las Vegas, Nev., U.S. Environmental Protection Agency, Environmental Monitoring Systems Laboratory, EPA/600/X-89/050, 100 p. Also available at http://www.karstwaters.org/files/gwkarstr.pdf

16. Wolfe, W.J.; Haugh, C.J.; Webbers, Ank; Diehl, T.H. Preliminary conceptual models of the occurrence, fate, and transport of chlorinated solvents in karst U.S. Geological Survey Water-Resources Investigations Report ; U.S. Geological Survey, 1997

17. Liedl, R. and Sauter, M.: Modelling of aquifer genesis and heat transport in karst system, Bulletin d'Hydrog´eology, 16, 185–200,1998

18. Adams, R. and Parkin, G. 2002. Development of a coupled surface – groundwater – pipe network model for the sustainable management of karstic groundwater. Env. Geol., 42, 4 513-517

19. Field, MS, 1997, Risk assessment methodology for karst aquifers: (2) Solute-transport modeling, *Environmental Monitoring and Assessment* 47, 23–37.

20. Field, MS 2002. The QTRACER2 Program for Tracer Break-through Curve Analysis For Tracer Tests in Karst Aquifers and Other Hydrologic Systems. U.S. Environmental Protection Agency, Office of Research and Development, EPA/600/R-02/001, 194 p

21. Lenore S Clesceri; Andrew D Eaton; Arnold E Greenberg; Mary Ann H Franson; (editors) *Standard Methods for the Examination of Water and Wastewater* : 19th edition supplement, (Prepared and published jointly by:) American Public Health Association, American Water Works Association, Water Environment Federation; Washington, DC, ©1996.

22. Gooseff, M. N., S. M. Wondzell, R. Haggerty, and J. Anderson, Comparing transient storage modeling and residence time distribution (RTD) analysis in geomorphically varied reaches in the Lookout Creek basin, Oregon, USA. Advances in Water Resources, 26, 925-93, 2003.

23. Danckwerts, P.V. Continuous Flow Systems. Distribution of Residence Times. *Chemical Engineering Science* **1952**, *50*, 3857-3866.

24. Shinnar, R.; Naor, P. Residence time distribution in systems with internal reflux. *Chemical Engineering Science* **1967**, *22*, 1369-1381.

25. Fetter, C. W.: 1992, *Contaminant Hydrogeology*, Macmillan Publishing Co. New York. 458 pp

26. Florea, L. J. and Carol M. Wicks, C. M.. "Solute Transport through Laboratory-Scale Karstic Aquifers" *Journal of Cave and Karst Studies* 63.2 (2001): 59-66.

27. Fogler, H. Scott (2006). *Elements of Chemical Reaction Engineering* (4th Edition ed.). Prentice Hall. ISBN 0-13-047394-4

28. Tom D. Byl and Roger Painter, Microbial Adaptations to Karst Aquifers with Contaminants, Tennessee AWRA,2009 Proceedings, Session 2C.9

29. Vogel, T.M. Handbook of bioremediation: Boca Raton, Fla. In Natural bioremediation of chlorinated solvents Norris, R.D.; Matthew, J.E., Eds.; Lewis Publishers, 1994; pp. 201-224.

30. White, WB. 2002. Karst hydrology: Recent developments and open questions. *Engineering Geology* 65 (2002) 85–105 1–21.

31. Bischoff, K.B., and Levenspiel, O., Fluid Dispersion: Generalization and Comparison of Mathematical Models: *Chemical Engineering Science*, **1962**,17, p. 245-255.

32. Tchobanoglous, G.; Burton, F.L.; Stensel, D.H. *Wastewater Engineering, Treatment and Reuse*, 4th ed.; McGraw Hill: City, State, USA, 2003; p. 279.

(54b) Prediction of Condenser Fouling in Thermoelectric Power Plant Cooling Systems

Iman Safari[1], Michael E. Walker[1], Javad Abbasian[1], Hamid Arastoopour[2], Ming-Kai Hsieh[3], Ranjani B. Theregowda[3], David A. Dzombak[3]and David C. Miller[4], (1)Chemical and Biological Engineering, Illinois Institute of Technology, Chicago, IL, (2)Department of Chemical and Biological Engineering, Wanger Institute for Sustainable Energy Research (WISER), Illinois Institute of Technology, Chicago, IL, (3)Department of Civil and Environmental Engineering, Carnegie Mellon University, Pittsburgh, PA, (4)National Energy Technology Laboratory

The thermoelectric power industry in the U.S. uses a large amount of fresh water. As available freshwater for use in thermoelectric power production becomes increasingly limited, use of nontraditional water sources is of growing interest. Utilization of nontraditional water, in cooling systems increases the potential for mineral precipitation on heat exchanger surfaces. In that regard, predicting the accelerated rate of scaling and fouling in condenser is crucial to evaluate the condenser performance. To achieve this goal, water chemistry should be incorporated in cooling system modeling and simulation.

This paper provides a model to estimate the rate of fouling in a recirculating cooling loop as a function of makeup water quality and operational conditions. Detail modeling of salts precipitation in condenser and volatile species desorption in cooling tower is presented. The model developed herein is used to perform case studies of various makeup water quality and treatment strategies to provide insight into the ramifications of degraded water use in the cooling loops of thermoelectric power plants.

(54c) A Combined Cost Model to Evaluate the Economic Impact of Degraded Water Use in Thermoelectric Cooling Loops

Michael E. Walker[1], Ranjani B. Theregowda[2], **Iman Safari**[1], Ming-Kai Hsieh[2], Javad Abbasian[1], Hamid Arastoopour[3], David A. Dzombak[2] and David C. Miller[4], (1)Chemical and Biological Engineering, Illinois Institute of Technology, Chicago, IL, (2)Department of Civil and Environmental Engineering, Carnegie Mellon University, Pittsburgh, PA, (3)Department of Chemical and Biological Engineering, Wanger Institute for Sustainable Energy Research (WISER), Illinois Institute of Technology, Chicago, IL, (4)National Energy Technology Laboratory

The thermoelectric power industry in the U.S. uses a large amount of freshwater. The large water demand is increasingly a problem, especially for new power plant development, as the availability of freshwater diminishes in the United States. Reusing non-traditional water sources, such as treated municipal wastewater, provides one option to mitigate freshwater usage in the thermoelectric power industry. This paper presents the development of an integrated costing model to assess the implications of degraded water use, including treatment requirements and the impact of increased tube scaling in the main condenser. A case study of various treatment scenarios is provided for three plants in regions of differing water availability, to provide insight into the ramifications and benefits of degraded water use as makeup to thermoelectric recirculating cooling loops.

(54d) Can Monochloramine Be Used to Control Biofouling in Recirculating Power Plant Cooling System Using Treated Municipal Wastewater?

Shih-Hsiang Chien[1], David A. Dzombak[2] and Radisav Vidic[1], (1)Department of Civil and Environmental Engineering, University of Pittsburgh, Pittsburgh, PA, (2)Department of Civil and Environmental Engineering, Carnegie Mellon University, Pittsburgh, PA

Biofouling, which involves biological growth in both suspended and sessile form, reduces both heat exchanger and cooling tower efficiency in a thermoelectric power plant. This problem is further aggravated when treated municipal effluent, which contains high concentration of organic matter and nutrients compared to freshwater, is used in a power plant recirculating cooling system as an alternative to freshwater. Although secondary treatment and nitrification remove large fraction of biodegradable components, significant biocide addition is still required to suppress biological activity in cooling systems employing treated municipal wastewater.

The key objective of this study was to evaluate the effectiveness of chlorine-based biocides (sodium hypochlorite, pre-formed monochloramine, and chlorine dioxide) and determine their optimal chemical treatment regimen when using secondary and tertiary treated municipal wastewater as power plant cooling makeup. These three biocides are widely used in water and wastewater disinfection but their effectiveness for biofouling control in highly concentrated (4-6 cycles of concentration, as in recirculating cooling systems) and aerated wastewater has not yet been established. Optimal chemical regimens for achieving successful biofouling control were determined by monitoring biocide usage and heterotrophic planktonic and sessile bacterial populations in both laboratory- and pilot-scale studies. In addition, the cost and degree of environmental hazard associated with each biocide when used under optimal conditions were determined using Life Cycle Costing and Life Cycle Assessment tools.

Laboratory experiments indicated that a pre-formed monochloramine (MCA) dose of 2 mg/L was required to achieve industry biofouling control criteria (below 10^4 CFU/mL for planktonic bacteria and below 10^4 CFU/cm^2 for sessile bacteria) at 40 ºC, while sodium hypochlorite and chlorine dioxide required slightly lower initial doses to achieve the same biocidal efficiency. However, pilot-scale studies revealed that less pre-formed monochloramine is needed to maintain sufficient residual and suppress biological growth. Although laboratory studies demonstrated the ability to form MCA in treated municipal wastewater through the reaction between added free chlorine and inorganic ammonia present in the treated wastewater, the addition of free chlorine in pilot-scale studies to achieve a total chlorine residual of 3-4 mg/L resulted in only 23% conversion to MCA. It was observed that 91% of inorganic ammonia in the incoming secondary effluent was stripped out in the cooling tower, leading to unstable in-situ MCA formation. A cost analysis for these biocides was performed based on the results of pilot scale experiments to provide relevant practical information for wastewater reuse. This analysis confirmed that the addition of pre-formed MCA for biocidal control in power plant recirculating cooling systems using treated municipal wastewater is effective and deployable at modest cost. Results of Life Cycle Assessment also suggested that pre-formed MCA has the least environmental and human health impact for this application.

A Statistical Approach to Prediction of Transmembrane Pressure in Membrane Bioreactors

Hiromasa Kaneko, Kimito Finatsu, The University of Tokyo, Tokyo, Japan

Introduction

In water treatment fields such as sewage treatment and industrial liquid waste treatment, the membrane bioreactor (MBR) has been widely used to purify wastewater for reuse [1,2]. MBRs combine biological treatment with membrane filtration. First, bacteria within activated sludge metabolize the organic pollutants and produce environmentally-acceptable metabolites, then a microfiltration (MF) or ultrafiltration membrane separates liquids from solids, i.e. the clean water from the sludge.

However, MBRs suffer from some practical difficulties. One of the critical difficulties is membrane fouling [3,4]. Membrane fouling is a phenomenon wherein foulants, such as activated sludge, sparingly-soluble compounds, high molecular weight solutes and colloids, absorb or deposit on the membrane surface and absorb onto and block the membrane pores. For example, in cases where the MBR is operated under constant-rate filtration, significant energy is required to achieve constant permeate flow rate due to membrane fouling.

To reduce membrane fouling, physical cleaning, for example, aeration, backwashing, or back-pulsing, is performed periodically during MBR operations. Furthermore, chemical cleaning must be carried out with chemical reagents after a given period of processing time, when the transmembrane pressure (TMP) exceeds a given value, because some foulants cannot be removed by physical cleaning and these residual foulants will prevent the recovery of membrane performance with time. Frequent chemical cleaning and replacement of membranes are both expensive.

Hence, to enable chemical cleaning to be performed at an appropriate time [5], membrane fouling must be predicted in the long term. When an MBR plant is operated under constant-rate filtration, this corresponds to prediction of the TMP. Moreover, to enable the distributed MBR systems mentioned above, each MBR must be operated automatically and controlled remotely. Therefore, the TMP must be predicted *a priori* and a schedule of chemical cleaning must be created in advance.

Geng et al. assumed that fouling that could not be reduced by aeration and backwashing was deep pore clogging and that an increase of TMP could be represented as an exponential function of processing time, then fitted parameters of the exponential equation with experimental data [6]. Kim et al. attempted to adapt changes of states in an MBR plant by recursively identifying the parameters of the exponential equation [7]. Additionally, Chen et al. focused on a critical flux [8], below which membrane fouling can be neglected in the short term, and a TMP jump [9], which is a rapid increase in TMP after a period of processing, and constructed two models that predicted the initial increase and rapid increase of TMP, respectively [10]. However, the predictive performance of these models for new data were not confirmed, although the fitting accuracy of the models was high. Furthermore, water quality variables were not considered in the models and hence it is conceivable that they cannot predict TMP well in other MBR plants and lack broad utility.

On the other hand, Meng et al. studied the relationships between membrane fouling resistance and water quality variables [11]. By investigating correlations among the water quality variables and membrane fouling resistance, the results suggested that there should be a direct influence of MLSS concentration, sludge particle size distribution and extracellular polymeric substances and an indirect influence of the other variables on membrane fouling. Thus, water quality variables have multiple effects on MBR fouling phenomena and the simultaneous consideration of these effects is necessary to predict membrane fouling resistance and TMP in the long term in various MBR plants.

Therefore, in this study, we use chemoinformatics [12] to construct statistical models to predict TMP from other MBR parameters, **X**, such as water quality variables and operating conditions. Chemoinformatics uses informatics methods to solve chemistry problems and is used to study a variety of topics, such as quantitative structure–property relationships, quantitative structure–activity relationships, reaction design, drug design, and process systems.

In this paper, we target MBR plants operated under constant-rate filtration, construct prediction models for the increase in TMP per unit time, ΔTMP, from **X**, and predict TMP using the models. Figure 1 shows the basic concept of this TMP prediction. First, a statistical model is constructed between **X** and ΔTMP by using databases of variables that have been measured in the past. Next, the constructed model predicts the increase in TMP from time n to time n+1, $\Delta TMP_{pred}(n+1)$, by inputting MBR parameters in new data **x**(n). The target value of TMP at time n+1, $TMP_{pred}(n+1)$, can be calculated by adding $\Delta TMP_{pred}(n+1)$ to TMP(n), which is measured previously.

When long term prediction of TMP is carried out, the constructed model predicts $\Delta TMP_{pred}(n+2)$, $\Delta TMP_{pred}(n+3)$, ... by inputting expected values or setting values of MBR parameters at times n+1, n+2, ..., respectively. We then use these predicted values as the baseline and can then predict $TMP_{pred}(n+2)$, $TMP_{pred}(n+3)$, ... by adding $\Delta TMP_{pred}(n+2)$, $\Delta TMP_{pred}(n+3)$, ... to $TMP_{pred}(n+1)$, $TMP_{pred}(n+2)$, ..., respectively.

In our study, two methods are used to construct regression models, the partial least-squares (PLS) [13] method and the support vector regression (SVR) [14] method. PLS models are based on linear regression and we can obtain semi-quantitative regression coefficients for each considered parameter to predict the increase in TMP. On the other hand, SVR models are based on nonlinear regression and are reported to have more predictive accuracy than that of neural networks in some cases.

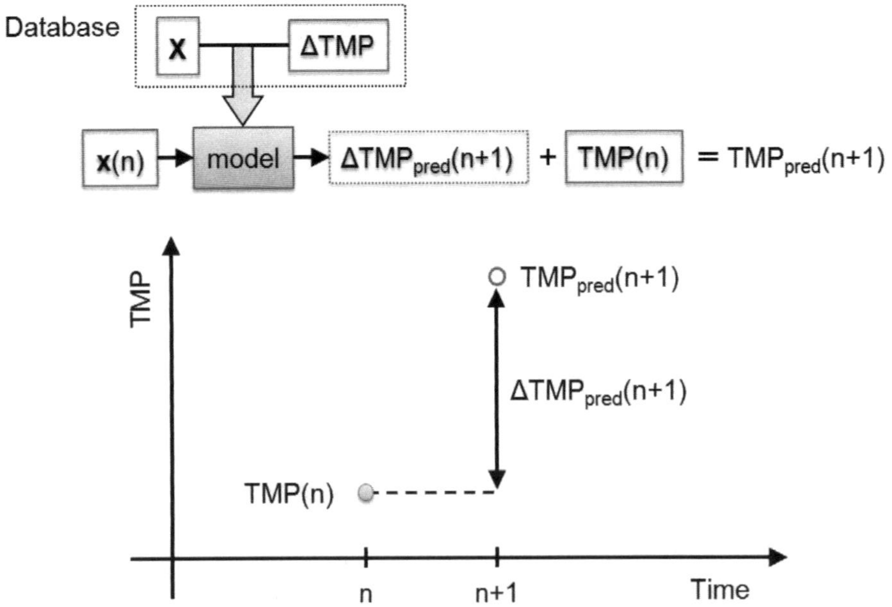

Figure 1. Basic concept of TMP prediction. **X** contains MBR parameters such as water quality variables and operating conditions of a training data set; ΔTMP is an increase in TMP for a training data set; n represents time; **x** represents MBR parameters of new data; ΔTMP_{pred} represents a predicted value of ΔTMP; and TMP_{pred} represents a predicted value of TMP.

Method

PLS

PLS is a method for relating explanatory variables, **X**, and an objective variable, **y**. Using a linear multivariate model, it goes beyond traditional regression methods in that it also models the structures of **X** and **y**. In PLS modeling, the covariance between **y** and the score vector $\mathbf{t}_i$ is maximized.
A PLS model consists of the following two equations:

$$\mathbf{X} = \mathbf{TP}' + \mathbf{E} \tag{1}$$

$$\mathbf{y} = \mathbf{Tq} + \mathbf{f} \tag{2}$$

where **T** is a score matrix, **P** is an **X**-loading matrix, **q** is a **y**-loading vector, **E** is a matrix of **X** residuals, and **f** is the vector of **y** residuals.
The PLS regression model is as follows:

$$\mathbf{y} = \mathbf{Xb} + \mathbf{const} \tag{3}$$

$$\mathbf{b} = \mathbf{W}(\mathbf{P}'\mathbf{W})^{-1}\mathbf{q} \tag{4}$$

where **W** is an **X**-weight matrix and **b** is a vector of regression coefficients. The number of components must be decided appropriately to construct a highly predictive model. r^2 and q^2 values are used as the measure of fit. The details of these statistics are shown in Appendix B. The optimum number of components is determined in this paper by the first local maximum of q^2.

SVR

The SVR method applies a support vector machine (SVM) to a regression analysis and can be used to construct non-linear models by applying a kernel trick as well as a SVM. The primal form of SVR can be shown to be the following optimization problem:
Minimize

$$\frac{1}{2}\|\mathbf{w}\|^2 + C\sum_i |y_i - f(\mathbf{x}_i)|_e \tag{5}$$

subject to

$$|y_i - f(\mathbf{x}_i)|_e = \max(0, |y_i - f(\mathbf{x}_i)| - e) \tag{6}$$

where y_i, and $\mathbf{x}_i$ are training data, **w** is a weight vector, e is a threshold, and C is a penalizing factor that controls the trade-off between a training error and a margin.
By minimization of Equation (5), we can construct a regression model that has a good balance between generalization capabilities and the ability to adapt to the training data. The kernel function in our application is a radial basis function:

$$K(x, x') = \exp\left(-\gamma |x - x'|^2\right) \tag{7}$$

where γ is a tuning parameter controlling the width of the kernel function. In this study, LIBSVM [15] is used as the machine learning software.

Case Study

We verified the usefulness of the proposed TMP prediction method with real industrial data obtained from operation of MBRs.

Data

The data were obtained from a MBR plant operated under constant-rate filtration. A brief summary of the target MBR plant is shown in Table 1. In this MBR plant, sewage water flows into two lines of the MBR, A and B. MBR system A comprises anoxic and aerobic tanks and B comprises anaerobic, anoxic and aerobic tanks. A membrane module is immersed in each aerobic tank. Flocculants are added in MBR A.

Values of process variables measured online, such as TMP and temperature, were averaged daily while those of other variables, measured offline, such as MLSS concentration and soluble total organic carbon (TOC), could be obtained only once every few days. Time plots of TMP for A and B are shown in Figure 3. At the points where values of TMP decreased rapidly in Figure 2(a) and (b), plant operators had carried out chemical membrane cleaning.

For training data in MBR A, we used data from monitoring that took place from 5th April to 12th September 2003, and for test data we used data from 13th September 2003 to 4th February 2004. The numbers of training and test data points were 119 and 96, respectively. For training data in MBR B, we used data from monitoring that took place from 2nd June to 17th October 2004, and for test data we used data from 18th October 2004 to 12th April 2005. The numbers of training and test data points were 126 and 153, respectively. Data measured during chemical membrane cleaning and shutdown of each MBR were eliminated in advance.

Table 1. Brief Summary of the Target MBR Plant.

Operation	Constant-rate filtration
Treatment object	Urban wastewater
Treatment scale	25–32 m^3/day × 2 lines (A and B)
Residence time	4.3–6.0 hours
MLSS concentration	8–20 g/l
Membrane	Immersion type flat sheet membrane (MF)
Material	Organic polymer
Pore diameter	~0.4 μm
Membrane area	40 m^2/line

Process Variables

We modeled the relationship between an increase of TMP over a period of one day, ΔTMP, and values of other process variables for the day, **X** (Figure 1). If a new value of **X** is obtained, we can estimate the corresponding ΔTMP and predict TMP for the next day by adding this to TMP for the day. Moreover, by using future values of **X**, we can predict future values of TMP. **X**-variables include temperature, inlet flow, variables relevant to water quality and so on, and the measurement frequency for these variables differ. Variables that are rarely measured have little information and therefore the variables measured at a frequency of less than 30% of the TMP measurement were deleted in advance. Table 2 shows the selected process variables. Values of **X** that were not measured were interpolated from measured data. We used a simple interpolation method, where the newest value was used as a missing value, and the Hermite interpolation method [16].

Results and Discussion

We constructed PLS and SVR models predicting TMP through the intermediary of ΔTMP, with training data. The results for MBR A are shown in Table 3. r^2 represents the fitting accuracy of the constructed models and q^2 represents the predictive accuracy of the constructed models. Values close to unity for both r^2 and q^2 are favorable. The low values of root mean squared error (*RMSE*) and $RMSE_{CV}$

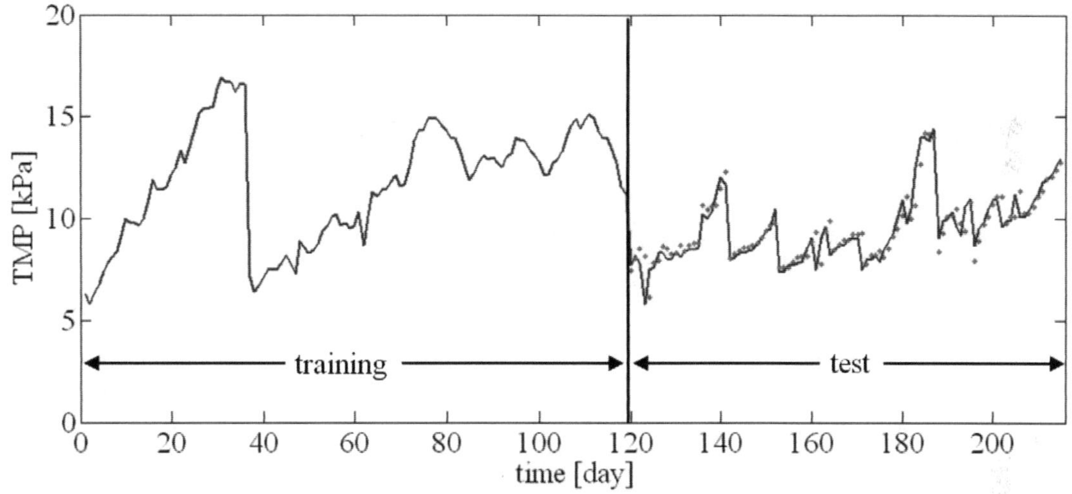

(a) MBR A

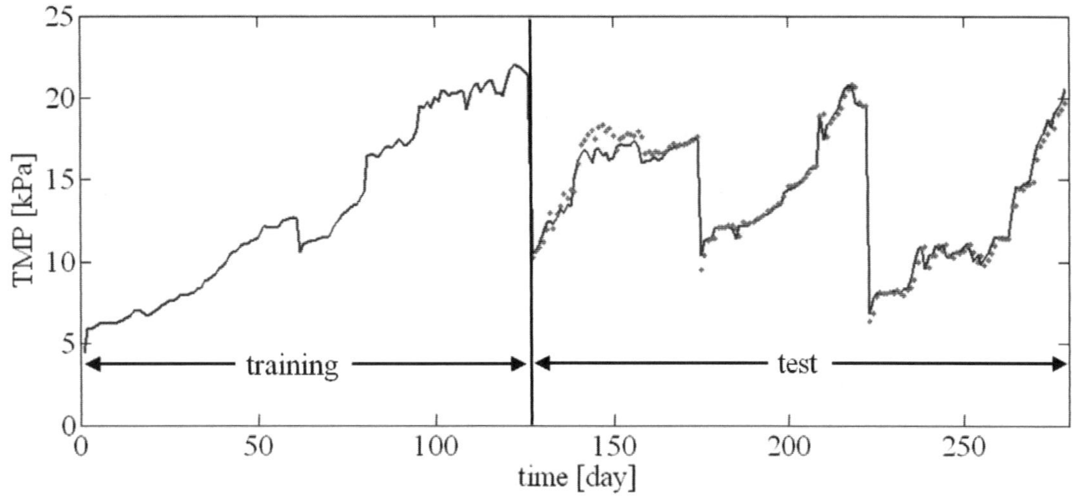

(b) MBR B

Figure 2. Time plot of TMP. The lines represent the observed values of TMP and the points represent the predicted values of TMP.

Table 2. Process Variables.

Line	Process variable
A	Water temperature Soluble TOC in aerobic reactor MLSS concentration in aerobic reactor
B	Water temperature Treatment flow rate Viscosity in aerobic reactor Filtration rate using a filter paper

indicated that calculation and validation error of the models were small. The predictive model could be constructed with only water temperature, soluble TOC in the aerobic reactor, and MLSS concentration in the aerobic reactor. In addition, there was little difference between r^2, *RMSE*, q^2 and $RMSE_{CV}$ values after applying the simple interpolation technique and those after the Hermite interpolation technique. While nonlinear functions are used in the Hermite interpolation technique, in this case the simple interpolation technique could satisfactorily deal with the data.

For the SVR method, interpolation techniques did not affect the results as much as when using the PLS method, as shown in Table 3.

The modeling results for MBR B are shown in Table 4. All the r^2 and q^2 values exceeded 0.9, reflecting that the TMP models could be constructed with high accuracy. The *RMSE* and $RMSE_{CV}$ values were small. For MBR B, there was little difference between the results using nine and four **X**-variables, similar to MBR A. We concluded that the predictive model could be constructed using water temperature, treatment flow rate, viscosity in the aerobic reactor, and the filtration rate through a filter paper.

The prediction results for MBRs A and B are shown in Tables 3 and 4, respectively. The r_{pred}^2 and the $RMSE_P$ are the r^2 and the *RMSE* for test data, respectively. The r_{pred}^2 value was low and the $RMSE_P$ value was exceptionally high when the data B and the Hermite interpolation are used to construct the TMP model with nine **X**-variables from Table 4. This is because the interval between the day when the variables measured with low frequency were finally measured and the final day of the test data were large,

Table 3. Modeling and Prediction Results for MBR of A.

		Modeling				Prediction	
Reg.[a]	Interp.[b]	r^2	*RMSE*	q^2	$RMSE_{CV}$	r_{pred}^2	$RMSE_P$
PLS	Simple	0.971	0.474	0.970	0.484	0.843	0.665
	Hermite	0.971	0.473	0.970	0.483	0.842	0.668
SVR	Simple	0.990	0.280	0.977	0.422	0.840	0.672
	Hermite	0.985	0.345	0.975	0.447	0.833	0.686

Table 4. Modeling and Prediction Results for MBR of B.

		Modeling				Prediction	
Reg.[a]	Interp.[b]	r^2	*RMSE*	q^2	$RMSE_{CV}$	r_{pred}^2	$RMSE_P$
PLS	Simple	0.997	0.284	0.997	0.291	0.981	0.477
	Hermite	0.997	0.284	0.997	0.291	0.981	0.480
SVR	Simple	0.997	0.285	0.996	0.328	0.983	0.447
	Hermite	1.00	1.27×10^{-4}	0.997	0.298	0.983	0.447

[a] The regression method.
[b] The interpolation technique.

and therefore the values interpolated by the Hermite interpolation became too large. Careful attention is required when applying Hermite interpolation to variables having low measurement frequency. However, even in this situation, prediction can be achieved by using simple interpolation, only **X**-variables with sufficiently high measurement frequencies, or the SVR method, as shown in Table 4.

The $RMSE_P$ values were low, though these are slightly higher than the *RMSE* and $RMSE_{CV}$ values of training data. Almost all the r_{pred}^2 values exceeded 0.8, indicating that the constructed models could predict TMP of test data with high accuracy. In Figure 2, the points represent the predicted values of TMP. As shown in Figure 2(a) and (b), the models could predict the values of TMP, following the observed TMP values with high accuracy. We can therefore conclude that highly predictive models for TMP from other process variables can be constructed by making full use of chemoinformatic methods.

Conclusion

To predict TMP in MBR plants, we proposed statistical models relating the increase in TMP and other process variables with a database of their historical values. The analysis of two sets of real industrial data showed that the measurement frequency of the process variables, interpolation techniques, and regression methods did not affect the modeling and prediction results. However, it is necessary to pay close attention to the use of Hermite interpolation for variables with low measurement frequency. A linear relationship between the increase of TMP and MBR parameters was indicated because there was little difference between the results of PLS and SVR approaches. Furthermore, the effects of the water quality variables and operating conditions on TMP were analyzed using standard regression coefficients obtained by the PLS method.

Variables measured at a low frequency were not be used because they have little information. However, we will be able to use the deleted variables if the frequency of their measurement increases or the variables are predicted by ASM models in the future. In addition, the dynamics of a MBR plant is important and therefore, the predictive accuracy of the prediction models will increase by incorporating the dynamics into the model.

In this paper, we predicted values of TMP over only a one day period. However, we plan to predict TMP values over longer terms. Of course, the constructed models can perform long-term prediction of TMP by inputting expected values or setting values of MBR parameters. However, an improvement of prediction accuracy will be achieved by combining statistical methods with the concepts of critical flux and TMP jump i.e. we will consider water quality, operating conditions, and the phenomena of adsorption and deposition of foulants on the membrane simultaneously.

Our results suggest that the statistical approach used in this study can be a practical tool for the analysis and prediction of TMP. In addition, we believe that by applying our method to process control, MBR plants could be operated effectively and stably.

Acknowledgments

The authors acknowledge the support of the Core Research for Evolutional Science and Technology (CREST) project 'Application of Integrated Intelligent Satellite System (IISS) to construct regional water resources utilization system' of the Japan Science and Technology Agency (JST) and the financial support of Japan Society for the Promotion of Science.

References

1. W.B. Yang, N. Cicek, J. Ilg, State-of-the-art of membrane bioreactors: Worldwide research and commercial applications in North America, J. Membr. Sci. 270 (2006) 201–211.
2. F. G. Meng, S.R. Chae, A. Drews, M. Kraume, H.S. Shin, F.L. Yang, Recent advances in membrane bioreactors (MBRs): Membrane fouling and membrane material, Water Res. 43 (2009)1489–1512.
3. P.L. Clech, V. Chen, T.A.G. Fane, Fouling in membrane bioreactors used in wastewater treatment, J. Membr. Sci. 48 (2008) 534–541.
4. M. Kraume, D. Wedi, J. Schaller, V. Iversen, A. Drews, Fouling in MBR: What use are lab investigations for full scale operation?, Desalination 236 (2009) 94–103.
5. M.J. Kim, B. Sankararao, C.K. Yoo, Determination of MBR fouling and chemical cleaning interval using statistical methods applied on dynamic index data, J. Membr. Sci. 375 (2011) 345–353.
6. Z. Geng, E.R. Hall, P.R. Berube, Membrane fouling mechanisms of a membrane enhanced biological phosphorus removal process, J. Membr. Sci. 296 (2007) 93–101.
7. M.J. Kim, B. Sankararao, C.K. Yoo, Determination of MBR fouling and chemical cleaning interval using statistical methods applied on dynamic index data, J. Membr. Sci. 375 (2011) 345–353.
8. R.W. Field, D. Wu, J.A. Howell, B.B. Gupta, Critical flux concept for microfiltration fouling, J. Membr. Sci. 100 (1995) 259–272.
9. B.D. Cho, A.G. Fane, Fouling transients in nominally sub-critical flux operation of a membrane bioreactor, J. Membr. Sci. 209 (2002) 391–403.
10. Y. Ye, V. Chen, A.G. Fane, Modeling long-term subcritical filtration of model EPS solutions, Desalination 191 (2006) 318–327.
11. F.G. Meng, H.M. Zhang, F.L. Yang, S.T. Zhang, Y.S. Li, X.W. Zhang, Identification of activated sludge properties affecting membrane fouling in submerged membrane bioreactors, Sep. Purif. Technol. 51 (2006) 95–103.
12. J. Gasteiger, T. Engel, Chemoinformatics – A Textbook, first ed., Wiley-VCH, Weinheim, 2003.
13. S. Wold, M. Sjöström, L. Eriksson, PLS-regression: a basic tool of chemometrics, Chemom. Intell. Lab. Syst. 58 (2001) 109–130.
14. V.N. Vapnik, The Nature of Statistical Learning Theory, second ed., Springer, New York, 1999.
15. C. C. Chang, C.J. Lin, LIBSVM: a Library for Support Vector Machines, 2001. Software available at http://www.csie.ntu.edu.tw/~cjlin/libsvm.
16. J.H. Bramble, S.R. Hilbert, Bounds for a class of linear functionals with applications to Hermite interpolation, Numer. Math. 16 (1971) 362–369.

BIOMASS TO OIL: FAST PYROLYSIS AND SUBCRITICAL HYDROTHERMAL LIQUEFACTION

Nicolas Doassans-Carrère, Laboratoire M2P2, Aix-en-Provence, France
Guillain Mauviel, LRGP, Nancy, France
Jean-Henry Ferrasse, Laboratoire M2P2, Aix-en-Provence, France
Jacques Lédé, LRGP, Nancy, France
Olivier Boutin, CEA, Marcoule, France

Abstract

The present paper deals with the comparison of two biomass-to-oil processes: fast pyrolysis and subcritical hydrothermal liquefaction. Using the same biomass (beech sawdust), fast pyrolysis was led thanks to a cyclone reactor (wall temperature between 870 and 1040 K) and subcritical hydrothermal liquefaction thanks to a 150-ml-batch-reactor (temperature between 420 and 600 K). Mass balances and phases analysis (ultimate analysis, higher heating value (HHV), Karl-Fischer, gas chromatographies, H^1 NMR) allow the comparison of both processes and the characterization of pyro-oils (heavy oils, light oils and aerosols), liq-oils (heavy oils and water soluble organics), solids and gas.

For both processes, there is a temperature for which a high conversion of biomass is reached with a minimum of gas mass yield. Up to 62.6 wt% of pyro-oil and 47.0 wt% of liq-oil, on dry basis, were obtained in this work, with oxygen contents and chemical compositions different from one to the other.

Introduction

The global energy demand is continuously increasing, due to the growth of both population and industrialization. The supply of CO_2-neutral energies has to become more important in the future, and bio-fuels from biomass may be a part of those energies.

Different processes have been developed to produce bio-oils from different biomass, among them fast pyrolysis and subcritical hydrothermal liquefaction are studied in this work. Fast pyrolysis can produce up to75%wt of bio-oil (pyro-oil), composed typically of 56%wt of C, 38%wt of O and 6%wt of H [1]. Subcritical hydrothermal liquefaction consists in thermal decomposition of biomass in water in high pressure and temperature condition but below its critical point (647 K and 22.1 MPa). Subcritical water is mainly used today for wet air oxidation [2]. Using subcritical hydrothermal liquefaction, 35%wt of bio-oil (liq-oil) can be reached. This liq-oil, less oxygenated than pyro-oil, is composed typically of 74%wt of C, 17%wt of O and 9%wt of H [1]. Those two types of bio-oils must be upgraded to obtain bio-diesel, thanks to catalytic hydrodeoxygenation (HDO) which reduces the proportion of oxygen.

The aim of this paper is a direct comparison of fast pyrolysis and subcritical hydrothermal liquefaction for the production of bio-oil. This comparison is based on mass balances and products characterization: bio-oils, solids and gas.

Experimental apparatus

In order to compare those processes, experimentations were led with the same biomass, beech sawdust.

Fast pyrolysis

Fast pyrolysis is achieved thanks to the cyclone reactor [3] which is a continuous reactor with an oil recovery system separating three fractions of pyro-oil (Figure 1). Up-stream of the cyclone, the installation is composed of the biomass (tank and screw conveyor and nitrogen (inert carrying gas) feeds, with a mass flowrate ratio between nitrogen and biomass equal to 7.8. Particles, carried by nitrogen, are pyrolysed along the cyclone wall which is heated by induction. A by-pass system allows preheating and equilibrium set-up before biomass feeding. Thanks to the separator properties of the cyclone, solids are isolated from gaseous products and recovered in the solids collector. Gaseous products, evacuating through the upper part of the cyclone, pass through three water cooled heat exchangers in series to recover the first fraction of oil in a collector, "heavy oils" (HO). Then the gas flows through refrigerated coils inside which the "light oils" (LO) fraction is recovered. In order to recover the last fraction, "aerosols" (AE), both electrostatic and membrane filters are used in series. At the exit, clean gases are finally sampled and analyzed by micro-GC.

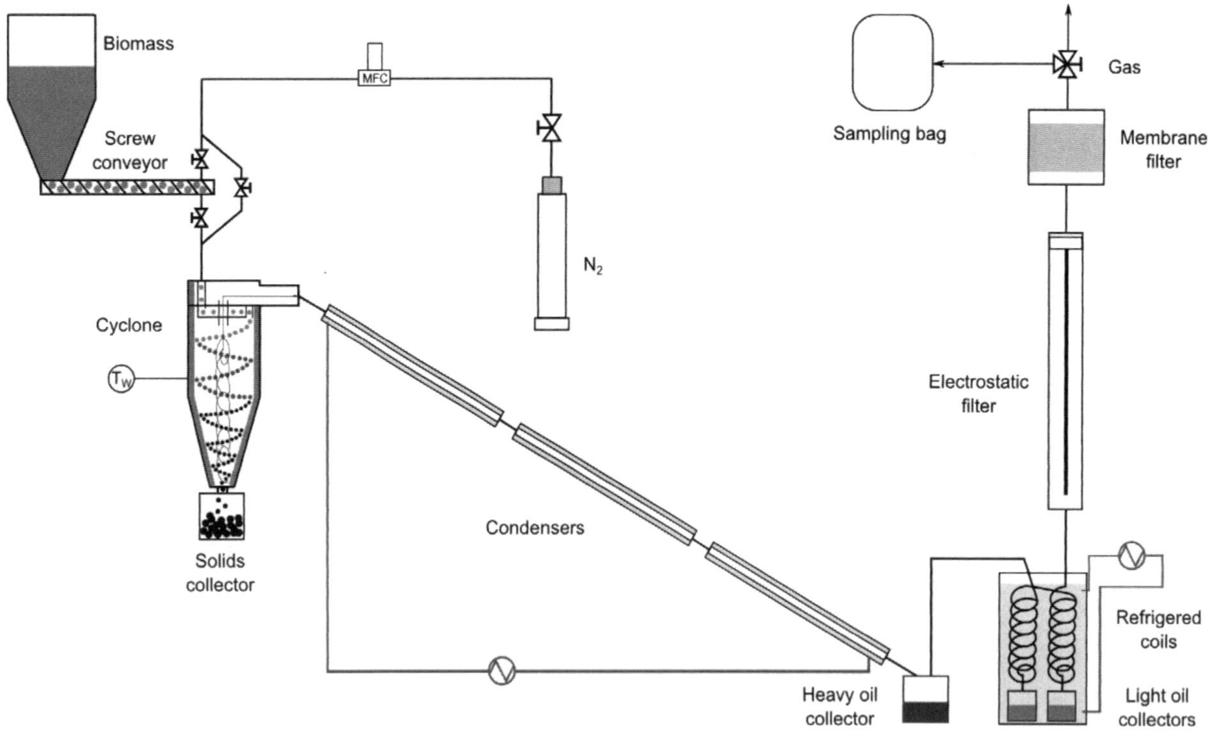

Figure 1. Fast pyrolysis cyclone reactor installation scheme

Subcritical hydrothermal liquefaction

Subcritical hydrothermal liquefaction is achieved in a 150 ml autoclave where biomass (6 g) is introduced prior to the catch of the reactor with the locking straps (Figure 2). Thanks to a volumetric pump, nitrogen or distilled water can be introduced into the autoclave at a desired pressure. After inerting the reactor with nitrogen and setting the initial working pressure (1 MPa), 100 ml of distilled water are added into the reactor, heated by an electric resistance. Once achieved, the working temperature is maintained at the experiment duration time, then the reactor is cooled down by a refrigerating system using air and then water. Once the reactor at the ambient temperature, gases are sampled and analyzed by GC-TCD. The reactor is opened in order to collect the products (liquids and solids). Based on [4], pure acetone is used to rinse the reactor and wash solids. After filtration, three phases are obtained: solids, aqueous phase and organic phase. Solids are placed into a drying oven. Evaporation at reduced pressure is used to remove acetone from the organic phase and water from the aqueous phase, in order to get respectively two oil fractions, "heavy oils" (HO) and "water soluble organics" (WSO).

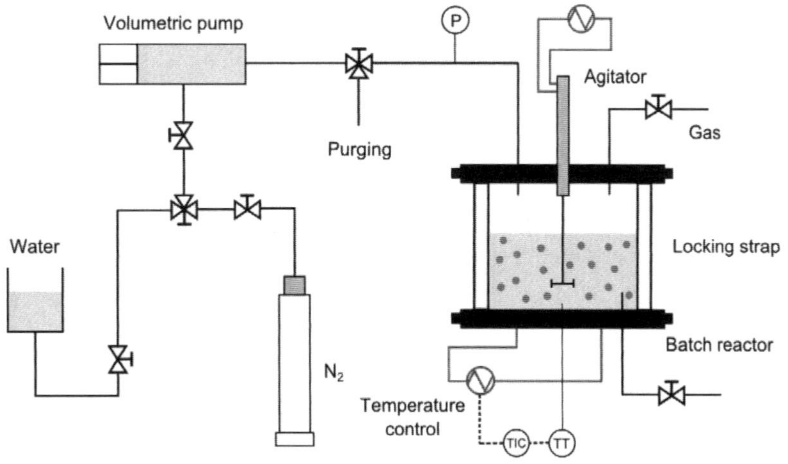

Figure 2. Subcritical hydrothermal liquefaction installation scheme

Results and discussion

Maximum of bio-oil mass yield

For both installations, mass balances are based upon mass measurement of solids and bio-oils, and evaluation of the gas composition. Thanks to the known amount of nitrogen used in those installations (mass flowrate for fast pyrolysis and initial pressure for subcritical hydrothermal liquefaction), the mass of produced gas can be determined from gas concentrations in the gaseous phase. Mass yield are defined from dried biomass, with a biomass moisture of 9 wt.%. It has to be noticed that for fast pyrolysis, after data reconciliation based upon atom balances, mass balances close at 100 % (the initial overall mass balances closures range between 83 and 100 %). No data reconciliation could be made for subcritical hydrothermal liquefaction, and the overall mass balances closures range between 64 and 83 %. This misclosure is often explained by a water production [5].

Figure 3 shows influence of temperature on mass balances for fast pyrolysis and subcritical hydrothermal liquefaction. For both processes, there is a temperature for which a maximum of bio-oil mass yield is reached. For fast pyrolysis, at 970 K, 62.6 wt% of pyro-oils are obtained with 25.7 wt% of gas and 11.7 wt% of solids. For subcritical hydrothermal liquefaction, at 573 K, 47.0 wt% of liq-oils are obtained with 5.5 wt% of gas and 17.8 wt% of solids. This maximum correspond to a high conversion of biomass coupled with a low gas mass yield. The partial conversion of biomass is revealed by ultimate analysis of solids recovered. In fast pyrolysis, the composition of solids recovered at 870 K ($CH_{1.26}O_{0.57}$) is closer to initial biomass one ($CH_{1.48}O_{0.72}$) than chars ones obtained at 1040 K

($CH_{0.24}O_{0.17}$). In subcritical hydrothermal liquefaction, until 528 K, the composition of solids ($CH_{1.50}O_{0.68}$) is similar to the biomass one, whereas from 573 K, solids recovered are closer to chars ($CH_{0.75}O_{0.24}$).

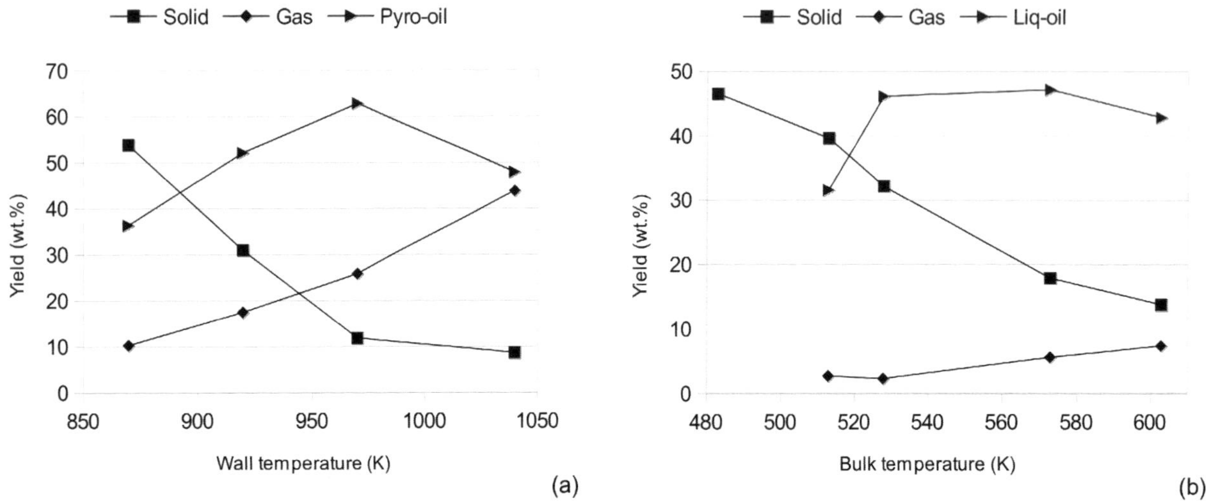

Figure 3. Mass yields evolutions according to temperature for (a) fast pyrolysis and (b) subcritical hydrothermal liquefaction

Products characterization

Apart from ultimate analysis made on solids, which have revealed partial biomass conversion, characterization of gas and bio-oils allows the direct comparison between both fast pyrolysis and subcritical hydrothermal liquefaction.

Gas from fast pyrolysis is a mixture of CO, CO_2, H_2, CH_4, C_2H_4 and others C_2-C_3, with a majority of CO, CO_2 and CH_4 (> 91 wt.%). At the opposite, gas from subcritical hydrothermal liquefaction is mainly composed by CO_2 (> 99 wt.%), with a maximum of 1 wt.% of H_2 at 603 K.

Pyro-oil is the most oxygenated bio-oils with composition of heavy oil, light oil and aerosols of, respectively, $CH_{3.34}O_{1.73}$, $CH_{4.08}O_{1.98}$ and $CH_{1.32}O_{0.46}$. Composition of liq-oil fractions (heavy oil and water soluble organics) are respectively $CH_{1.11}O_{0.43}$ and $CH_{1.65}O_{0.78}$. This difference of oxygen content is partly explained by the water content of pyro-oil. In fact, contrary to liq-oil free of water (consequence of evaporation steps), 40 wt.%, 36.5 wt.% and 6.5 wt.% of, respectively, heavy oil, light oil and aerosols is water.

Gas chromatographies and H^1 NMR confirm this oxygen content difference. Using the same analytical apparatus, a direct comparison is possible between pyro-oil and liq-oil. Those bio-oils have some compounds in common, like phenol, guaiacol or acetic acid, but major compounds are different. For example, aldehydes or carbohydrates like levoglusan, major pyro-oil oxygenated compounds, are absent in liq-oil. At the opposite, hydrogen from alkanes or aromatics, are in higher proportion in liq-oil, indicating longer-chain compounds (and therefore less oxygenated) or higher quantity of compounds from those families.

Finally, ultimate analysis and HHV-measurement, made on solids and fractions of pyro-oils and liq-oils, confirm the use of a correlation Dulong-like between both of those properties [6]. Indeed, liq-oils have higher HHV than pyro-oils, but with a lower mass yield. Energetic balances are close to mass balances with up to 59.5 % and 51.5 % of initial biomass HHV are recovered, respectively, in pyro-oils and liq-oils.

Conclusions

Fast pyrolysis and direct liquefaction of beech sawdust has been studied in the production of bio-oils, through mass balances and products characterization. For both processes, a maximum of bio-oils mass yield is observed which corresponds to a high conversion of biomass coupled with a minimum gas mass yield.

The higher oxygen content of pyro-oils than liq-oils, revealed by ultimate analysis, is explained by the water content and different compositions, shown by GC-MS/FID and H1 NMR analysis.

For future work, those results are used as a basis for energy analysis which must be achieved to ensure the viability of the two processes for the production of bio-diesel, including HDO step. This step is, in particular, simulated thanks to bio-oils analysis like oxygen content and compounds families.

References

1. Huber, G. W., Iborra, S., and Corma, A. (2006). Synthesis of transportation fuels from biomass : Chemistry, catalysts, and engineering. *Chemical Reviews*, 106(9) :4044–4098.
2. Lefèvre, S., Boutin, O., Ferrasse, JH., Malleret, L., Faucherand, R. and Viand, A. (2011). Thermodynamic and kinetic study of phenol degradation by a non-catalytic wet air oxidation process. *Chemosphere*, 84 (9) :1208-15.
3. Lédé, J., Broust, F., Ndiaye, F.-T. and Ferrer, M. (2007). Properties of bio-oils produced by biomass fastpyrolysis in a cyclone reactor. *Fuel, 86*, 1800-1810
4. Sun, P., Heng, M., Sun, S.-H., and Chen, J. (2011). Analysis of liquid and solid products from liquefaction ofpaulownia in hot-compressed water. *Energy Conversion and Management*, 52(2) :924 – 933.
5. Cheng, S. D'cruz, I., Wang, M., Leitch, M., and Xu, CC (2010). Highly efficient liquefaction of woody biomass in hot-compressed alcolhol-water co-solvents. *Energy & Fuels*, 24(9):4659-67
6. Channiwala, S. and Parikh, P. (2002). A unified correlation for estimating hhv of solid, liquid and gaseous fuels. *Fuel*, 81(8):1051

(62b) Pyrolysis Oil and Its Applications

Rajeswar R. Gattupalli, Process Design Development, UOP LLC, A Honeywell Company, Des Plaines, IL, Stanley Frey, Renewable Energy & Chemicals, UOP LLC, Des Plaines, IL and Lance Baird, UOP LLC, A Honeywell Company, Des Plaines, IL

Fuels obtained from biomasses are becoming a valid alternative to the use of fossil fuels, especially in the light of stringent environmental constraints.Fast pyrolysis of lignocellulosic biomass produces a renewable liquid fuel called pyrolysis oil that has a variety of applications as a potential replacement for fossil fuels.One aspect of pyrolysis oil that has been addressed to promote its broader use is improving the stability of pyrolysis oil with respect to viscosity increase over time, making long range transportation and storage possible. Other properties of pyrolysis oil present challenges during its use in standard equipment constructed for petroleum derived fuels. Significant effort has been spent on research and development to improve the stability of pyrolysis oil and for its use in burners and other energy applications.

This presentation is divided into two parts - in the first part, we will give a brief background on pyrolysis oil and its stability concerns, the second part will focus on the rapid thermal processing or RTP process for fast thermal conversion of wood and/or other biomass to high yields of liquid fuels and the various applications of stabilized RTP green fuel.

Catalytic performance of Mn-promoted Ni/MgO catalysts in CO_2 reforming of toluene used as model compound of tar produced from biomass: Influence of preparation methods

Meng Kong, Jin-hua Fei*,Xiao-ming Zheng*

Institute of catalysis, Xixi Campus, Zhejiang University, Hangzhou, P. R. China, 310028

Abstract

Catalytic performance of Mn-promoted Ni/MgO catalysts prepared by different methods was investigated in CO_2 reforming of toluene. By XRD, H_2-TPR and TEM analysis and reforming tests, catalysts made by either conventional impregnation or precipitation methods in which $Mn(NO_3)_2 \cdot 4H_2O$ was used as the start precursor behaved similarly in catalytic performance, Mg_6MnO_8phase was present in all the calcined catalysts which caused worse activities in reforming tests. In contrast, by using a mechanic milling method in which MnO was initially added by physical milling with MgO as a support and then the calcined support was impregnated with nickel nitrate, the main crystalline phase of Mn species was α-Mn_2O_3 which worked as a support, after reduction, Ni can distribute well on both MnO and MgO, in this case, more Ni particles can be obtained because the interaction between Ni and MnO was less strong than that with MgO, therefore, a better performance than Ni/MgO and other catalysts can be yielded.

Keywords: Tar, Reforming, Ni/MgO, Mg_6MnO_8, α-Mn_2O_3

Introduction

Tar is an inevitable by-product during biomass gasification which can pollute environment, block process pipes and decrease calorific value.[1-3]Reforming of tar compounds to syngas is an efficient method to improve the overall energy recovery during gasification while minimizing environmental pollution.[2]

Many researchers have used steam as an oxidizing agent for reforming tar to produce H_2-rich syngas. However, the concentration of steam in the rough production gases from biomass gasification is much less than the concentration of CO_2 which can range from 10 to 30 vol. %.[3,4]Therefore, CO_2 could be a better choice than H_2O for reforming the tar to boost the effective conversion of carbon and minimize the release of CO_2 into the atmosphere by providing more syngas. In our study, toluene was selected as a tar model compound because it represented one of the major tar species 错误!未定义书签。, after comparing some supported catalysts, Ni/MgO was recognized as an efficient catalyst in CO_2 reforming of toluene, 80% of toluene was transformed to syngas at 600℃with a CO_2/toluene mole ratio of 14.[5]In order to improve the catalytic performance especially at a lower reaction temperature, experimental study and Co-Ni bimetallic catalysts have been investigated.

In this contribution, MnO was used as a promoter, because it showed good capacities in stabilizing Ni particles and eliminating coke in Ni/Al_2O_3catalyst for dry-reforming of methane.[6,7] However, the conventional impregnation and precipitation methods were not suitable in Ni-MnO-MgO catalyst system. Therefore, in this paper, the influence of preparation method of MnO-promoted Ni/MgO catalyst in CO_2 reforming of toluene was investigated, which was helpful for readers to understand the relationship between catalytic performance and catalysts' structure.

2. Experimental Section

2.1. Catalyst preparation

Ni/MgO, Ni/MnO and Ni-MnO-MgO catalysts were prepared by the following methods. In all catalysts the loading of Ni was controlled at 5% and MnO was 10% by weight, in Ni/MnO, the loading of Ni was also 5%, the rest part was MnO.

Ni/MgO was prepared by impregnating with Ni $(NO_3)_2 \cdot 6H_2O$ aqueous solution and MgO overnight, and then dried at 120℃ for 6h, finally calcined at 700℃ for 6h.

Ni-MnO/MgO-CI catalyst was made by a co-impregnation method. MgO was impregnated initially with Ni $(NO_3)_2 \cdot 6H_2O$ and 50 wt.% Mn $(NO_3)_2$ aqueous solution overnight, then followed the above steps.

Ni/MnO-MgO-SI catalyst was made by a sequential impregnation method. MnO/MgO-IM support was prepared first by the impregnation method as same as Ni/MgO, then the modified MnO/MgO support was impregnated with Ni $(NO_3)_2 \cdot 6H_2O$ solution, dried and calcined at the same conditions above.

Ni/MnO-MgO-PI catalyst was made by a precipitation-impregnation method. MnO/MgO-PR support was first prepared by the precipitation method in which50 wt.% Mn $(NO_3)_2$aqueous solution was used as precursor, and Na_2CO_3 was the precipitation agent, and the PH was controlled at 10. After dring and calcination, the modified support was impregnated with nickel nitrate solution as Ni/MnO-MgO-SI did.

Ni/MnO-MgO-MI catalyst was also made by an impregnation method. First, the support of MnO/MgO-MM was prepared by mechanic milling MnO_2 with MgO uniformly, and then calcined at 700℃ for 6h, this support was finially used to prepare Ni-based catalyst by an impregnation method as above.

Ni/MnO_2 catalyst was prepared by co-precipitation with aqueous solution of Ni $(NO_3)_2 \cdot 6H_2O$ and Mn $(NO_3)_2$ and Na_2CO_3 as precipitation agent. Ni/MnO_2 was not made by impregnation, because MnO_2 has a badly worse hydrophilic capacity and it is very hard to squeeze to a suitable size of 0.3-0.45mm..

2.2. Catalyst characterization

The crystalline phases contained in the samples were examined by powder X-ray diffraction system (Rigaku, D/MAX 2500/PC) using Cu Ka radiation (45 kV, 40 mA).

The TEM experiments were performed by using JEM-2010(JEOL Ltd.) TEM instrument operated at 200 kV. An X-ray analyzer for EDS is incorporated into the instrument for elemental analysis under STEM mode. The sample was first dispersed in ethanol and supported on Lacey-formvar carbon on a 200 mesh Cu grid before the TEM images were recorded.

2.3. Catalyst activity test

CO_2 reforming of toluene was carried out at atmospheric pressure in a tubular quartz fluidized-bed reactor having an inside diameter of 9.8 mm. The catalyst (0.5 ml, 0.3-0.45 mm) was first reduced at 700 °C in H_2 for 60 min and then tested for 450 min at 570 °C. For testing, liquid toluene was supplied at a rate of 0.3 ml/h byan HPLC pump (ASI, 501) into a constant temperature evaporator (120°C) where a gas mixture composed of 285 ml/min Ar and 14.5 ml/min CO_2 (atmospheric pressure) was introduced at the same time. The flow rates of CO_2 and Ar were measured by mass flowmeters (Brooks, 5850E).

The effluent gases were analyzed with an on-line gas chromatograph (Agilent, GC-6820) equipped with a TDX-101 column and a thermal conductivity detector (TCD) to determine the H_2, CO, and CO_2 concentrations (No CH_4 was detected). Simultaneously, toluene and other aromatic products (just less benzene) were determined with an HP-5 capillary column and a flame ionization detector (FID).

3. Results and discussion

3.1 Characterizations of catalysts

From the XRD patterns of supports used in this study(Fig. 1-A),it was obvious that both Mn/MgO-IM and Mn/MgO-PR supports had distinct diffraction peaks of Mg_6MnO_8 which derived from the rock salt structure of MgO by replacing one-eighth of the Mg^{2+} ions with Mn^{4+} ions and one-eighth with vacancies,[8]where Mn^{4+} ions appeared at the support surface and reacted with MgO to form Mg_6MnO_8,[9] also, due to this formation, MnO could disperse by one single layer on MgO surface.[10] In XRD spectra, there were some diffraction peaks (ca. 42.9°and 62.3°) overlapped together which indicated that both Mg_6MnO_8 and MgO were present on Mn/MgO-IM and Mn/MgO-PR supports. The size of Mg_6MnO_8 particles in either Mn/MgO-IM or Mn/MgO-PI was similar at 14-16 nm (Mg_6MnO_8 (311), $2\theta = 35.4°$). However, for MnO/MgO-MM support, besides the characterized peaks of MgO, the dominant phase of Mn species was α-Mn_2O_3 with a particle size of 32 nm (Mn_2O_3 (211), $2\theta = 32.9°$) which came from calcination of MnO_2.[11]

In XRD patterns (Fig. 1-B), a physical mixture composed of 50wt.%NiO and 50wt.%MgO without calcination exhibited evident difference of NiO and MgO signals in the range of 70-80°, so this region can be used to determine NiO and MgO.[12] However, in Ni/MgO catalyst, no NiO phase was detected, and the diffraction peaks were close to pure MgO, which indicated NiO dispersed well on the support of MgO owning to solid solution structure between NiO and MgO.[12] In Ni/MnO catalyst prepared by co-precipitation, there were also no NiO peaks besides α-Mn_2O_3 with a particle size of 58nm (Mn_2O_3 (211), $2\theta = 33.1°$), which suggested that Ni species also had a good dispersion.

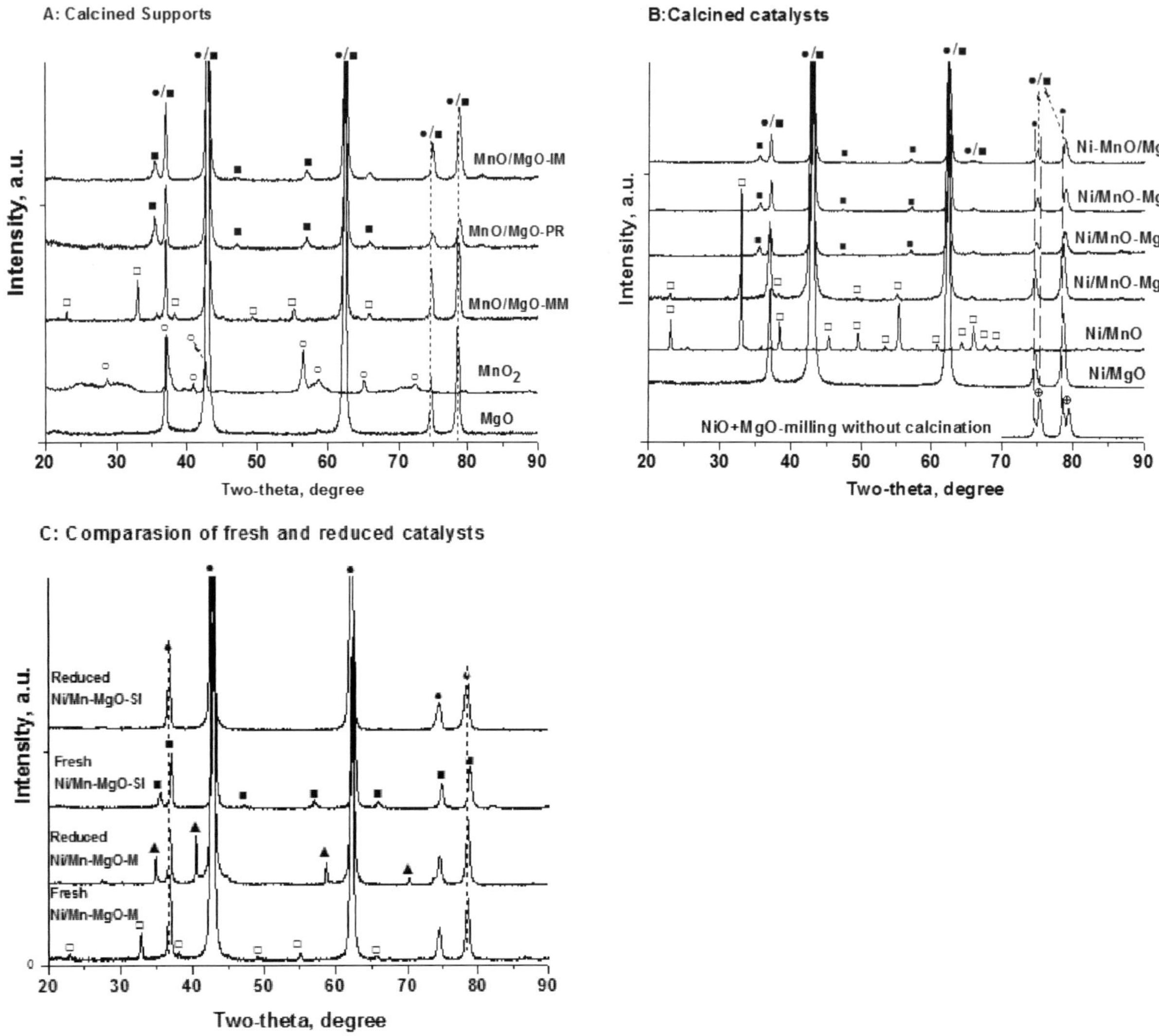

Fig. 1 XRD patterns of (A) Calcined supports, (B) Calcined catalysts, (C) Reduced catalysts. Symbols: □Mn_2O_3, ○MnO_2, ▲MnO,⊕NiO, ●MgO/or Ni-Mg-O, ■Mg_6MnO_8.

For the catalysts made by $Mn(NO_3)_2$ as the start precursor (SI, CI, PI), the crystalline phases of Mn and Mg species in XRD patterns did not change obviously in which $Mg_6MnO_8$and MgO were the main materials. But because the catalysts were calcined at 700℃ for 6h again, the particle size of Mg_6MnO_8 was increased slightly from 14-15nm to 18-20nm. Of particular interest in fig.1-B was that there were no NiO peaks present on catalysts, which might because NiO was covered byMg_6MnO_8 phase. As for Ni/MnO-MgO-MI catalyst, α-$Mn_2O_3$was also the main Mn species with a particle size of 36nm, no NiO and other manganese-contained phases.

In sum, Mg_6MnO_8 presented on all the catalysts with $Mn(NO_3)_2$ as the start precursor, whereas, Mn_2O_3 was the main phase of manganese on Ni/MnO-MgO-MI catalyst. So Ni/MnO-MgO-SI can be selected as a representativecatalyst to compare with NiO/Mn-MgO-MI.

When these two fresh samples were reduced at 700°C for 1h in H_2, the main states of Mn changed (Fig. 1-C). It was clear that Mg_6MnO_8 disappeared in the reduced Ni/MnO-MgO-SI catalyst without other Mn species, which suggested that the reduced Mn species would have a good dispersion on MgO surface. However, Mn_2O_3 was reduced to MnO on Ni/MnO-MgO-MI after reduction.

TEM pictures gave very different morphology of the refuced catalysts in Fig. 2. Fig. 2-A was the reduced Ni/MnO-MgO-SI, it was clear there were some squares on the MgO surface which was not showed in Fig. 2-B of the reduced Ni/MnO-MgO-MI. The different images derived from the different Mn species definitely. Noted that, in the EDS analysis of reduced Ni/MnO-MgO-MI, the Mn element seemed absent, that was because MnO has a high density which easily deposited in ethanol before TEM characterization.

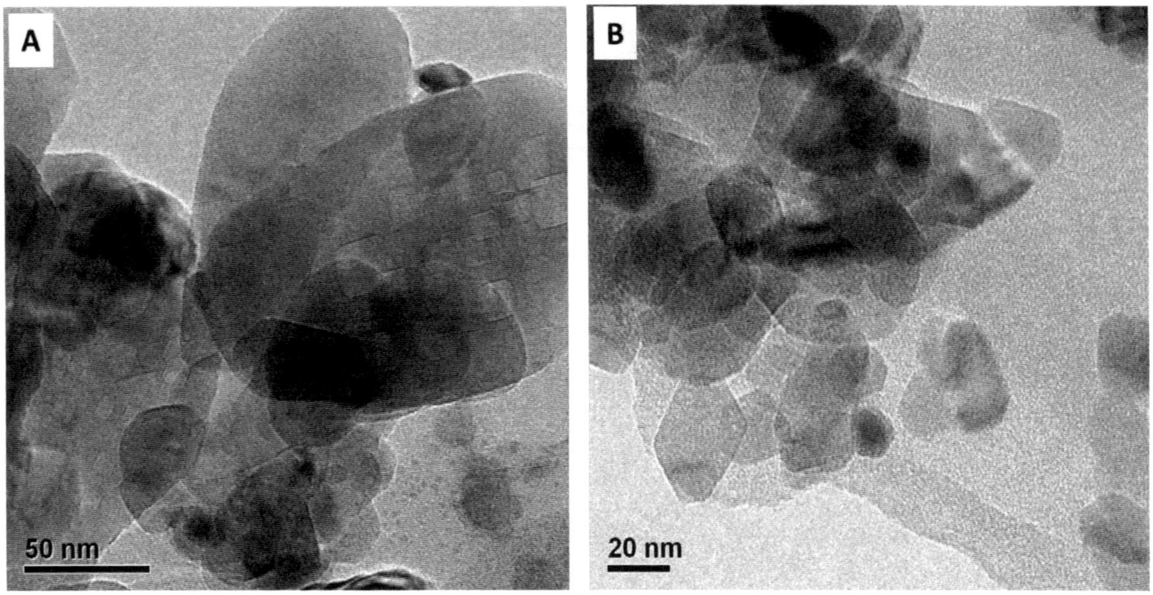

Fig. 2 TEM pictures of(A) reduced Ni/MnO-MgO-SI (B) reduced Ni/MnO-MgO-MI.

3.2 Activity of the employed catalysts in CO_2 reforming of toluene

The performances of catalysts mentioned above were drawn in Fig. 3. As (is) seen from these curves, over Ni/MgO catalyst without any promotes, the initial conversion of toluene can reach to 72.5% (data at 30min), comparably, it will be just 37% when just MnO severed as the support. However, for the Ni-MnO-MgO catalysts with $Mn(NO_3)_2$ as the precursor, the conversions of toluene were no more than 20%, among them, Ni/MnO-MgO-PI had a relatively higher activity with 22.5% of toluene converted at initial time, but it decreased quickly to less than 10% after 5h test. Interestingly, by using MnO_2 as the precursor, a highest and stable performance with 81% of toluene converted can be obtained over Ni/Mn-MgO-MI catalyst. For Ni-MnO-MgO with $Mn(NO_3)_2$ as the precursor, the total conversion of toluene were no more than 10%, including （that） 1.5%-2.5% of toluene was converted into benzene, they had a poor capacity for deep cracking of benzene. However, the proportion for toluene to benzene was also 0.8 % over Ni/MgO, 0.4% over Ni/Mn-MgO-MI, it can be considered that the existence of Mn promotes the conversion of toluene.

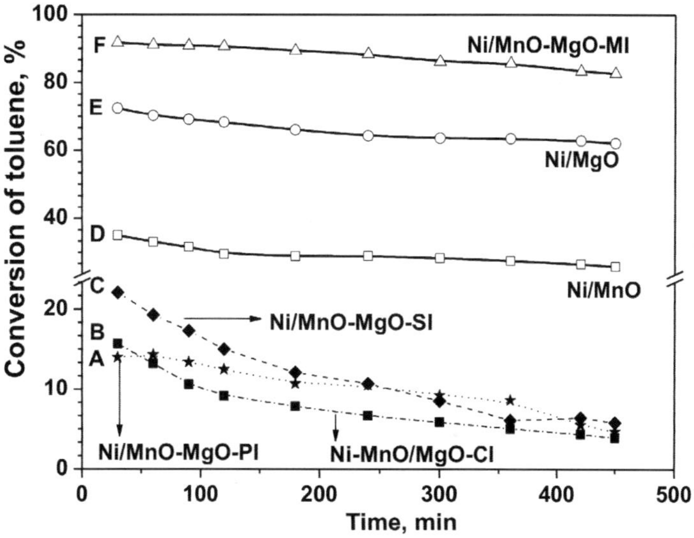

Fig. 3. The conversion of toluene over the catalysts on stream. A, Ni-Mn/MgO-PI; B, Ni/Mn-MgO-CI; C, Ni/Mn-MgO-SI; D, Ni/MnO; E, Ni/MgO; F, Ni/Mn-MgO-MI. Reforming conditions: 0.5ml catalyst, 0.005ml/min toluene, 14.5ml/min CO_2, 285ml/min Ar, GHSV= 36000 h^{-1}, temperature was 570℃.

Conclusion

Mn-modified Ni/MgO was investigated for CO_2 reforming of toluene. It was found that the preparation method played a crucial role in catalytic performance. Using $Mn(NO_3)_2$ as the precursor, Mg_6MnO_8 formed during calcination, which made the catalysts obtain a worse activity, however, using MnO_2 as precursor, the interaction between Mn species and MgO became weak, MnO can work as a support to distribute Ni particles, which benefited the reforming reaction.

Acknowledgements

This research work was supported by the Hi-Tech Research and Development Program of China (2007 AA 05Z415) and the Natural Science Foundation of Zhejiang Province (Y407113).

Reference

[1] Coll, R., Salvado, J., Farriol, X., Montane, D., Fuel Process. Technol., 2001, 74, 19-31.

[2] Zhang, R.Q., Brown, R.C., Suby, A., Cummer, K., Energy Conversion and Management. 2004, 45, 995-1014.

[3] Garcı´a. L., Salvador, M.L., Arauzo, J., Bilbao, R., Energy Fuels., 1999, 13, 851-859

[4] Li, X.T., Grace, J.R., Lim, C.J., Watkinson, A.P., Chen, H.P., Kim, J.R., Biomass Bioenergy. 2004, 26,171-193

[5] Kong, M., Fei,J.H., Wang, S., Lu, W., Zheng, X.M., Bioresour. Technol. 2011, 102, 2004-2008.

[6] Nandini, A., Pant, K.K., Dhingra, S.C., Appl. Catal. A: Gen. 2005; 290: 166-174

[7] Seok, S.H., Choi, S.H., Park, E.D., Han, S.H., Lee, J.S., J. Catal. 2002, 209, 6–15.

[8] Taguchi, H., Nagao, M., J. Mater. Sci. Lett. 1991,10, 658 - 659

[9] Guilliat, I.F., Brett, N.H., J. Chem. Soc. Faraday Trans. 1972, 68, 429.

[10] Yao, J.S., Meng, M., Luo, J.Y., Zha, Y.Q., Xie, Y.N., Hu, T.D., Liu, T., J. Mol. Catal. (China), 2006, 20, 300-305

[11] Ardizzone, S., Bianchi, C.L., Tirelli. D., J. Electroanal. Chem. 1997, 425, 19-23.

[12] Ruckenstein, E., Hu, Y.H., Appl. Catal. A: Gen. 1999,183, 85-92.

(62d) Direct Inter-Processes Integration for Energy Efficiency Optimization and GHG Emissions Reduction in Oil Refining

Mahmoud Bahy Noureldin, Process and Control Systems, Saudi Aramco, Dhahran, 31311, Saudi Arabia and Faisal AlMousa, Process and Control Systems Department, Saudi Aramco, Dhahran, Saudi Arabia

Industrial communities are invited to conserve energy resources not only to give a chance to pre-developing countries and future generations to grow and prosper but also to preserve the environment. Energy efficiency optimization is a quick answer to energy-based GHG emissions reduction. It enables us get two targets with one arrow. From GHG emissions outlook, energy efficiency optimization applications in industry can help us get out tens of thousands of cars abstractly from our roads and practically from our air every year.

Since the emanation of the pinch technology and its evolution to pinch analysis technique and the adaptation of MINLP techniques for process synthesis of mega complexes, the direct inter-processes integration has been considered impractical. For inter-processes "energy" integration many arguments against the direct inter-processes integration are certainly valid.

Plants are normally have different start up and shut down times; the processes can work at partial loads; the processes can have seasonal changes in its operating conditions; utility systems, heaters, coolers and HEN capital will not be reduced and may be even increased due to differences in processes/plants schedule and operation philosophy; the disturbance in one process plant can propagate to another one if they are mass and/or energy integrated; the distance-time/velocity lags in process will affect the controllability of integrated processes; the geographical distances among processes will cost us energy in pumping or compression and capital in piping; plants' safety might be impacted due to the travel of a fluid from one "hazardous" area classification to another; the fear of leakage and so on. Therefore, direct inter-processes integration while is beneficial to energy conservation and the GHG emissions reduction as will be demonstrated in the paper, is still to date almost ignored in the industrial community. Due to those concerns most of the current methods for inter-process integration are indirect and conducted using buffer systems. Buffer systems are either steam system, boiler feed water, hot water system, hot oil system or sometimes combinations of them. The industrial community does agree that, direct integration approach in inter-process integration (between several plants) is efficient and can render more saving in energy consumption and energy-based greenhouse gas emissions beyond the maximum possible using intra-integration only. In this paper, we demonstrated that huge potential for energy consumption and GHG emissions reduction in oil refining can be attained through smart integration among processes. The paper has two folds of results. First, while we agree with the validity of some of the arguments against the direct inter-processes/plants integration we believe and will demonstrate that we still have room for improvement in energy efficiency enhancement and energy-based GHG emissions reduction. We will identify at the targeting phase best possible scenarios for inter-processes integration and the most cost effective solution(s) now and in the future via "Plants' Smart Matching", without the discouraging solutions which require many inter-connections among plants as sometime reported in literature. Second, the intuition that mandates sophisticated inter-plants integration to achieve good benefits, the more you integrate among the processes (new mega sites consists of integrated refining, chemical, petrochemical and plastics plants), the better you save in energy consumption is not always true. The paper presents those two results and demonstrates the benefit that could be obtained from the direct inter-processes integration through two industrial case studies.

(62e) Energy Efficient Control Strategy to Operate Post- Combustion CO2 Capture in Response to Power Plant Load Reduction

Sepideh Ziaii, Gary T. Rochelle and Thomas F. Edgar, Department of Chemical Engineering, The University of Texas at Austin, Austin, TX

Energy Efficient Control Strategy to Operate Post- Combustion CO_2Capture in Response to Power Plant Load Reduction

Increasing concern about global warming has motivated the development CO_2capture and sequestration as a potential option for reducing emissions from PC and IGCC plants. One of the mature technologies that can capture CO_2emissions from a coal-fired power plant is amine-based absorption/stripping, which has the advantage of being added to an existing power plant. However, this technology is a very energy intensive, which can reduce the efficiency of power plants by 30%.

In thistechnology the aqueous amine solutioncountercurrently$_2$from the flue gas by exothermic chemical reactions and significantly reduces the CO_2concentration in the gas stream exiting the absorber. The CO_2-rich amine solution from bottom of the absorber is sent through a cross exchanger and directed to the stripper. The stripper temperatureis maintained$_2$reaction and liberate CO_2. The lean solution leaving the bottom of the stripper iscooled and $_2$. The concentrated CO_2stream from the top of stripperis compressed

Due to high-energy demand that is mostly attributed to solvent regeneration and CO_2compression, many research efforts seek to find the most energy efficient steady sate design . However there are few studies havebeen performed on minimizing energy consumption during possible operational scenarios for an existing capture plant. This study has simulated

This work incorporated models of MEA absorption/stripping process, CO2flowsheet in the framework of Aspen Custom Modeler. Absorber and stripper are packed columns modeled as multi-segment components with time variant mass and energy balance equations. Each segmentis modeledwas modeledare modeled

During boiler load variation in power plants, so many process variables are subject to change. One of the important variables influencing downstream capture plant operation is the property of flue gas entering the absorber. Based on discussion in literature on reduced load operation of coal-fired power plants, itcould be assumed that as load reduces, the rate of flue gas is reduced accordingly while the composition and temperature are constant. The other issue resulting from power plant load change is related to thesteam rate1st

Regarding energy efficient operation, the simulation results demonstrated that a variable speed compressor is much more beneficial than a constant speed one. Adjusting compressor speed reduces lost work by pushing the compressor away from surge condition and avoids additional energy use due to recycling gas in the compressor. In addition, as total steam rateis reduced in power cycle, the reboiler steam pressure and flow rate is reduced and consequently CO22heat duty equivalent work; however, the temperature should not exceed 1202is limited by maximum compressor speed, which is typically 120% of rated speed.

Another potential processvariable which can be manipulated to minimize reboiler lost work and save the compressor from surging is solvent circulation rate. At fixed reboiler temperature, there is an optimum solvent rate that minimizes lost work and maximizes CO2is set by compressor and pump operability. The minimum flowis always setis operated

Based on minimization of total lost work by compressor speed and solvent circulation rate performed in the presence of operational constraints associated with pumps, compressor and solvent thermal degradation, this study gives a practical optimum CO_2removal over 40%-100% operating load range. For this range of load change in an MEA plant designed for 90% removal, the plant can remove up to 94% of inlet CO_2. The practical minimum removal over that load range is about 85%, where the reboiler temperature reaches its maximum allowable value and the lost work is at a minimum.

Reducing the load of power plant that reduces extracted steam rate in the capture results pushes the compressor operating curve to the surge limit such that at minimum load (40%) , adjusting solvent rate and compressor speed no longer moves the operating curve away from

Based on energy minimization in response to power plant load reduction, two control strategies were found to operate the capture close to minimum energy path at each desiredremoval .

1. Ratio control between reboiler steam rate and CO_2rate in rich solution or
2. Controlling lean loading to vary proportional to the CO_2removal change.

In the future we plan to observe and compare dynamic performance of these two control strategies.

(62f) Study on Cascade Refrigerant and Mixed-Refrigerant Systems for Ethylene Plants

Ha Dinh, Jian Zhang and Qiang Xu, Dan F. Smith Department of Chemical Engineering, Lamar University, Beaumont, TX

Refrigeration system holds an important role in most chemical/petrochemical processes. A refrigerant system transfers heat from low-temperature sources to high-temperature sinks through vapor-compression cycles at the expense of mechanical work, magnetism, laser or other means. In many industrial processes, the refrigeration systems are employed for chilling and freezing the final liquid/solid products or liquefying the gas intermediates/products for the downstream cryogenic separation. Optimal synthesis of these systems is attractive not only because of their economic benefits but also because they incorporate most of energy transfer operations in widespread use in the chemical industry. The traditional cascade refrigeration system (CRS) used in ethylene plants includes multiple refrigerants and multiple temperature and pressure levels. Each component subsystem includes a compression section, a condensing section and several temperature levels with sub-coolers, expansion valves, evaporators and flash drums. Recently, mixed-refrigerant system (MRS) has shown some advantages comparing to the single-component one. A MRS uses a mixture as the refrigerant instead of several pure components as in conventional cascade refrigeration systems. This system operates within smaller temperature differences at the lower limit; therefore, it leads to a smaller increase in entropy and consequently a smaller loss in energy. Besides, an MRS also requires a simpler unit configuration and fewer maintenance problems.

In this study, both types of refrigeration systems used in ethylene plants are studied based on exergy analysis, a powerful method to explore system energy loss below ambient temperature. The procedure involves three consecutive steps: Thermodynamic Analysis, Synthesis Model Development and Solving, and Solution Validation and Evaluation. Rigorous steady-state models of both refrigeration systems are firstly constructed and employed to generate Exergy-Temperature (B-T) chart in order to analyze their thermodynamics and energy distributions. An exergy-based mixed-integer linear programming (MINLP) model is developed to maximize the energy efficiency of each system, where multiple refrigerants with multiple recycling loops are simultaneously addressed and all of cooling demands are fulfilled. The optimal solution is subsequently examined by conducting rigorous simulation to check their feasibility and consistency. Economic evaluation is also carried out from the calculated capital cost of new equipments and annual operating cost. The efficacy of the developed methodology is demonstrated by two case studies: a cascade refrigeration system and a mixed-refrigerant system used in two ethylene plants.

(63a) Preliminary Technical Feasibility Study Into Effect of Withdrawing Spent Caustic Stream From Existing Activated Sludge Treatment Plant

Janit Pongpuak[1], Wiwut Tanthapanichakoon[2], Vasimon Ruanglek[1], Nakarin Mongkolsiri[1] and Surachate Chalothorn[1], (1)SCG Chemicals, Central R& D, Technology Centre, Bangkok, Thailand, (2)Department of Chemical Engineering, Tokyo Institute of Technology, Japan

Our olefins plant produces olefins (ethylene, propylene, butylenes) as well as aromatics (Benzene, Toluene, and Xylene, or BTX) by thermally cracking naphtha in the presence of dilution steam. The plant employs an activated sludge (A/S) wastewater treatment plant to treat a combination of the following three major wastewater streams: the spent caustic stream (SCS) from the cracked gases scrubbing unit, the blowdown stream (DSG) from the dilution steam generator used to produce steam for the thermal cracker, and the dilution water stream (DWS) consisting of rainwater, equipment cleaning water, fire alarm testing and etc. As part of our waste minimization policy, a new plan is afoot to divert the SCS completely to a caustic soda recovery plant, thereby not only significantly reducing the total influent flow to the A/S plant but also changing the wastewater characteristics. This leads to a necessity to conduct a technical feasibility study into the corresponding effects on the performance of the A/S plant. In practice it is also necessary to know how the SCS flow diversion should be implemented in order to ensure smooth withdrawal of SCS during the transition period.

Simultaneously, the feasibility of total elimination of the DWS and furthermore the possible inclusion of wastewater from a nearby HDPE plant in the near future are to be investigated. To smoothly accomplish the above-mentioned series of planned changes, a series of preliminary experimental studies at the lab-scale level (using batch reactors and continuous reactors) have been carried out to study the A/S process performance under different scenarios. Based on the experimental data, a dynamic math model of the lab-scale A/S process is also developed and substantiated. In the next step the extension of the math model to the existing A/S plant will be carried out whilst taking into the difference in residence time distribution in the lab-scale and full-scale aeration tanks. Then the validated model will be used to predict the A/S plant performance in response to the planned withdrawal of SCS and DWS streams as well as the subsequent inclusion of the HDPE plant stream. What will be reported here is the lab-scale experimental results and the corresponding math model substantiated by these experimental results.

(63b) Cleaner Chemical Industry Through HEN Advanced Retrofit Approach

Mahmoud Bahy Noureldin, Process and Control Systems, Saudi Aramco, Dhahran, 31311, Saudi Arabia, Mohammad Alibrahim, Process and Control systems Department, Saudi Aramco, Dhahran, Saudi Arabia, Mana Al-Owaidah, Process and Control Systems, Saudi Aramco, Dhahran, Saudi Arabia and Ali AlQahtani, Process and Control systems, Saudi Aramco, Dhahran, Saudi Arabia

The economics of industrial production, the limitations of global energy supply, and the realities of environmental conservation are an enduring concern for all industries. Wherever you turn, there's another entreaty to save energy, reduce carbon emissions and protect the environment for posterity.

Industrial plant's heat exchangers network (HEN) shall always be ready to face a high level of operational changes along its lifetime. These changes are short-term, such as process disturbances; uncertainty in feed stock conditions and product demand, and long-term such as the need to process more raw materials that warrants the debottlenecking of the facility including the HEN to increase its capacity. Nowadays, and since late seventies of the last century, another important frequent change is happening due to the continuous escalation in energy prices in a rate that is higher than plant equipment cost, also warrants continual modification of the facility's HEN to enhance the plant energy efficiency along its lifetime (sometimes reaching up to 30 years). In such cases, the HEN retrofit objective/task is to produce a practically implementable cost effective HEN design modification that satisfies the new process objective and its new operating constraints. There are many possible modifications for an existing HEN to retrofit the original design to the new objective. It can include a combination of all possible process operating and design conditions modifications, existing HEN topological/structural modifications, and existing HEN unit design modifications and parametric modifications (such as heat transfer enhancement to improve U) as well. In huge industrial facilities such problem can be solved using pinch technology or MINLP , where we have an objective to minimize total cost of the existing HEN modifications and the rest of the plant's process units design and operation modifications minus the realized energy cost saving. There are some known drawbacks in the existing methods. In one hand the most widely used to date pinch method and its modifications is non-systematic and involves heuristics that produce solutions whose quality depends entirely on the designer experience in applying the heuristics. For instance, determining the location of the pinch point for an assumed best globalDT_min, (which globalDT_min should be used) is a little arbitrary. As the value of assumed best globalDT_min changes, the location of the pinch point changes and therefore the assessment of which unit to transfer heat across the pinch also changes. The same also applies upon process changes that result in same output. The retrofit is a one package solution that has no consideration for yearly step-by-step modifications and practicality for future retrofit projects. The mathematical programming-based methods on the other hand have the drawback of providing little or almost no scope for user interaction. Nowadays, it is possible to solve larger problems than before but still not on the scale needed in many industrial mega sites. The problem of the HEN retrofit is NP-hard, and including all possible process design modifications, different types and configurations of heat exchanger units and so on in the superstructure is impractical and will make the problem harder, intractable and impossible to solve. In addition, the MINLP model cost objective function needs comprehensive data base for the model solution to be realistic with information/data that is normally not available during conceptual phase of the retrofit project, especially in de-centralized engineering environment where projects cost control depends entirely on rigorous cost calculation that always comes after the retrofit project basic engineering and is carried out at different departments in most of the companies. Besides neither practical nor logical to assume with high fidelity all possible retrofit solutions that can be implemented for a given plot plan of a facility and rigorously develop cost correlation to all of it before running the mathematical program especially when there is always more than one option to carry out a designated modification such as the increase of the network surface area. Neither one of these methods (pinch technology, network pinch and mathematical programming) can rigorously assess the existing network; rigorously target for the solution before the go ahead of the retrofit project for practically attainable waste heat recovery and/or minimum number of units that need to be added to reach desired energy consumption targets or part of it, and to systematically renders solutions that can be implemented on phases with a guarantee that the solution implemented today in phase one will not become an obstacle, or need to be demolished, for a tomorrow retrofit solution to be implemented at later time in the future on phases two, three and so on along the plant's lifetime. The paper exhibits new elements, in the representation and assessment of the existing HEN studied for retrofit and in the systematic finding of the retrofit solutions that is evolutionary in implementation and revolutionary in consideration. It presents the concept of plant life-time retrofit-ability that addresses the identification of any possible contradiction between now and future retrofits to render opportunities of continuing retrofit projects to achieve ultimate target. The method is illustrated with two industrial case studies for NGL and oil refining plants.

(63c) Energetic Self Sufficiency: Theory and Application

Vasilios Manousiouthakis, Chemical & Biomolecular Engineering Department, UCLA, Los Angeles, CA and Jorge Pena Lopez, Chemical and Biomolecular Engineering Department, University of California, Los Angeles, Los Angeles, CA

Abstract

The concept of energetic self-sufficiency, for an open thermodynamic system, is first presented and its application to hydrogen production is discussed.

Let Ω be a steady-state open thermodynamic system with inlets in S_I; outlets in S_O; participating chemical species in S_C; participating elements in S_E; heat transfer rates $Q_j$with j in $S_Q$entering ($Q_j>0$) or exiting ($Q_j<0$) the system at temperatures $T_j$respectively, and shaft work rates W_j, j in $S_W$entering (consumed by) the system ($W_j>0$) or exiting (produced by) the system ($W_j<0$). This system has to satisfy the overall mass, element, energy, and entropy conservation laws. These latter two laws are often referred to as the 1st and 2nd laws of thermodynamics, which at steady-state require that the energy entering the system must equal the energy exiting the system, and that the total entropy change of both the surroundings and the system is greater than or equal to zero. Then

*Definition.*The system Ω is called energetically self-sufficient iff no heat is transferred from the surroundings to the system, heat is possibly transferred from the system to the surroundings ($Q_j<0$, j in S_Q) at the uniform surroundings temperature T_0, no shaft work is utilized (consumed) by the system, and possibly shaft work is generated (produced) by the system ($W_j<0$, j in S_W).

The considered application of this concept involves the thermodynamic assessment of hydrogen production from natural gas so that no carbon dioxide is emitted from the system. The concept of thermodynamic self-sufficiency is used to establish rigorous limits on the amount of hydrogen that can be produced.

(63d) Impact of Flaring Emissions on Regional Air Quality Associated with An Ethylene Plant Start-up

Jian Zhang, Qiang Xu and Thomas C. Ho, Dan F. Smith Department of Chemical Engineering, Lamar University, Beaumont, TX

Flaring is a common industrial practice designed to eliminate the accumulation of un-reacted feed gases or off-specification product gases through stack burning to ease the strain on a process unit especially during plant start-ups, plant shut-downs and operational up-sets. Although the practice is necessary and crucial to chemical plant safety, excessive flaring represents the loss of precious feed materials and also results in the generation of extra amounts of undesired emissions including VOCs and NO_X, both of which are ozone precursors.

In this study, the effects of flaring emissions associated with an ethylene plant start-up with various start-up scenarios on the increment of regional ozone concentrations were investigated through model simulations. In the practice, the dynamic process simulations associated with an ethylene plant start-up were performed first. The obtained flaring emissions were then incorporated into a CAMx air quality simulation model to simulate the effect of these extra VOCs and NO_Xemissions on the regional ozone concentrations. From the the CAMx simulations, the effects of start-up starting time, plant location, and episode day on ozone increments associated with various start-up scenarios were investigated and scientific insights were also provided. Finally, the environmentally-preferred start-up schedule is identified as the one that avoids having the peak flaring emissions occur near the time of the highest solar intensity to minimize the impact of the start-up process on the regional ozone concentrations.

(63e) Phytoremediation of A Wetland SOIL, by TWO METHODS In Bogota Colombia

Ingrid Rivera[1,2], Sergio Pedraza[1] and Gabriel Camargo[3], (1)Environmental Engineering, Universidad Libre, Bogota, Colombia, (2)Universidad Libre, Bogota, Colombia, (3)Ingenieria Mecanica, Universidad Libre, Bogota, Colombia

In this project, the efficiency of two phytoremediation methods for soil (phytoextraction and phytostabilization) was investigated to remove heavy metals such as Pb, Cu and Ni. The place for this research was a wetland call "La conejera".

In this wetland the pollution levels has been growing due the urbanization impact in Colombia. The first step in the project was physicochemical characterization, followed by the selection of the phytoremediation process. The methods chosen to remove Pb, Cu and Ni, were phytoextraction and phytostabilization. The ability and effectiveness to remove the heavy metals was evaluated.

For the phytoextraction process, the mustard (sinapis Alba) and barley (*Hordeum vulgare)*were the plants selected. The mustard was used due its adaptability in wet soils, faster germination and growth. The barley was used due resistance to extreme weather conditions. The results obtained shown that all metals were remove by the barley and only the Copper when the mustard was used.

In the phytostabilization method, the plant used was Ryegrass perenne (*lolium perenne L.*). The results obtained shown that all metals was retired for this plant.

2012 AIChE Spring Meeting &
Global Congress on Process on Process Safety
April 1-5, 2012 - Houston, TX

Search of Suitable Halon 1301-Like Fire Suppressants Based on Flame Inhibition Mechanism Understanding

C. Osorio[1], E. Petersen[2] and M.S. Mannan[1]

Mary Kay O'Connor Process Safety Center, Artie McFerrin Department of Chemical Engineering, MS-3122, Texas A&M University, College Station, Texas USA
[2]Department of Mechanical Engineering, Texas A&M University, College Station, Texas USA
Corresponding author: epetersen@tamu.edu

ABSTRACT

Halon 1301 (CF_3Br) had been widely used as a fire extinguisher for many years due to its unique properties including high extinguishing efficiency at low concentrations, high dispersion capabilities, and storage stability, among others. Unfortunately, CF_3Br and other halogenated fire suppressants have been restricted by the Montreal Protocol due to their high ozone depleting potential. This restriction has led to the necessity to find cleaner alternatives with similar, Halon 1301-like capabilities. Despite all the scientific efforts made during the last few decades, no single suitable Halon 1301 substitute has been found. A more likely approach to overcome this issue is to study a variety of novel choices that can be used on specific situations. Identification of these alternatives requires a systematic methodology based on the understanding of the flame inhibition mechanisms that could address this search in promising directions. To this end, the present work presents a methodology to investigate fundamental phenomena involved in the combustion inhibition mechanism of CF_3Br by using detailed chemical kinetics which have been experimentally validated. Specialized combustion equipment, including a high-pressure shock tube and a flame speed bomb, are used to obtain important parameters including ignition delay time, OH* time histories, and flame speeds which are also employed as metrics for model validation. In particular, laminar flame speed is considerably important since it represents a fundamental property of the reacting mixture and provides a tool for deeper flame inhibition analysis to evaluate thermal and chemical inhibition contributions and saturation effects. Sensitivity analysis is performed to generate valuable information regarding key elementary reactions that limit and control flame inhibition behavior. Halon 1301 may be unique in its ability to modify both the chemical kinetics and thermal properties of a reacting system. With a fundamental understanding of the chemical and thermal advantages of CF_3Br as a fire suppressant, the same fundamental tools can be employed to test for similar attributes in candidate compounds. This approach provides a guide for identification of novel, chemically active inhibitors which behave similar to CF_3Br but can be significantly less harmful to the environment.

1. INTRODUCTION

Halon 1301 (CF_3Br) had been widely used for many years due to its unique properties including high extinguishing efficiency at low concentrations, high dispersion capabilities, and storage stability, among others. Unfortunately, CF_3Br and other halogenated fire suppressants have been restricted by the Montreal Protocol due to their high ozone depleting potential [1-3]. It led to the necessity to find cleaner alternatives with similar, Halon 1301-like capabilities. Despite all the scientific efforts made during the last decades, researchers have not been able to find a single, suitable Halon 1301 substitute. Identification of likely alternatives requires a systematic methodology based on flame inhibition mechanistic understanding that can address this search in promising directions.

To this end, the present work presents a methodology to investigate the effect of fire suppressants. This paper shows part of an overall research effort that started with the examination of CF_3Br. Therefore, potential candidates can be tested and compared with Halon 1301, which is used as reference.

2. FLAME INHIBITION ANALYSIS: METHODOLOGY

The proposed methodology, shown in Figure 1, is applied to investigate fundamental phenomena involved in the combustion inhibition mechanism of different flame inhibitors, starting with CF_3Br. Then, similar properties can be sought in other fire suppressant candidates/blends using similar analytical procedure. To accomplish this, model formulations and experimental designs are integrated in order to examine the chemical kinetics and thermal effects of fire suppressants on different hydrocarbons, especially methane, ethane and propane. Comparison between numerical and experimental combustion parameters, such as ignition delay times, OH* time histories, and flame speed are used as metrics for mechanism validation. In particular, laminar flame speed, considered a fundamental property of the reacting mixture in premixed flame, will be used for flame inhibitor characterization, but also to examine the thermal and kinetic inhibition mechanism of a specific substance or blend. In addition, identification of the key elementary reactions and pathways that most affect inhibition mechanism are determined through sensitivity analysis.

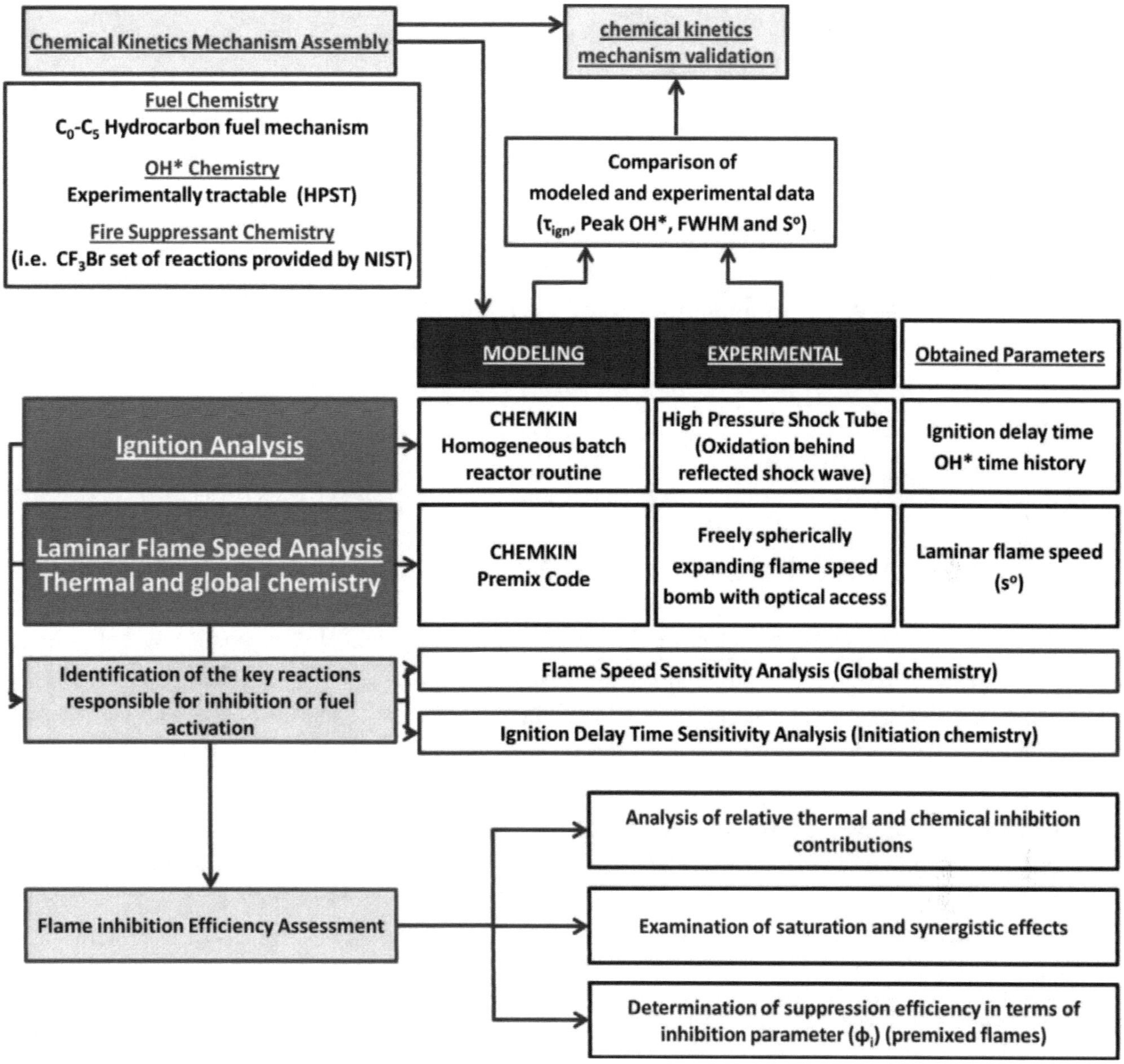

Figure 1. Methodology for flame inhibition analysis

2.1 Modeling

Modeling predictions are conducted using CHEMKIN. A detailed chemical kinetics mechanism was assembled from three existing sets of chemical reactions: (1) a C_0-C_5 hydrocarbon mechanism from NUI-Galway [4], (2), the Hall and Petersen OH* sub mechanism [5], and (3) Halon 1301 chemistry which have been provided by National Institute of Standards and Technology (NIST) [6]. Use of this model provides novel information regarding fire suppressant effects on a wide range of fuels (C_0-H_5), as well as OH* chemistry that can be directly compared with experimental ignition results. The complete mechanism, used for Halon 1301 study, includes 333 species involved in 1980 reactions.

2.2 Experimental

Examination of fuel ignition using a High Pressure Shock Tube: Effect of CF_3Br on the fuel ignition is analyzed using a High Pressure Shock Tube (HPST). The facility is provided with driven and driver sections that have ID's of 15.24 cm and 7.62 cm, respectively, both with a 1.27-cm wall thickness, constructed entirely of stainless steel 304 (Fig. 2). The two sections are separated by a diaphragm that ruptures depending on the pressure ratio between the two sections. The reactant mixture is contained at the driven section (low pressure). It will ignite by the action of a shock wave produced after the diaphragm ruptures. In these experiments, fuel oxidation is reached behind reflected shock waves where zero dimensional conditions are achieved. Reaction progress is monitored using sidewall emission measurements of OH* chemiluminescence at 310 n. Then, ignition delay time (τ_{ign}) can be determined by analyzing the Pressure/OH* emission profiles obtained in each experiment (Figure 3). A more detailed description of this facility and technique is provided by Aul et al. [7].

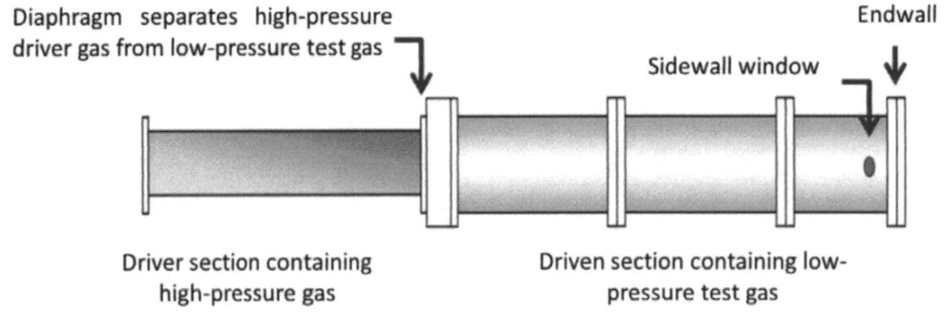

Figure 2. Primary sections and diaphragm station of a simple shock tube. [7].

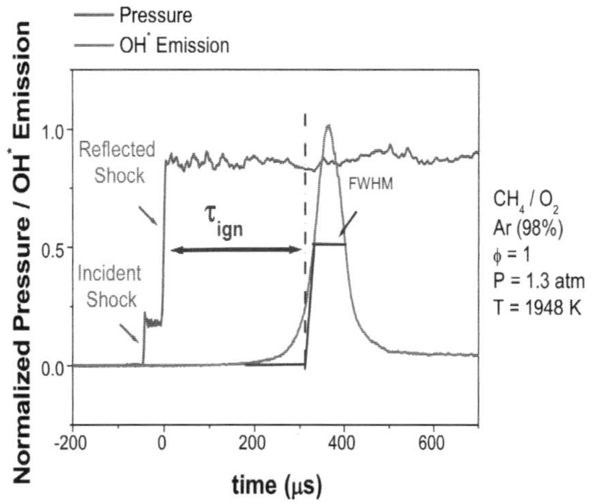

Figure 3. Typical pressure and OH* measurements from the shock-tube experiment

Laminar Freely Propagating Flame Speed: Evaluation of the laminar flame speed can be made by using different techniques, however, most of them involved flame curvature, non-uniform upstream flow, and heat transfer and other effects that makes difficult their modeling [8]. The method used in this study examines freely propagating flames, minimizing complexity of the system (e.g. minimum heat losses). This facility, described in more detailed by de Vries et al. [9], consists of a aluminum alloy constant-volume cylindrical vessel, with an inner diameter of 30.5 cm and length of 35.6 cm. Direct observation of the flame allows tracking the changes of flame radii as a function of time (Figure 4). For this radii tracking, a high-speed camera (up to 20,000 frames per second) is used as element of the Z-type schlieren optical diagnostic technique.

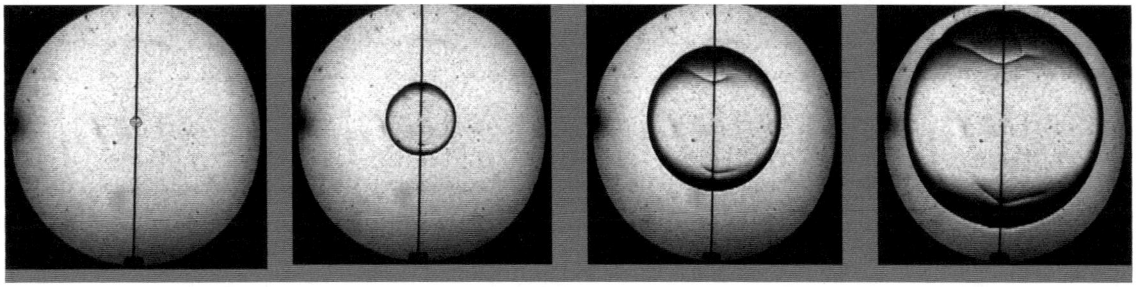

Figure 4. Example of experimental images of an expanding flame.

3. RESULTS

Ignition results: Different mixtures of fuel (methane, ethane and propane), oxygen, and Halon 1301, highly diluted in argon (98% Ar by volume) were evaluated. Experiments were conducted at different temperatures (1250 - 2250 K) and equivalence ratios at atmospheric pressure. Figure 5 shows the effect of CF_3Br on ignition delay time for stoichiometric mixtures. Two different aspects are examined: 1) the effect of CF_3Br on methane, ethane, and propane combustion, and 2) the agreement between experiment and model. Ignition Delay Time results show that CF_3Br speeds up ignition in all methane mixtures tested. On the other hand, CF_3Br acts by retarding the ignition in those mixtures containing ethane and propane. In most of these cases, the model predicts the experimental behavior; but in general, it underestimates the resulting effect, indicating that some improvements can still be made.

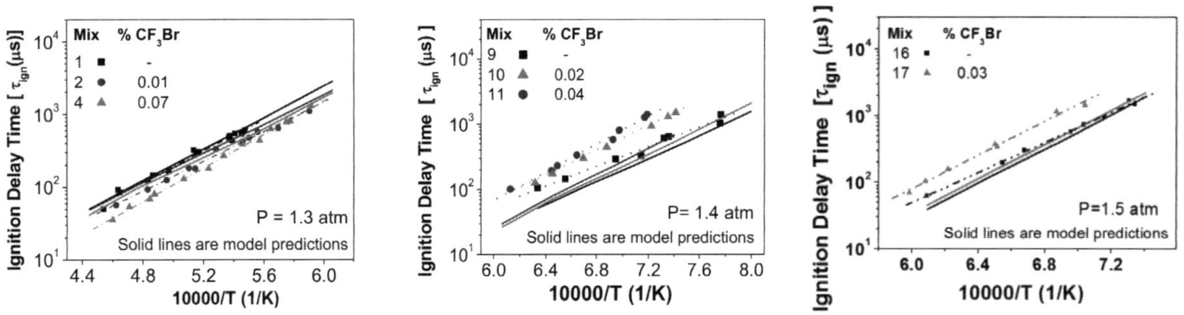

Figure 5. Ignition Delay Time results comparing data to kinetics model for (a) Methane ϕ=1.0, (b) Ethane ϕ=1.0, and (c) Propane ϕ=1.0.

Laminar flame speed results: The effect of CF_3Br on methane, ethane and propane flames was assessed by adding two different concentrations of 0.5% and 1% of Halon 1301 to different fuel/air mixtures, at different stoichiometric ratios from 0.7 to 1.3, at ambient temperature. Figure 6 illustrates the results obtained from experimental and modeling calculations. In all the cases, CF_3Br shows to present an inhibitor effect by decreasing the laminar flame speed. Modeled data foretells same experimental trend, however, improvement in the kinetic mechanism can be made for better predictions.

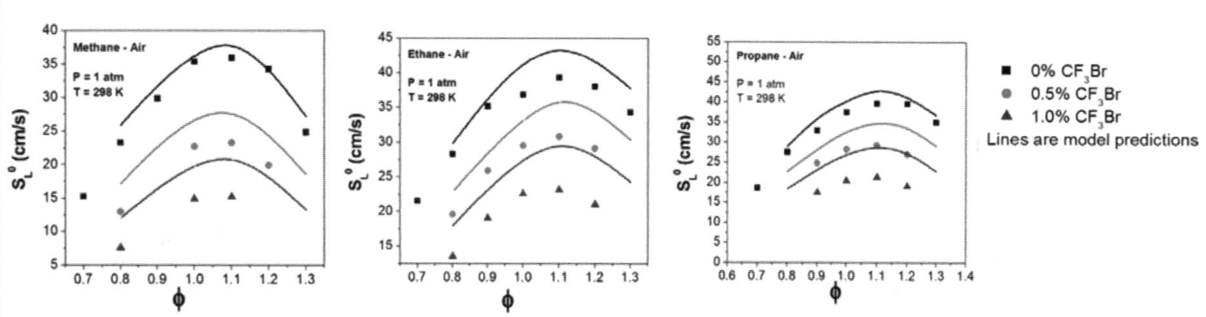

(a) Methane – Air (b) Ethane – Air (c) Propane - Air

Figure 6. Experimental and modeled Flame Speed of different fuel-air mixtures at different CF_3Br concentrations.

<u>Model improvement and identification of key elementary reactions</u>: To improve the model, it was necessary to identify the key elementary reactions that most affect the laminar flame speed in a determined system. Among the reactions that are activated by CF_3Br (reactions involving Br or F), the reaction (R1): Br + HCO => HBr + CO presented the highest negative sensitivity coefficient, for all the systems analyzed. Considering this result, the R1 pre-exponential factor (A) was modified by trial and error. Results illustrated in Figure 7, shows an improvement of methane flame speed predictions by changing the rate of R1 by five Similar results were found for ethane and propane. These results are not aimed to be used as the final and more precise model predictions, but rather show the methodology that can be used for chemical kinetic mechanism improvement.

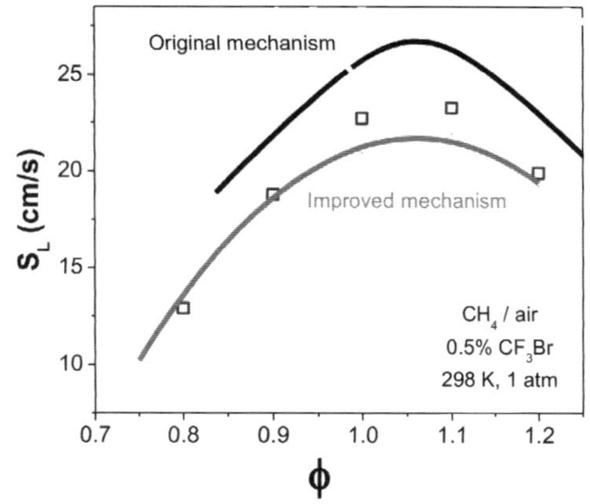

Figure 7. Comparison between experimental (blue points) with the original (black line) and improved modeled (red line) flame speed data for methane mixtures (0.5% CF_3Br, 1 atm, 298K).

Flame inhibition efficiency assessment: Analysis of the relative physical (thermal) and chemical (kinetic) inhibition contribution of CF_3Br was conducted. First, it was necessary to determine the laminar flame speed reduction due to the global effect (physical + chemical) in terms of (S/So) global. For this calculation, the flame speeds with Halon (S) and without Halon (So) were calculated using the mechanism containing F and Br chemistry. Then, an inhibition parameter can be obtained considering the following expression [6], where X_{O2} and X_{CF3Br} are the concentrations of oxygen and Halon 1301, respectively.

$$\Phi = \frac{X_{O2}}{X_{CF3Br}} ln\left(\frac{S_L}{S_o}\right)^{-1} \quad \text{Equation 1}$$

Table 1. Relative chemical and physical inhibition contributions of CF_3Br on methane, ethane and propane stoichiometric flames.

Inhibition Parameter	Methane	Ethane	Propane
Φ_{global}	17.7	9.6	10.8
$\Phi_{physical}$	2.1	2.0	1.6
$\Phi_{chemical}$	15.6	7.6	9.2

Then, the reduction of flame speed by physical means was calculated by isolating the F and Br chemistry from the chemical kinetics mechanism, and calculating the flame speed at this condition. The chemical contribution was obtained by subtracting the physical from the global effect. Results (Table 1) show that CF_3Br behaves as a fire suppressant mainly by its chemical mechanisms. The results, in agreement with previous authors [10-13], demonstrates the capability of CF_3Br to reduce H-radical concentrations in the flame reaction zone, eventually decreasing the rate of the chain branching reaction.

4. SUMMARY AND CONCLUSIONS

Despite the effort conducted toward finding suitable fire suppressants, there is still a need of novel candidates with properties similar to CF_3Br. This search should be based on flame inhibition understanding. Based on this desire, the present work aims to improve the understanding of the Halon 1301 mechanism as a fire inhibitor by employing an analytical methodology capable of examining both thermal and chemical effects on hydrocarbon flames. Similar analysis can be used for evaluating suitable substitutes using Halon 1301 as a reference.

The proposed analytical methodology is based on detailed chemical kinetics mechanisms which can be validated experimentally. The fuel kinetics mechanism employed in this work (C_0-C_5) allows extending the examination of a fire suppressant's effect to more-practical fuels. Specialized combustion equipment, including a high-pressure shock tube and a flame speed bomb were used to obtain important parameters including ignition delay time, OH* time histories, and laminar flame speed. These parameters serve as metrics for model validation and will be tools for deeper flame inhibition mechanism analysis. Sensitivity analysis was performed to obtain valuable information regarding key elementary reactions and pathways. This approach helps to understand the limiting and controlling flame inhibition behavior, and can also be used as a tool for model improvement as well.

5. REFERENCES

[1] J.S. Daniel, S. Solomon, R.W. Portmann, R.R. Garcia, J. Geophys. Res., 104 (1999) 23871-23880.
[2] UNEP, in: U.N.E.P. (UNEP). (Ed.), Montreal, Canada, 1987.
[3] in: U.S. Code (Ed.) VI, 1990.
[4] N. Donato, C. Aul, E. Petersen, C. Zinner, H. Curran, G. Bourque, Journal of Engineering for Gas Turbines and Power, 132 (2010) 051502.
[5] D. Healy, N.S. Donato, C.J. Aul, E.L. Petersen, C.M. Zinner, G. Bourque, H.J. Curran, Combustion and Flame, 157 (2010) 1526-1539.
[6] T. Noto, V. Babushok, A. Hamins, W. Tsang, Combustion and Flame, 112 (1998) 14.
[7] C.J. Aul, DeVries, J. and Petersen, E. L in: Eastern States Fall Technical Meeting of the Combustion Institute, Charlottesville, VA, 2007.
[8] J. De Vries, in: Mechanical Engineering, Texas A&M University, College Station, 2009, pp. 186.
[9] J. De Vries, W. Lowry, Z. Serinyel, H.J. Curran, E.L. Petersen, Fuel, 90 (2011) 8.
[10] C.K. Westbrook, Combustion Science and Technology, 34 (1983) 201-225.
[11] T. Noto, V. Babushok, J. Burgess, A. Hamins, W. Tsang, A.W. Miziolek, Proceedings of the Combustion Institute, 26 (1996) 1377-1383.
[12] V. Babushok, D. Burgess, W. Tsang, in: Halon Options Tenchical Working Conference, 1994, pp. p. 205.
[13] C.R. Casias, J. Thomas McKinnon, Proceedings of the Combustion Institute, 27 (1998) 2731-2739.

OPTICAL FIBER REFLECTANCE PROBE FOR DETECTION OF SUPERCRITICAL TRANSITION

Yujian Sun, Boung Wook Lee, Milorad P. Dudukovic
Chemical Reaction Engineering Laboratory (CREL),
Department of Energy, Environmental and Chemical Engineering,
Washington University in St. Louis, St. Louis, Missouri 63130

Abstract

An optical fiber reflectance probe is developed for the fast, in-situ detection of supercritical transition in multiphase and multicomponent systems. Experiments with two binary mixtures methanol-cyclohexane and CO_2-methanol are conducted to demonstrate the capability of the reflectance probe to capture the supercritical transition in both liquid-liquid mixture and gas expanded liquids. The current probe can withstand pressures up to 140 bars and temperatures of 120 °C, which suggests its applicability in many practical processes.

Keywords

Supercritical transition, reflectance probe, methanol-cyclohexane, CO_2 expanded liquids

1. Introduction

Supercritical fluids, at temperature and pressure above their critical points at which the phase boundaries cease to exist, are suitable substitutes for organic solvents in many processes. Specifically, carbon dioxide expanded liquids (CXLs) are promising as alternative media for performing separations, extractions, reactions, and other applications [1-7]. Although the critical phase transition of a mixture from subcritical to supercritical state can be observed by the naked eyes in vessels with fully transparent walls, that transition is hard to detect in most practical reactor vessels with steel walls unless expensive view cells are provided. High temperatures and high pressures (HTHP) are typically employed in mixed autoclaves for processes where the above transition is expected to occur at some operating conditions. Thus, it is desirable to develop an approach for in-situ detection of the critical phase transition.

Making use of the fact that a transition from subcritical to supercritical state coincides with the change of the fluid from heterogeneous opalescent to homogeneous transparent state, we are able to design and manufacture an optical fiber probe to detect the reflected light which travels a certain known distance in the mixture and therefore carries information related to the transition [8]. We illustrate here the changes in the light signal captured during in-situ detection in the methanol-cyclohexane (liquid-liquid) mixture as the system temperature is raised from below to above critical temperature at fixed pressure. We also illustrate the signal changes with pressure in the CO_2-methanol (gas-liquid) system at different temperatures. These examples illustrate the applicability of the optical fiber probe in the detection of critical phase transition for multiple types of mixtures.

2. Experimental

2.1. Probe Setup

The optical probe system used in this work consists of three optical fibers (I, II and III), as shown in Figure 1. Fiber III (ø1035 μm) is coupled closely with fibers I and II (ø630 μm), which are connected to the laser source (Thorlabs SIFC635PM) and the light detector (Thorlabs PDA36A), respectively. The other end of fiber III, with a small piece of mirror attached about 1~2mm away, is immersed in the mixture being studied. In this way, light coming from the laser source can be transmitted to the mixture through fiber I and III, and then reflected back by the mirror to fiber III and sent back to the light detector through fiber II. The light signal can be converted to an analog signal by the detector, which can then be transferred to the data acquisition system (PowerDAQ PD-NBC-16) and recorded by the computer.

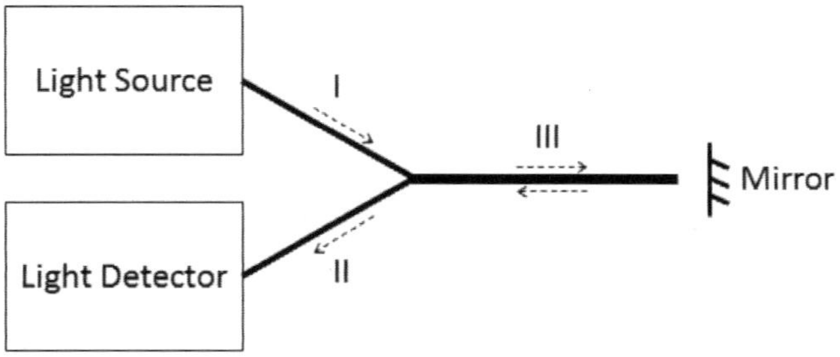

Figure 1. Schematic for the optical fiber probe system

2.2. Rationale

The light signal acquired by the detector consists of two parts, one reflected by the mirror and the other scattered by the mixture. The mirror is attached sufficiently close to the probe end that the reflected part dominates in the total signal. The "noisy" part of the signal due to contribution of scattered light is negligible compared to the intensity of the reflected light from the mirror. That is why we selected the name of reflectance probe.

When a mixture is at subcritical state, there exist distinct interface boundaries between the two immiscible distinctly different phases. Under sufficient stirring, the interface boundaries are broken down into small pieces and distributed randomly throughout the heterogeneous mixture, resulting in strong fluctuation of light signals. During supercritical transition, when interphase boundaries cease to exist, the mixture turns to homogeneous transparent state, thus the fluctuation of the light signal is greatly reduced at the same stirring rate. This disappearance of the fluctuations distinguishes the supercritical state from the subcritical state, and helps in determining the critical conditions.

2.3. Procedure

For methanol-cyclohexane system, as it is stirred and gradually heated up, the temperature change is recorded by video at the same time with data acquisition of the light signal. When the temperature exceeds the critical temperature for the specific composition, there is a signal change which coincides with the transition. From the data obtained, we can find the exact time of this signal change, and since we have a record of the temperature with time we obtain the critical temperature.

For CO_2-methanol system, the experimental setup is shown in Figure 2. CO_2 is pumped into the 3-in. inner diameter (ID), 1-L autoclave. Pressure is controlled via a Tescom 4000 back-pressure regulator and Validyne pressure transducers. Temperature control is achieved using Omegalux heating tape wrapped around the autoclave, an Omega thermocouple inserted into the vessel's thermowell, and an Omega CN132 temperature controller [9]. The end of the optical reflectance probe is immerged in the liquid (methanol) to detect the supercritical transition. As CO_2 is gradually pumped in, the pressure increases and exceeds the supercritical pressure at specific temperature, which results in a signal change that can be captured and analyzed.

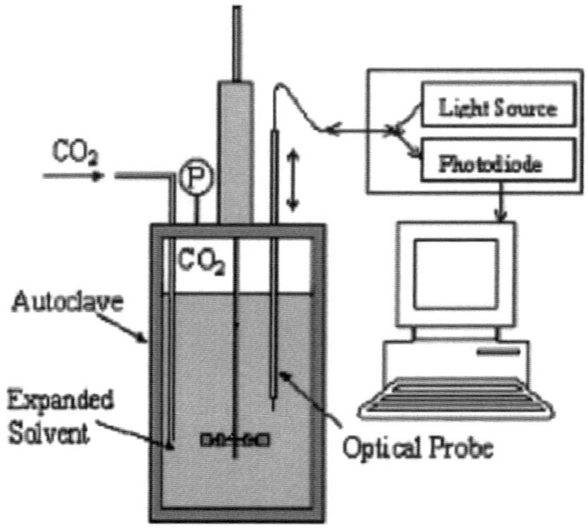

Figure 2. Experimental setup for the autoclave

3. Results and Discussion

3.1. Methanol-Cyclohexane (L-L)

As the liquid mixture of methanol and cyclohexane is heated up to reach its critical temperature, the stirred mixture turns to homogeneous supercritical state, and thus a light signal change can be clearly captured. Figure 3 shows in-situ data collected with 50% mole fraction of methanol. In the same way, critical temperatures of methanol-cyclohexane mixtures are determined at different compositions, and all agree well with previously obtained data [10], as shown in Figure 4.

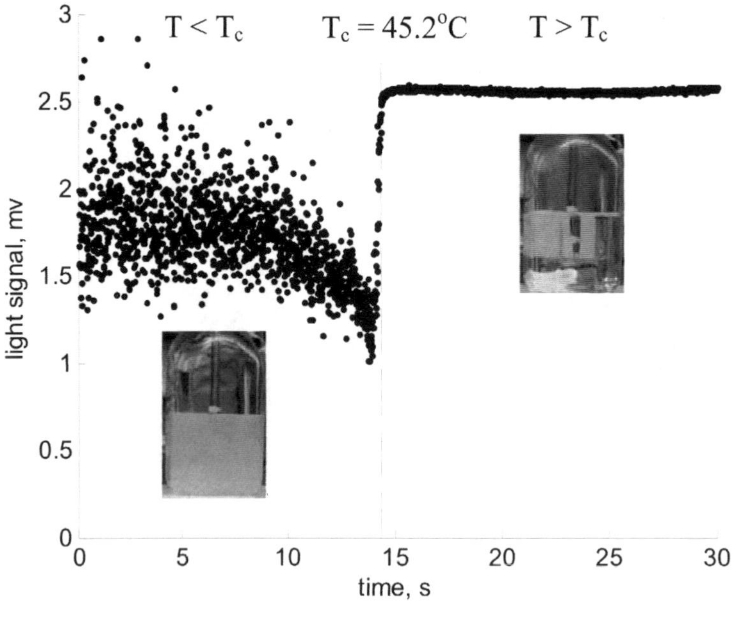

Figure 3. In-situ light signal for methanol-cyclohexane mixture during supercritical transition ($X_{methanol}$ = 50%; Sampling Frequency = 100 Hz)

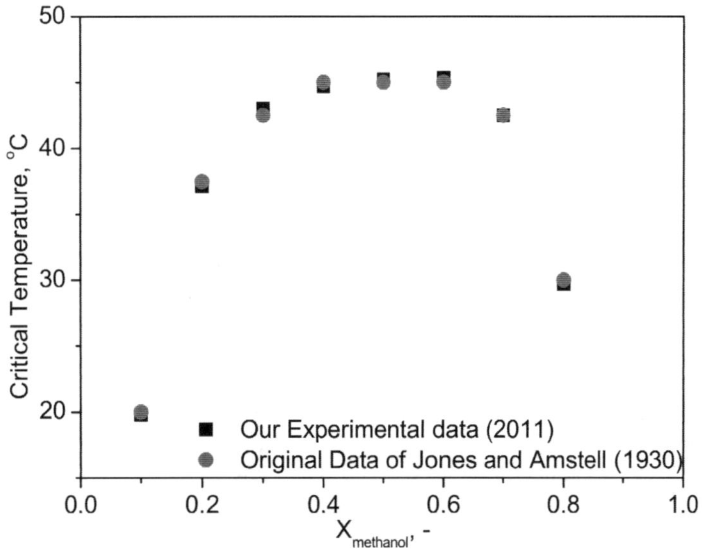

Figure 4. Critical temperatures for methanol-cyclohexane at different compositions

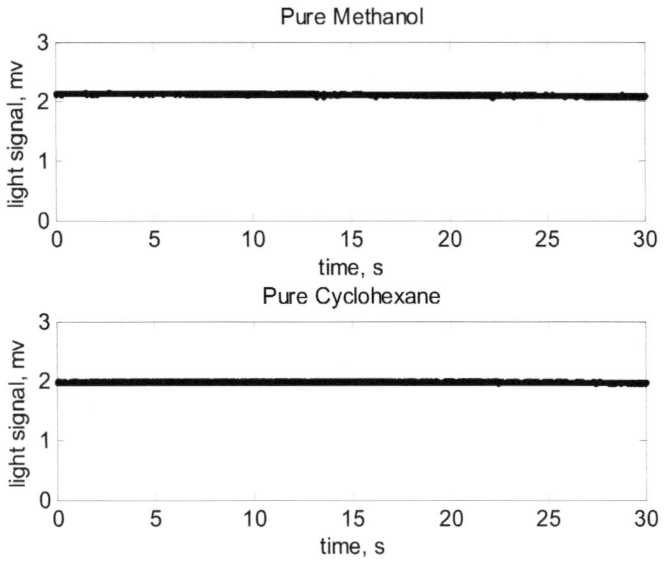

Figure 5. Light signals for pure methanol and pure cyclohexane during heating up

As seen above, at transition point we always find an abrupt rise in the signal. The reason for this change is as follows. In the subcritical methanol-cyclohexane mixture, the heterogeneity results in opalescence as shown in the left photo in Figure 3. This opalescence increases the scattering in the liquid, but it greatly reduces the reflectance since it blocks the path of reflected light and scatters it also. During supercritical transition, the mixture turns from heterogeneous opalescent to homogeneous transparent state, thus the reflected part of the light is enhanced while the scattered part is reduced. Since reflectance dominates in the signal, the entire signal increases sharply at transition.

There is another interesting phenomenon worth noting. The signal intensity continuously and gradually goes down until it reaches a minimum before rising up sharply during transition. Since the same phenomenon regarding detected light intensity cannot be observed for pure methanol and pure cyclohexane, as shown in Figure 5, this has to be related to interactions between the two phases. It seems to indicate that interphase boundaries increase in number during heating up which is responsible for blocking more reflected light and resulting in decreased total signal intensity. Whether that implies smaller diameters of dispersed phase but large numbers requires more study.

3.2. CO_2-Methanol (G-L)

Critical pressures for CO_2-methanol at different temperatures are determined similarly by the same reflectance probe. Figure 6 shows a set of data collected at 25.2 °C. We can still see the disappearance of light intensity fluctuations during supercritical transition. However, the abrupt rise in the signal at transition, which was evident in Figure 3 for the methanol-cyclohexane mixture, is clearly missing in Figure 6 for the carbon dioxide-methanol mixture. The reason might be that this initially heterogeneous gas liquid mixture consisting of small bubbles dispersed in the liquid has different optical properties opalescence character than the methanol-cyclohexane mixture. Our probe can withstand high pressures as shown in Figure 7, which suggests its applicability in many processes.

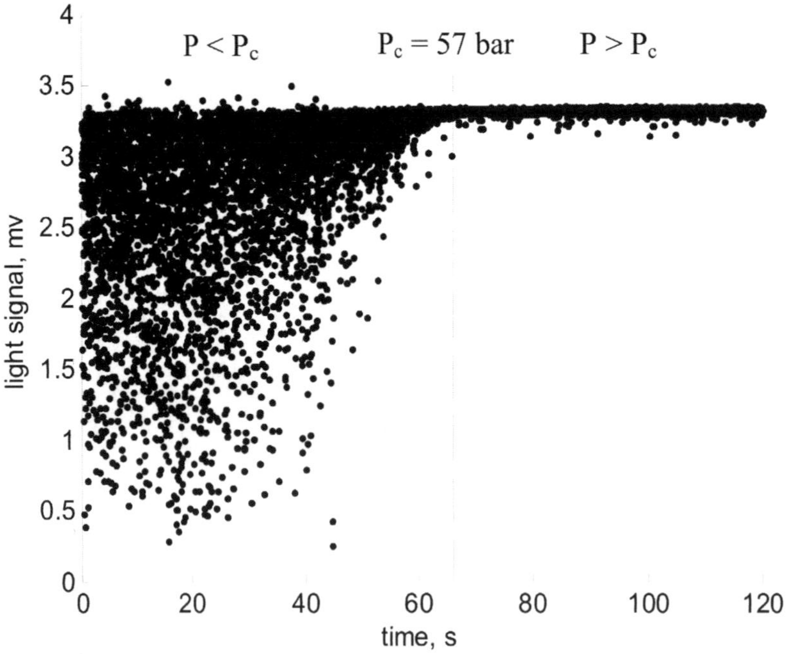

Figure 6. In-situ light signal for CO_2-methanol mixture during supercritical transition (T = 25.2 °C; Sampling Frequency = 100 Hz)

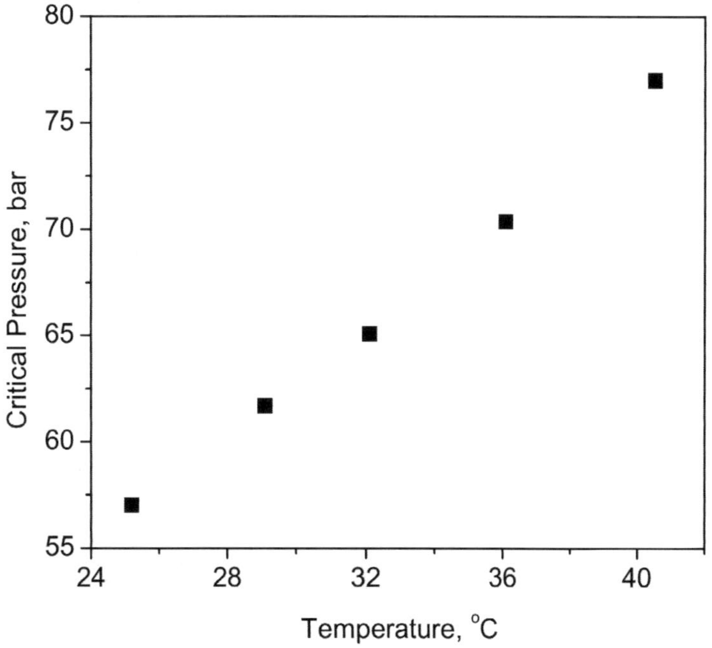

Figure 7. Critical pressures for CO_2-methanol at different temperatures

4. Conclusion

A reflectance probe is developed in this work to detect the reflected light signal change during supercritical transitions of both methanol-cyclohexane and CO_2-methanol system. This signal change can be coupled to the temperature or pressure recorder to determine the critical condition at interest.

ACKNOWLEDGMENTS

This work was supported in part by the grant from the National Science Foundation (CBET 0933780).

References

1. Adams D. (2000), “Clean and green…but are they mean?” Nature, 407, pp. 938-940.
2. Subramaniam B. and McHugh M. (1986), "Reactions in Supercritical Fluids-A Review", Ind. Eng. Chem. Process Res. Dev., 25, pp. 1-12.
3. Savage P. E., Gopalan S., Mizan T. I., Martino C. J. and Brock E. E. (1995), "Reactions at supercritical conditions: Applications and fundamentals", AIChE J., 41, pp. 1723-1778.
4. Jessop P. G. and Subramanian B. (2007), “Gas-Expanded Liquids”, Chemical Reviews, 107, pp. 2666-2694.
5. McHugh M. and Krukonis V. J. (1994), “Supercritical Fluid Extraction”, Butterworth-Heinemann, Boston.
6. Jessop P. G., Ikariya T. and Noyori R. (1999), "Homogeneous Catalysis in Supercritical Fluids", Chem. Rev., 99, pp. 475-494.

7. Wei M., Musie G. T., Busch D. H. and Subramaniam S. (2002), "CO_2-expanded solvents: unique and versatile media for performing homogeneous catalytic reactions", J. Am. Chem. Soc., 124, pp. 2513-2517.
8. Mueller S. G. (2009), "Optical Measurements in Gas-Liquid Stirred Tanks", Doctorate Dissertation, Washington University in St. Louis.
9. Mueller S. G., Werber J. R., Al-Dahhan M. H. and Dudukovic M. P. (2007), "Using a Fiber-Optic Probe for the Measurement of Volumetric Expansion of Liquids", Ind. Eng. Chem. Res., 46, pp. 4330-4334.
10. Jones D. C. and Amstell S. (1930), "The Critical Solution Temperature of the System Methyl Alcohol-Cyclohexane as a Means of detecting and estimating water in Methyl Alcohol", J. Chem. Soc., pp. 1316-1323.

() Musings on Management

Frank van Lier, The Lubrizol Corporation, Wickliffe, OH

It really is all about the people!

Hiring good people.

Retaining them.

Adjusting for different personal styles, cultures, encouraging diversity, embracing different perspectives.

Motivating people to work together in successful global teams.

Achieving top notch results that are better than any individual can achieve alone.

Maintaining the focus.

Getting the results.

Management is all about the excitement managers gets from having their people and teams grow and succeed.

(77a) Cultural and Technological Forces That Are Changing the Knowledge Management Game

Cindy Hubert, APQC, Houston, TX

With constantly changing technologies and work force demographics, developing innovative approaches to help your organization's critical knowledge flow is becoming increasingly essential to business advancement. Hubert will discuss these key points, including the steps to building and maturing knowledge management capabilities, generational concerns in knowledge transfer and retention, social networking on the Web, and other significant questions raised by an evolving business landscape.

(81a) Thermodynamic-Analysis-Based Design for Liquefied Natural Gas Receiving Terminal

Meiqian Wang, Jian Zhang and Qiang Xu, Dan F. Smith Department of Chemical Engineering, Lamar University, Beaumont, TX

Natural gas (NG) now has become the most promising alternate fuel energy. Recently, the estimation about large quantity of technically recoverable unconventional NG, such as shale gas, has been announced by the Energy Information Administration (EIA), which draws more attention on the development of natural gas. In the work, we developed a novel configuration for liquefied natural gas (LNG) receiving terminal, integrated with LNG regasification process, shale gas liquefaction and NGL recovery process. A great deal of cold energy from LNG regasification has been recovered through shale gas liquefaction and NGL fractionators' condensers.

The flexibility of the newly designed receiving terminal has also been addressed under different conditions, which include the changes of received LNG quantity and composition, transported shale gas quantity and composition, and heating value specified at local marketing place, etc. Moreover, heat duty integration has been involved in this design, where thermodynamic theory is employed to analyze and fundamentally evaluate the performance. All of the conditions together with thermodynamic and economic performance are integrated in a mixed-integer nonlinear programming (MINLP) model. This model is solved using optimization tool GAMS and validated via rigorous simulations in ASPEN Plus.

(81b) Design Approaches for Unconventional Gas Processing

John Harness, Technical Service, UOP, Des plaines, IL

Design Approaches for Unconventional Gas Processing

Membrane systems are well suited to remove CO2 and water to pipeline specifications although many people still consider membranes to be an "unconventional" technology. When applied to processing unconventional gas such as North American shale gas, membrane systems can offer distinct advantages.

This presentation will discuss the technology within membrane systems and present a case study from a gas plant that has been using membranes to process shale gas for the past 18 years. The presentation includes discussions on membrane theory and applications as well as discussion of a cost and benefit analysis comparing the use of membranes with Amine systems in unconventional shale gas applications.

UCSRP: A Flexible Sulfur Removal Process for Sweetening Natural Gas

Arun Basu, Howard Meyer, Dennis Leppin, Jim Zhou and Ajay Makkuni

Gas Technology Institute

ABSTRACT

With the recent activities in the area of unconventional natural gas (such as, Shale Gas) in the U.S., the need for a low-cost sulfur removal technology for processing relatively small as well as large CO_2-rich natural gas streams has increased significantly. In a conventional amine-based natural gas (NG) processing plant, the H_2S-rich acid gas stream from the stripper unit is treated in a Claus process for the production of elemental sulfur (the gas-phase reaction is $2H_2S + SO_2 = 3S + 2H_2O$). The Claus process has specific inherent limitations including (i) the need for a tail-gas clean-up unit (e.g., Shell Claus Off-gas Treatment: SCOT) to achieve the required H_2S recovery efficiency, (ii) inability to handle low sulfur production capacities (typically, for less than 20 tons/day sulfur), and (iii) the need for an additional CO_2-separation step prior to the Claus step if the gas stream from the Amine unit contains relatively high CO_2 (typically, if the CO_2 /H_2S ratio is greater than 2.3).

Gas Technology Institute (GTI) has been working with the University of California, Berkeley, for further development of their UCSRP (University of California Sulfur Recovery Process) technology. In this process, the Claus reaction is conducted in liquid phase at about 250-300 ^{o}F (and at any pressure) to produce liquid sulfur in the presence of an organic solvent and a homogeneous catalyst. A gas mixture consisting of a part of the H_2S-rich feed gas plus the unreacted H_2S stream from the UCSRP reactor is combusted with air in a furnace to form SO_2 (plus a small amount of sulfur) which is recovered in an absorber/stripper unit for recycle to the UCSRP reactor. As the overall sulfur recovery is 99.9+%, there is no need for a tail-gas clean-up unit. The UCSRP process can be scaled down to sulfur capacities significantly lower than 20 LT/day. For CO_2-rich feed streams from an amine plant, the gas can be processed directly in a UCSRP unit without a separate CO_2 removal unit. For direct processing of natural gas with relatively low CO_2 levels that already meet pipeline specifications, the UCSRP process will replace Amine/Claus-SCOT units; for higher CO_2 levels in feed natural gas, the exit gas from the UCSRP reactor will be treated to remove CO_2.without the complexity inherent when H_2S is also present in the stream.

For the processing of a H_2S-rich gas stream from an amine plant stripper with sulfur production at about 76 LT/day, Kellogg, Brown and Root Inc. had estimated the capital cost for an UCSRP plant to be about 60% and the operating costs to be about 68% of those for a three-stage Claus/SCOT plant.

INTRODUCTION

With the recent discoveries of relatively large quantities of unconventional natural gas (e.g., Shale Gas), the U.S. natural gas industry is interested in sulfur recovery processes that offer lower cost with high sulfur removal selectivity. Various amine-based technologies are commercially used to remove sulfur compounds from high-pressure natural gas feed as an acid gas which then requires a sulfur-recovery process. The modified Claus process coupled with a tail-gas treatment (TGT) process such as the SCOT (Shell Claus Off-gas Treating) process are typically used to recover elemental sulfur and produce a dischargeable plant tail-gas. Some plants produce sulfuric acid rather than elemental sulfur.

Gas Technology Institute (GTI) has been working with the University of California, Berkeley, for further development in a number of varying applications of their UCSRP (University of California Sulfur Recovery Process) technology for removal of sulfur compounds from various feedstock including natural gas and coal-derived synthesis gas.[1,2,3,4] In the UCSRP process, the Claus reaction ($2H_2S + SO_2 = 3S + 2H_2O$) is conducted in liquid phase (typically, using an organic solvent such as Diethylene Glycol Methyl Ether (DGM)) at about 250-300 °F (and at any pressure) to produce liquid sulfur. A homogeneous catalyst (e.g., 3- Pyridinemethanol), which does not dissolve in liquid sulfur, is added to the solvent in small quantities. At these conditions, as liquid sulfur is essentially immiscible in DGM solvent and over twice as dense, it is removed as a separate product from the reactor column. In this paper, we have (i) presented comparative capital and operating costs data (based on a prior study by Kellogg, Brown and Root Inc.[1]) for the UCSRP process to a conventional Claus/SCOT sulfur recovery system, and (ii) outlined key advantages of the UCSRP process for the sweetening of natural gas.

PROCESS OVERVIEW

Claus

The Claus process produces elemental sulfur by the partial oxidation of H_2S to SO_2 followed by the reaction of uncombusted H_2S with SO_2. The key reactions are:

$$2\,H_2S + 3O_2 = 2SO_2 + 2\,H_2O \qquad (1)$$
$$2\,H_2S + SO_2 = 3S + 2\,H_2O \qquad (2)$$

The overall reaction may be expressed as: $2\,H_2S + O_2 = 2S + 2\,H_2O$ (3)

In the original unmodified Claus process, the reaction was carried out in a single reactor (at about 400-650 °F) over a catalyst followed by a sulfur condenser.[2] With only one reactor, the sulfur recovery was limited to 75-90%. Most modern sulfur recovery plants use the modified Claus process, where the reaction is carried out in two steps. In the acid gas burner, 1/3 of the total H_2S in the feed gas is totally oxidized (using air) to SO_2. In the next step, the unburned H_2S and the SO_2 formed react in the reaction furnace without catalyst. This reaction proceeds rapidly in the high-temperature reaction furnace, and leads to a conversion of about 60% of the H_2S to sulfur vapor. The gas leaving the furnace is cooled to condense sulfur and then reheated for processing in a catalytic converter. Most modified Claus plants have two to four catalytic converters, each with its own condenser. Each succeeding converter operates at a lower temperature. The three-converter system typically provides a sulfur yield up to 98%. The gas leaving the final catalytic converter is typically sent to a Tail Gas Cleanup Unit (TGCU), such as a Shell SCOT unit. A schematic of the modified Claus process is shown in Figure 1.

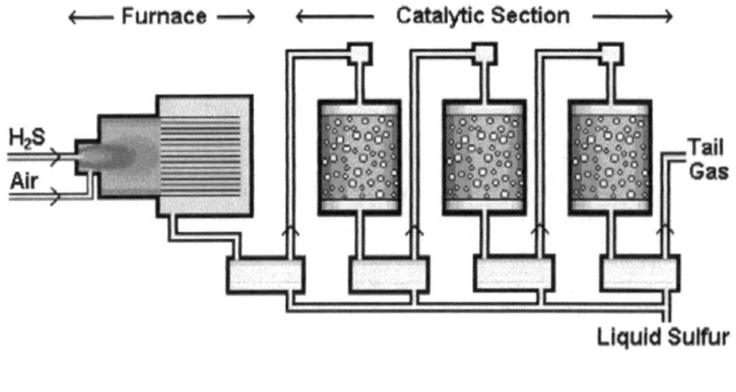

Figure 1: Schematic of the Modified Claus Process[5]

Several variations of the basic Claus process have been developed to handle acid gases containing different levels of H_2S. If the acid gas feed is relatively lean (less than 30 mol% H_2S on a dry basis), and if all of the acid gas is sent to the burner, the low level of H_2S would result in poor flame stability due to relatively low reaction furnace temperatures. Therefore, a fraction of the acid gas feed bypasses the burner to achieve suitable furnace temperatures necessary for flame stability. This configuration is referred to as the "Split-flow Claus". When the acid gas feed is sufficiently rich in H_2S (i.e., greater than 30-35 mol%), all of the acid gas is processed in the burner. This process configuration is known as the "Straight-through Claus".

A few key inherent problems of the modified Claus process are:

- As the furnace burns very fuel-rich, chances of soot formation increase significantly due to the conversion of hydrocarbons to COS and CS_2; this requires precise control of air flow to the furnace.
- Higher levels of CO_2 in the feed natural gas would significantly increase overall capital and operating costs of sulfur removal. As an example, if the CO_2/H_2S ratio is more than about 2.3, the off-gas from an amine stripper would have to be treated in an expensive CO_2 separation unit before the off-gas can be processed in a Claus plant.

SCOT

A schematic of a typical SCOT unit is shown in Figure 2.[2] The process was developed by Shell, and introduced in the early seventies as a process option for improving the efficiency of a Claus unit. By using the SCOT process, the sulfur recovery of a Claus plant can be increased to about 95-99.9%.

The SCOT process has two primary sections: (1) a hydrogenation/hydrolysis section, followed by water quench and (2) a selective amine gas treating section. In the first section, the tail gas from the Claus plant is heated to about 480-570 °F and then processed with H_2 (or a reducing gas) over a cobalt molybdate catalyst for converting all sulfur compounds (including SO_2,S, COS, CS_2) to H_2S. The effluent from the reactor is cooled to about 350 °F in a waste-heat exchanger (WHE) and the WHE effluent is further cooled by direct water quench. In the second section, the H_2S product stream is processed in an amine-based (e.g., MDEA or DIPA) absorber to recover most of the H_2S. The absorber overhead is usually incinerated (if the H_2S level is higher than 100 ppmv; using catalytic or thermal processes) to convert residual H_2S to SO_2 before discharge to the atmosphere. The H_2S-rich amine solution is processed in a regenerator to recover most of the H_2S which is recycled to the front end of the modified Claus unit. A schematic of the SCOT process is shown in Figure 2.

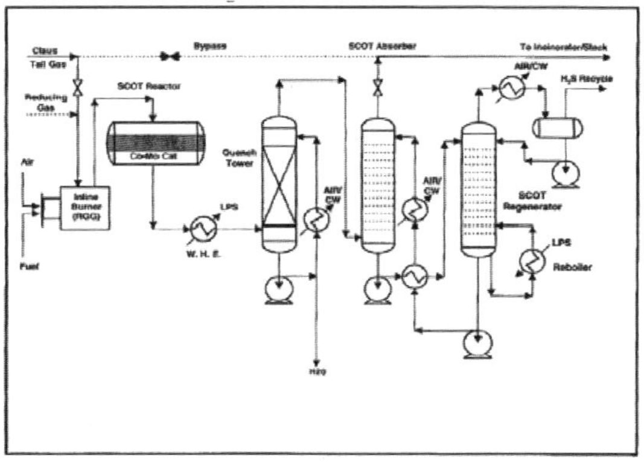

Figure 2: Schematic of the SCOT Process[2]

Although an integrated Claus/SCOT type process can meet current environmental regulations for sulfur emissions in the U.S., a SCOT-type tail-gas treatment process is quite expensive (typically costing as much as the modified Claus plant).[1]

UCSRP as a Replacement of the Claus/SCOT Processes

In the UCSRP concept (shown schematically in Figure 3), an organic solvent (e.g., DGM) containing a small quantity of a homogeneous catalyst is used to perform the Claus reaction in liquid phase at about 250-300 °F; this ensures that the reactor temperature is above the melting point (247 °F) of sulfur and below the polymerization temperature (310 °F) of elemental sulfur. The H_2S-rich acid gas is fed to the UCSRP reactor along with recirculated solvent plus SO_2 generated from combusting the reactor off-gases and part of the H_2S-rich feed with air in a furnace. As the liquid sulfur product is essentially immiscible and insoluble in DGM solvent and over twice the density as DGM, it is removed as a separate product from the reactor column thus shifting the chemical equilibrium toward sulfur formation.

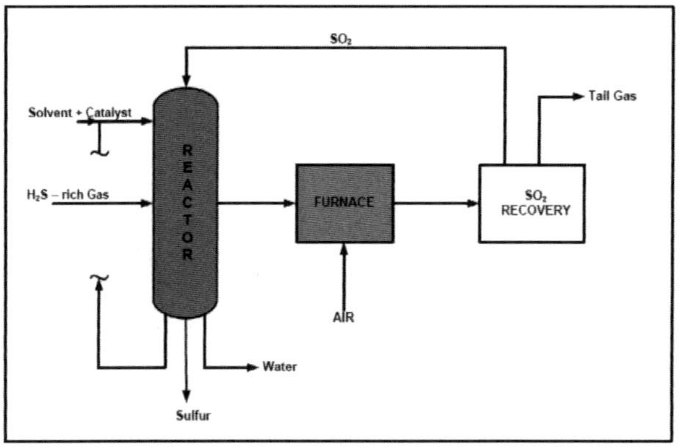

Figure 3: Schematic of the UCSRP Process

GTI is actively working with Prof. Scott Lynn (University of California, Berkeley), the inventor of the UCSRP process, to further improve the UCSRP technology, especially for high pressure (up to about 800 psig; the UCSRP-HP option) applications including sulfur removal from coal-derived syngas.[3,4] In the context of NG sweetening, GTI has demonstrated:

- The production of high-quality sulfur product (99.2% pure elemental sulfur) without any secondary treatment.
- Specific catalyst and solvent systems are stable under operating conditions based on a 1,000 hour test.
- Carbon steel is an acceptable material of construction for the UCSRP reactor

A few key advantages of the UCSRP process over the Claus process are: (i) the control of process stoichiometry is significantly simplified as the reactor is located before the furnace, (ii) slightly air-lean combustion precludes SO_3 and NO_x formation, (iii) hydrocarbons are burned primarily to CO_2 and water, (iv) relatively higher CO_2/H_2S ratios in the feed acid gas are tolerated, and (v) overall sulfur recovery is improved significantly.

- Cost Advantages of the UCSRP Process Over and Claus/SCOT: Scoping Economics Study by Kellogg Brown & Root, Inc.[1,2]

For sweetening of natural gas, KBR Inc. had conducted a scoping economics study to compare cost advantages of the UCSRP concept over the Claus/SCOT option in treating H_2S-rich gas from a conventional amine plant for the production of 76 long tons/day of sulfur.+1,2 A process flow diagram of the UCSRP process, as defined in the KBR study, is shown in Figure 4.

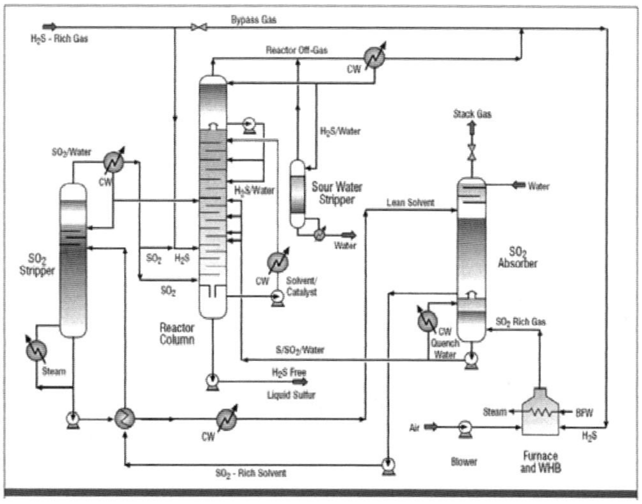

Figure 4: The KBR Study--Process Flow Diagram for the UCSRP Sulfur Recovery Process (for H_2S-rich Gas from an Amine Plant)[1]

As shown in Figure 4, the H_2S-rich gas stream enters the UCSRP unit at about 120 °F and 1.7 bar, typical of a Claus plant feed.[1] The reactor resembles a gas absorption column. A small fraction of the SO_2-rich gas stream enters the reactor column below the bottom bubble-cap tray. The remainder of the SO_2-rich gas together with about 75% of the H_2S-rich gas stream enters a tray or two higher and H_2S is in substantial excess at all points above. The remainder of the inlet H_2S bypasses the reactor and is sent to the furnace. The reactor is kept nearly isothermal by injecting water into the solvent flowing between trays to absorb the heat of reaction (from the Claus reaction) by vaporizing. The sulfur forms a second liquid phase. The solvent, which is recycled in a circulation loop from the bottom to the upper part of the reactor column, is saturated with elemental sulfur. The overhead vapor from the reactor is scrubbed with a water reflux to remove solvent vapor. A part of the water separated from the reactor off-gas is pumped back to the reactor as reflux. The net condensed water is sent to a sour-water stripper where dissolved H_2S and CO_2 are stripped overhead. The unreacted H_2S leaves the condenser and is mixed with the bypassed acid-gas feed and enters the furnace where it is burned with slightly sub-stoichiometric air to form SO_2. The hot combustion gas, containing SO_2, S, CO_2, H_2O and N_2 is cooled in a waste-heat boiler and is then fed to the SO_2 absorber where lean solvent from the SO_2 stripper is fed near the top. At the bottom of the absorber a water quench cools the combustion gas from the furnace to near ambient temperature, condensing a large fraction of the water vapors plus the sulfur vapor. The net dilute slurry of colloidal sulfur plus small amounts of higher sulfoxy acids is sent to the reactor column. The SO_2-rich solvent from the absorber is pumped to the SO_2 stripper to recover SO_2 for recycle to the UCSRP reactor.

As shown in Table 1, for the KBR study, two different H_2S/CO_2 feed gas ratios were evaluated:

Table 1: Feed Gas Composition (Design Basis: 76 LTPD Sulfur)[2]

	67/32 H_2S/CO_2	30/70 H_2S/CO_2
CH_4, mol%	0.2	0.2
H_2S, mol%	62.6	27.8
CO_2, mol%	30.2	64.9
H_2O, mol%	7	7
Feed Rate, kmol/hr	151.8	350.7
Claus Plant Configuration	Straight through flow	Split flow

Economic Results

- According to the KBR study[2], the relative capital cost for the UCSRP option is about 61% of a conventional Claus/SCOT unit (Figure 5) for a feed gas with 67% H_2S and 33% CO_2; for a feed gas with 30% H_2S and 70% CO_2, the estimate for the UCSRP option was about 60% of the Claus/SCOT plant.

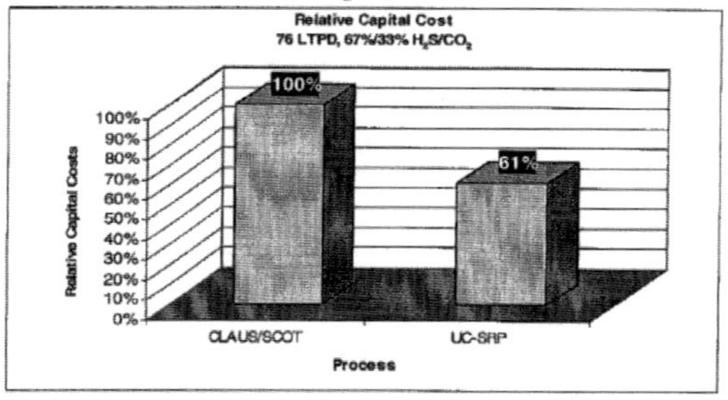

Figure 5: Relative Capital Cost – Claus/SCOT vs. UCSRP (Feed Gas with 67% H_2S, 33% CO_2)[2]

The relative Operating plus Maintenance costs for the UCSRP option were estimated at about (1) 68% of the Claus/SCOT design 67% H_2S/33% CO_2 Case, and (2) 67% of the Claus/SCOT design for the 30% H_2S/70% CO_2 Case.[2]

- Cost Comparison for a 935 LTD Sulfur Recovery plant[2]

As noted in Reference 2, another consulting firm performed an economic study for estimating relative costs for a 935 LTD sulfur recovery plant using the conventional Claus/SCOT and the UCSRP process. The relative capital cost for the UCSRP option was estimated to be about 62% of a Claus/SCOT system.

UCSRP as a Replacement of the Claus/SCOT Processes for a CO_2-rich Acid Gas Stream

If CO_2 level in a sour natural gas feedstock exceeds pipeline specifications (typically, less than 2 vol.%), an amine-type gas treating process is typically used to remove both H_2S and CO_2. In general, if CO_2/H_2S molar ratio is less than about 2.3, the off-gas from an amine stripper can be treated directly in a Claus/SCOT plant.[2] However, if this ratio is greater than 2.3, it may be necessary to carry out an expensive separation of the CO_2 prior to the Claus step. On the contrary, for an UCSRP unit, such a CO_2-rich off-gas from an "amine stripper" can be processed directly.

UCSRP as a Replacement of the Amine/Claus/SCOT Processes for a Natural Gas Stream

For direct sweetening of high-pressure natural gas feedstock, the high-pressure version of the UCSRP process would replace conventional amine plants integrated with a Claus/SCOT unit. A schematic of a two-stage UCSRP-HP concept, currently being evaluated at GTI for lab-scale demonstration, for high-pressure operations is shown in Figure 6; in this example, coal-derived syngas at about 800 psig is being processed to recover sulfur.[4] In this concept, the first UCSRP stage would be operated under "slightly excess SO_2" mode at the reactor outlet while the second stage would be operated under an "slightly excess H_2S mode". Typical sulfur balances for this example (in Figure 6) are shown in Table 2. In this example, the total SO_2 in the solvent plus syngas at the Reactor-1 outlet is about 9.7% of the SO_2 feed to this reactor; the H_2S content in the total solvent plus syngas flow at the Reactor-2 outlet is about 8.4% of H_2S in the syngas feed to the second reactor. For this example, the total syngas feed to the UCSRP reactors contains about 6,200 ppmv H_2S; about 1.34 vol.% of the feed bypasses the first reactor for processing in the second reactor. Based on GTI's design estimates, the exit syngas (sent to a CO_2 removal step) from the high-pressure separator (where Stream-11 is the inlet stream) would contain about 8 ppmv of H_2S plus COS.

Table 2: Stream Data for the Example Shown in Figure 6 (Basis: 1,000 lb-moles of total Syngas to the UCSRP Plant)[4]

Stream #	Stream Description	Lb-mole
1	Syngas to R-1[i]	986.6
2	Syngas to R-2[i]	13.4[ii]
3	SO_2 to R-1	3.1
4	Sulfur product	~ 6.2
5	Sulfur feed to Furnace	3.1
6	SO_2 from Furnace	3.1
7	SO_2 in solvent from R-1	0.26
8	SO_2 in syngas from R-1	0.04
9	H_2S in solvent from R-2	0.004
10	O_2 to Furnace	3.1
11	H_2S in syngas from R-2	0.003

i. Water vapor content: 1 mol%
ii. H_2S level: 0.083 lb-mole

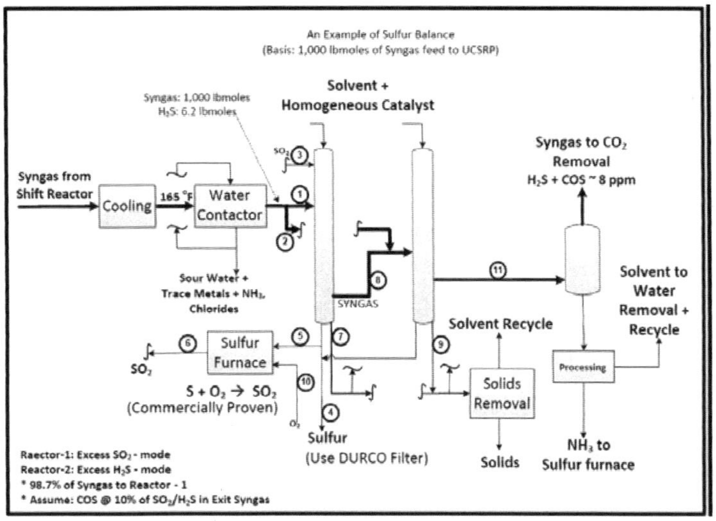

Figure 6: GTI's Two-stage UCSRP-HP Process Concept[4]

For direct sweetening of natural gas, an UCSRP-HP unit can replace an integrated Amine/Claus-SCOT plant. As shown conceptually in Figure 7, the USCRP-HP based plant designs would be relatively much simpler in handling various natural gas feedstocks with wide variations in (i) H_2S and CO_2 levels, and (ii) total gas flow rates.

As shown, if the CO_2/H_2S molar ratio is relatively high (~ greater than 2.3) in the H_2S-rich gas from an amine plant, we would need to first separate CO_2 before the gas can be processed in an integrated Claus/SCOT facility. For smaller plants (e.g., sulfur capacity less than 20 LT/days), we would need a "Liquid Redox" type process (and not a Claus/SCOT unit) to treat H_2S-rich gas (with CO_2/H_2S ratio of less than 2.3) from an amine plant. For the UCSRP-HP option, we can quite conveniently scale down to smaller sulfur capacities.[2]

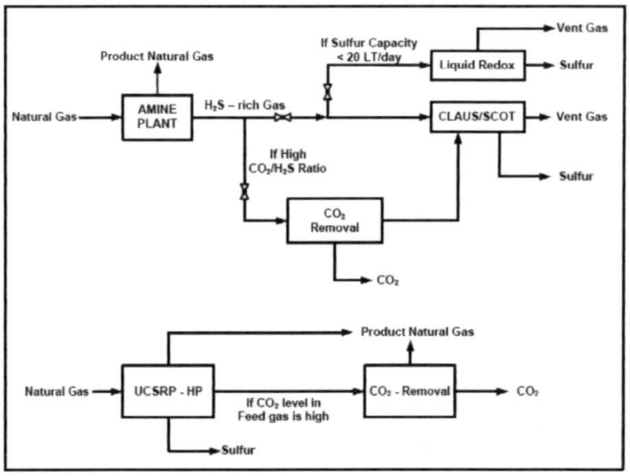

Figure 7: Relative Simplicity of an UCSRP-HP Process in Sweetening CO_2-rich Natural Gas

CONCLUSIONS

In summary, our key conclusions are:

- In comparison to the conventional Claus/SCOT process, UCSRP can reduce CAPEX by about 35-40% and OPEX by about 30-35%.
- In comparison to conventional Amine/Claus/SCOT for sweetening of natural gas feedstock, UCSRP-HP will reduce CAPEX/OPEX even more, especially for CO_2-rich NG streams.

References

1. Scot Lynn et al., "New Approach to Sulfur Removal Could Reduce Costs," GasTIPS, Winter 2002
2. Scot Lynn et al., "The University of California Berkeley's Sulfur Recovery Process: Claus Revisited," Paper presented at the 9th Gas Research Institute Sulfur Recovery Conference, October 24-27, 1999, San Antonio, Texas
3. S. J. Zhou, et al., "Economic Evaluation of the UCSRP-HP Process in IGCC Applications", Paper presented at the Clearwater Coal Conference, June 6, 2011.
4. Dennis Leppin, et al., "Economic Evaluation of the UCSRP-HP Process in IGCC Applications", SULPHUR 2011 International Conference & Exhibition, Houston, USA, November 6-10, 2011.
5. http://www.nelliott.demon.co.uk/company/claus.html

Refining Energy Utilization and CO_2 Emissions Modeling

John J. Marano, JM Energy Consulting, Gibsonia, PA

William R. Morrow III, Jayant Sathaye, Eric Masanet, Tengfang Xu, Lawrence Berkley National Laboratory, CA

Abstract

Petroleum refining consumes a significant amount of energy to convert crude oil into transportation fuels and other products. Based on U.S. Department of Energy statistics, this industry accounted for roughly 9% of U.S. industrial energy usage and 14% of industrial CO_2 emissions in 2010. Each refinery is configured differently, processing varying quality crudes into a variety of specification products; thus, no two refineries are the same. Process-specific energy use and CO_2 emissions for petroleum refineries are not publically available, and may not even be adequately or accurately measured; making it difficult to understand current or to predict future refining energy requirements and emissions. A tool for estimating individual process energy requirements and emissions as a function of varying feedstocks, products and processing would be useful to policy makers and researchers focused on improving energy efficiency and reducing emissions. JM Energy Consulting has developed such a modeling tool to support a Lawrence Berkley National Laboratory (LBNL) project aimed at estimating costs and benefits of energy-efficiency improvements in the U.S. refining sector. In this paper, we present our methodology and some preliminary results from this modeling effort.

Introduction

Sources of CO_2 emissions from petroleum refining include: on-site combustion of fuels to provide process heating and to generate steam and electricity for use within the refinery; the production of hydrogen used in refinery operations; and coke burn-off from catalyst regeneration (primarily in fluid catalytic cracking) [1]. Secondary, but sometimes significant, refinery CO_2 sources include flaring of off-gases during start-up, shut-down and process up-sets; biotreatment plants, and other fugitive emissions [2]. Figure 1 shows a typical breakdown of the major CO_2 refinery sources. The majority of CO_2 emitted from refining derives from fuel combustion. Interestingly, the distribution of emissions, as well as, total emissions are similar regardless of whether the refinery employs fluid catalytic cracking (FCC) or hydrocracking for the conversion of heavy gas oil to light product (*i.e.* fuel and power requirements are about the same, carbon is rejected either in the FCC regenerator or the hydrogen production plant required for hydrocracking).

Approaches for reducing refinery emissions may be categorized as follows [1,3]:

- Energy Efficiency Improvements – reduction of specific-energy required to manufacture product by minimizing energy losses
- Process Yield Improvements – reduction of specific-energy required to manufacture product by improving process yields

- Fuels and Utilities Switching – substitution of low-carbon fuels and other utilities for existing high-carbon ones
- Crude Oil and Other Feedstock Switching – substitution of light-sweet crudes for heavy-sour crudes, or introduction of renewable bio-oils as feedstocks
- Product Switching – substitution of drop-in biofuels as blendstocks, reducing refinery crude oil throughput
- Product Slate Optimization – adjustment of product slate to reduce processing requirements and emissions
- Carbon Capture and Sequestration – CO_2 removal from flue gas or syngas used to produce refinery hydrogen, followed by geological, oceanic or terrestrial storage in perpetuity

Some of the options identified above are broader than normally encountered in refinery specific studies aimed at identifying short term operating strategies, and would require changes in existing fuel markets outside of refiners' control. In order to get a complete and accurate picture, life-cycle emissions outside of the refinery must also be considered when making choices among options, since it is total emissions that are important.

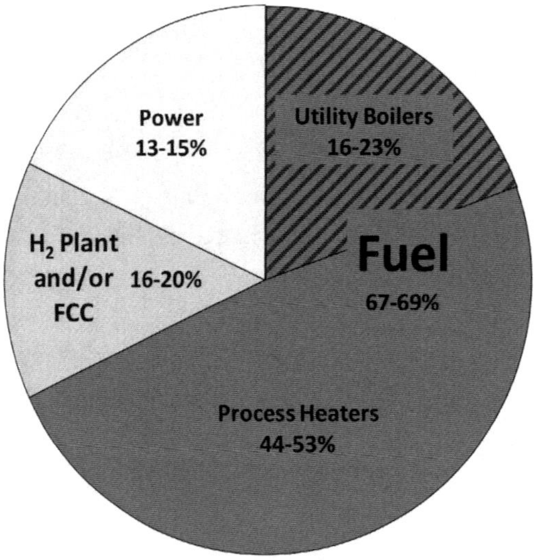

Figure 1. Typical Refinery CO_2 Emissions Distribution [4]

Adoption of efficient process technologies is an important approach to reducing CO_2 emissions, in particular those associated with combustion. In many cases, implementing energy efficiency measures is among the most cost-effective approaches that any refiner can make, improving productivity while reducing emissions. Therefore, careful analysis of the options and costs associated with efficiency measures is required to establish sound carbon policies addressing global climate change, and is the primary focus of LBNL's current petroleum refining sector analysis for the U.S. Environmental Protection Agency. The analysis is aimed at identifying efficiency-related CO_2 abatement measures and developing energy and CO_2 emissions abatement supply curves for the U.S. refining industry. A refinery model has been developed for this purpose that being general, also has the potential to be used to examine the costs and benefits of the other CO_2 mitigation options listed above. This model is described in more detail below.

Methodology - Background

Lawrence Berkley National Laboratory [5] and others [6-8] have been active in identifying possible measures for improving the efficiency of petroleum refining operations. However, it has been difficult to quantify the full impact of implementing these measures industry wide. One approach for doing this, attempts to quantify energy use and emissions at the process plant level within the refinery using average performance parameters for the individual process plants. This approach has been successful in establishing nominal baselines for energy use and emissions (*e.g.* [9]) for the refining industry as a whole. On the other hand, the approach has been less than definitive when it comes to quantifying potential benefits and costs from applying energy efficiency improvement measures [10]. This shortcoming results from the absence of predictive capabilities in the approach; as well as, issues associated with quantifying the inherent variability that exists between individual refineries.

Petroleum refineries in the U.S. are complex industrial facilities involving multiple processing plants configured both in parallel and series. The component plants are integrated through utility systems, which supply fuel, steam, cooling water, and electricity to the various processes. Individual operations within these plants are also extensively integrated. Due to the high degree of energy integration, both at the refinery and plant level, efficiency improvements to a single operation may have implications throughout the refinery. Efficiency improvements are in general not additive, with the possibility of the effectiveness of later modifications being dampened by those implemented earlier (*e.g.*, a 10% reduction in fuel consumption based on a revamped heat exchanger network design and a 5% reduction due to less furnace fouling does not necessarily result in a 15% total reduction). For this reason, it is essential to examine efficiency improvements in refining at the process level using a bottom-up, predictive approach to estimate energy usage on an operation-by-operation basis. A number of excellent case studies based on this approach can be found in the open literature [11-13]; however, these do not systematically address the state-of-play within the industry as a whole, and also require very sophisticated and detailed analyses (*e.g.*, heat exchanger network pinch analysis).

A predictive approach requires a certain degree of engineering to be incorporated into the analysis that may not be necessary when considering simpler manufacturing facilities. Technical descriptions of the various processes are required that relate process feed and operating conditions to process product streams and utility requirements on a step-by-step basis. There is precedence for this type of analysis within the refining industry for determining economically optimal refinery operation using Linear Programming (LP) models. A number of articles have appeared over the years that describe use of refinery LP models to examine possible CO_2 emissions reductions [1,4,14,15]. However, problems exist with standard approaches to refinery LP modeling that must be addressed when analyzing efficiency improvements.

Traditional LPs are formulated as an optimization of a refinery economic objective function and a system of constraints expressed as equalities and inequalities related to material and utility balances, along with operating and product quality specifications. For real refineries, models are very large with many variables being considered for optimization, which may be performed at a refinery on a day-by-day basis. Accuracy is achieved through the use of measured plant data or rigorous front-end process simulations (*i.e.* the LP model must be tuned-up). For the present application, interest is not in determining absolute economic optima, but rather *in sequentially organizing the cost hierarchy of a series of energy efficiency measures that might be applied across the industry*. Simplicity and flexibility is essential, and the primary concern is that the model mirrors correct refinery responses to changes of interest [16]. For this application, a detailed LP model may miss the big picture by focusing too much on the details (*e.g.*, multi-tiered product specifications, elaborate process models, multiple crude oils). This response to non-routine situations is sometimes referred to as model "constraint gridlock".

In addition, LP formulations are typically limited to the present or the near future, have difficulty handling capital investment decisions, and mass and energy balances may not be closed. The focus here is on short to midterm (twenty-year horizon) investments and operational changes that can improve efficiency, with attention being given to state-of-the-art and possible next generation technologies. For the latter in particular, costs are not well known and are dependent on the outcomes of on-going research, development and demonstration projects. When examining solutions to long range problems, accurately estimating production volumes may be more important than maintaining rigorous product specifications. There is a considerable degree of uncertainty with product quality requirements and other assumptions when looking as much as twenty years out into the future.

Special attention must be given to mass and energy balances. Early refinery LP models were neither mass nor energy balanced. Material balances were formulated on a volume basis (*e.g.*, barrels per day), and a total mass balance for the refinery model was not achieved without adjustments to stream densities. This situation has been mostly corrected in advanced LP models in use today; however, compositional balances for carbon and hydrogen are still not adequately represented in most LP models. Efforts to date have focused on methods to close these balances around the refinery model; however, these appear to be by means of forcing closure through arbitrary adjustments to specific stream compositions. Such methods are not adequate when one of the primary functions of the model is the prediction of CO_2 emissions. The best example of where this situation cannot be tolerated is when compositions of fuel gas streams are approximated by an externally specified and fixed fuel composition, leading to poor estimates of CO_2 emissions [1]. Likewise, early models and those still in use today perform only utility balances, that is the amount of steam, power, etc. used is equal to that generated or purchased. Once again, this is not adequate when considering issues directly related to energy efficiency. It is necessary to explicitly estimate energy use and losses throughout the refinery, instead of inferring them from utility balances. In order to do this, the energy content of refinery feed, product and internal streams must be known or predicted, and process-to-process heat exchange must be explicitly considered.

Methodology – Approach

To address the issues described above, standard methods for building refinery LP models have been adapted for the present application of generating refinery energy-usage abatement supply curves. A combination of empirical data correlations and semi-rigorous engineering models has been used to develop reduced-form refinery models to estimate refinery product yields and energy requirements. In many respects, these reduced-form models are similar to what might be used in a more traditional LP model applying "delta-based" sub-models. The refinery model developed has the following attributes:

- The model is formulated to maintain total mass and energy balance, as well as elemental balances for carbon, hydrogen, nitrogen and sulfur. This has been achieved by reformulating individual process yields data on a weight carbon basis (*e.g.*, ton of carbon in naphtha product per ton of carbon in feedstock). Hydrogen contents of process streams are based on correlated changes between the hydrogen-to-carbon ratio of feed and product streams, as are stream energy content (heating values). Nitrogen and sulfur removal efficiencies and distributions to liquid products are explicit input to the individual conversion processes represented in the model.
- Relatively simple sub-models are employed for conversion processes, relating yields and utility use to at most a few essential feed-quality and operating-severity parameters. Intermediate streams within the refinery are not aggregated (pooled) within the model, and streams of distinctly different character employ different yield correlations (*e.g.*, vacuum gas oil and coker gas oil FCC feeds). Utilities usage and emissions are disaggregated at the process level, and as much as possible, estimated at the operational level (*e.g.*, for the atmospheric crude unit, steam usage is estimated for individual column product draws, instead of being specified as a fixed proportion to crude feed rate).
- In general, individual product specifications are met by refining and not blending. What this means is the model does not optimize intermediate-stream pathways through the refinery, which are only varied when considering alternative crude oils and refining configurations (*e.g.*, the amount of light straight-run naphtha blended vs. that upgraded is input; whereas, the amount of gasoil sent either to hydro or cat cracking is calculated to achieve the desired gasoline-to-diesel ratio of the product slate specified). Sufficient processing severity is specified within the model in order to infer that product specifications are met. Individual refinery configurations are based on a statistical analysis of U.S. refinery process capacity, and an elementally-balanced slate of fourteen crude oil types is used to select a blend of crudes meeting the desired characteristics for any given refinery analysis.

Input to the refinery model includes production volumes for key products, in most cases these are transportation fuels: gasoline, jet, diesel, and marine fuel oil. Predicted outputs are crude and intermediate stream flows. Refinery energy usage and emissions are estimated on a process-by-process basis. The model was developed in MS Excel®; calculations are performed in the following sequence:

1) The refinery carbon balance is solved with Solver® using carbon-based yield correlations for individual processes;

2) Based on correlations, the hydrogen contents of all process streams are calculated, and the refinery hydrogen balance is closed by purchasing hydrogen or reforming natural gas on-site;
3) Electricity, fuel, steam and cooling water requirements for individual process operations are estimated, and utility balances for the refinery are closed by calculating the amounts of supplemental fuel (typically natural gas), electricity and make-up water required. Utilities may either be purchased or generated on-site;
4) Process energy usage and losses are calculated to achieve an overall refinery energy balance, and refinery CO_2 emissions are calculated;
5) The energy flows for the base refinery simulation are combined with a database of cost and performance efficiency-measures and used to sequentially rank the measures based on their return on investment.

The above procedure ensures that the refinery model is elementally and energy balanced without the need for ad hoc adjustments. While the models and methods used are less rigorous than those employed in sophisticated process simulators, they contain a more detailed engineering description than found in standard LP refinery models.

Sample Results

To demonstrate the robustness of the method described above, refinery CO_2 emissions were estimated and allocated to the various refinery intermediate streams that are blended into gasoline and distillate fuels. Only the results relating to distillate products are discussed here. The refinery was modeled as a "notional" refinery set up to approximate a typical U.S. Gulf Coast refining configuration.

Figure 2 compares the embodied refinery CO_2 emissions and combustion emissions for the following streams that may be blended into diesel fuel: straight-run kerosene, hydrocracker kerosene, straight-run light-gasoil, hydrocracker light gasoil, cat-cracker light-gasoil produced from straight-run vacuum gasoil, cat-cracker light-gasoil from heavy coker gasoil, and coker light-gasoil. The coker-derived streams are produced from delayed coking of vacuum resid. Other than the hydrocracker product streams, all of the distillate streams have been hydrotreated to meet product sulfur specifications. Also shown are the emissions for the final diesel product blend. The first thing to note is that the combustion emissions for all the diesel blendstocks are almost identical, varying by not more than 2% from the blended product value.

The embodied emissions are reported as a percentage of the combustion emissions. These are significantly less than the combustions emissions, ranging from a low of slightly less than 2% for hydrotreated straight-run kerosene to a high of slightly greater than 20% for the hydrotreated, cat gasoil produced from heavy coker gasoil. It is clear from Figure 2 that embodied CO_2 emissions are directly related to the extent and severity of processing that a stream experiences within the refinery. In particular, refinery conversion operations such as hydrocracking, cat cracking, and coking produce more emissions than hydrotreating operations alone. The embodied emissions for the diesel blend are about 9% of combustion emissions.

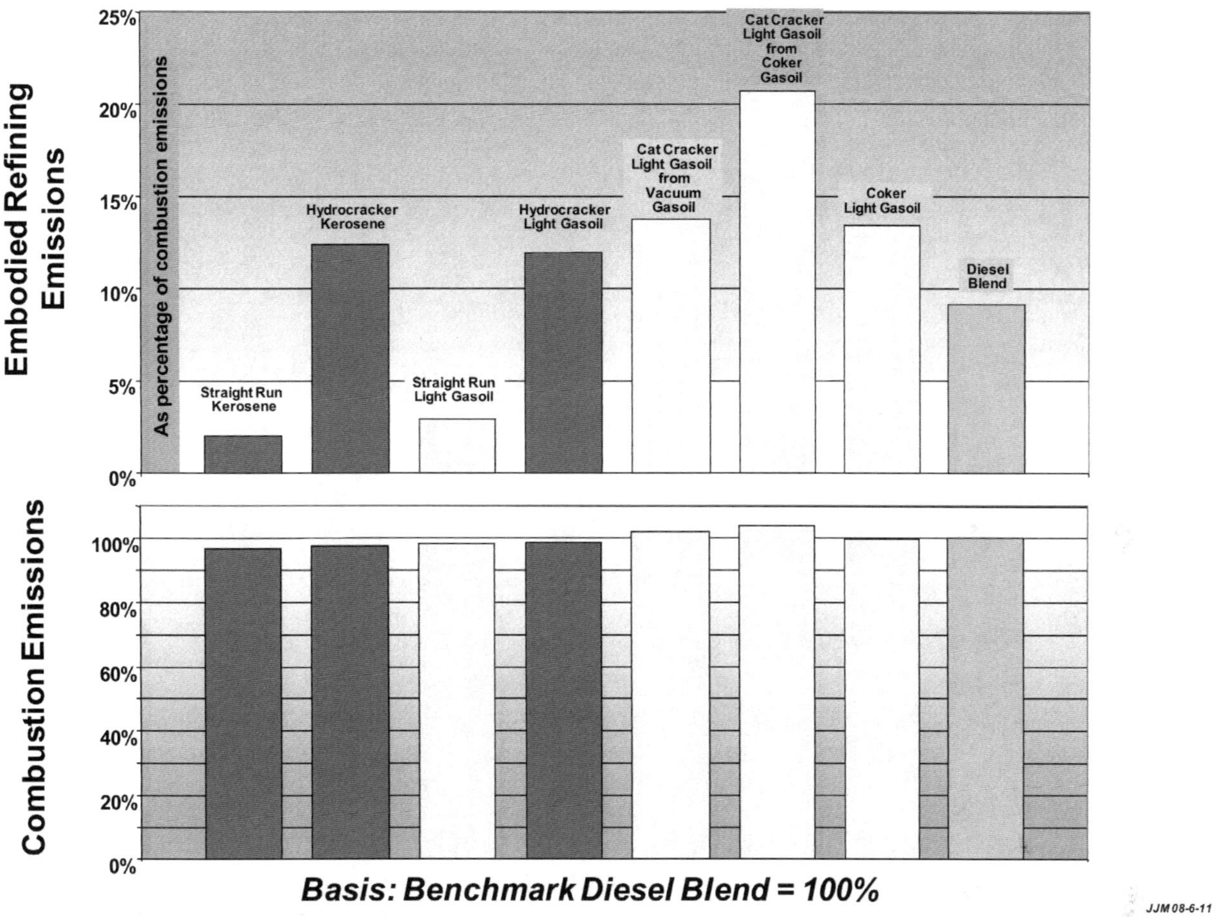

Figure 2. Relative CO_2 Emissions of Various Diesel Blendstocks [17]

Figure 3, compares CO_2 emissions for blended diesel produced from the refining of two different quality feedstocks, high-sulfur, light and high-sulfur, heavy crude oil. Embodied emissions from the heavy crude are about 20% higher than those from the light crude. This result is due more to the final distribution of distillate streams in the blend than to specific variations in the severity of refining seen by the individual blendstocks. This can be seen from the second row of pie charts in Figure 3. The heavy crude contains significantly less straight-run distillate, with low embodied emissions, than the light crude (47 vs. 56% of the diesel pool), and produces more FCC light gasoil (21 vs. 15%) due to the heavier crude's higher content of vacuum gasoil. Interestingly, the contribution of the coker to embodied emissions is relatively constant for both crude oils (45+14=59% vs. 39+21=60%).

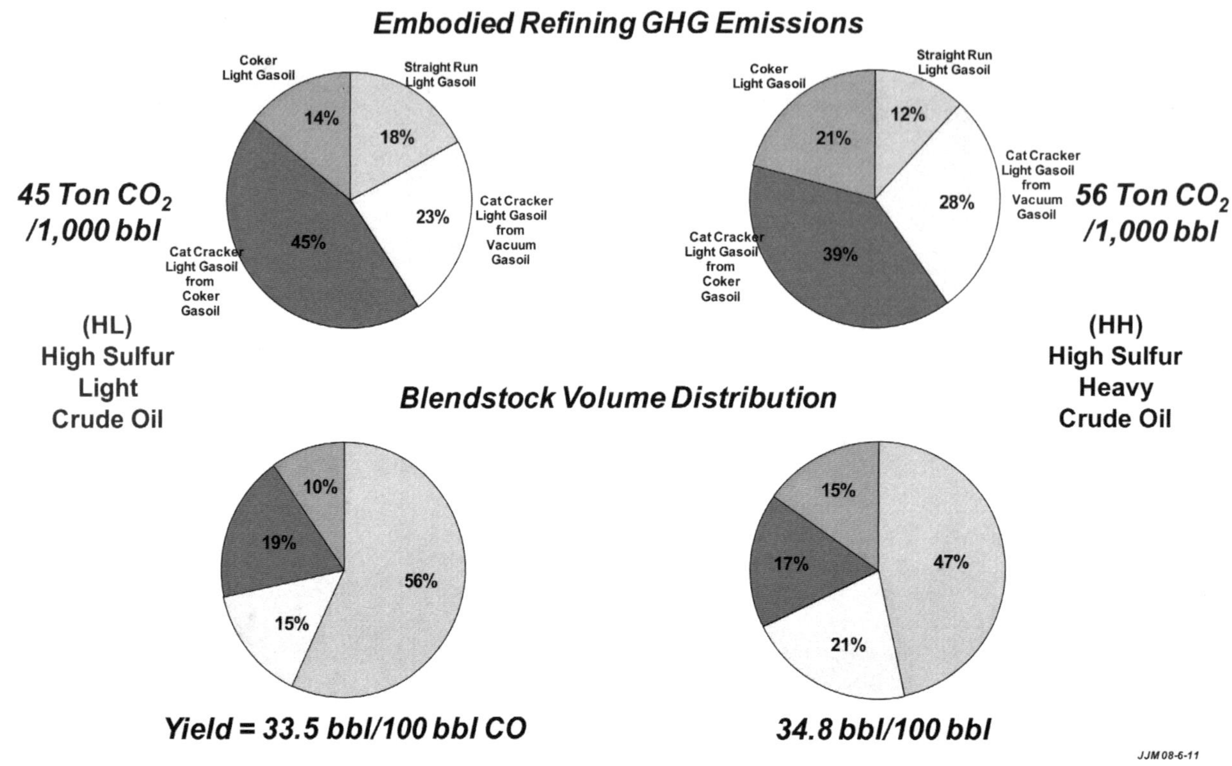

Figure 3. Comparison of CO_2 Emissions for Diesel Production from Light & Heavy Crudes [17]

Status and Next Steps

JM Energy and LBNL are continuing work to establish supply curves for refinery energy-usage and CO_2 emissions abatement. The current focus in on developing a database of energy abatement measures, efficiency and costs that can be applied to the refinery model results. In addition, a methodology for handling the variation between U.S. refineries is under development. The major challenge here, is establishing a baseline for the existing industry, since individual refineries do not publically report their energy use; though, there are now regulations in place requiring individual refineries to report their CO_2 emissions. A question still to be addressed is: How widely have any given efficiency measures been adopted by refiners? A clue to answering this question may be available from examining historical energy usage data for the industry, which clearly shows an improvement in efficiency over the last ten years. Data compiled on energy use will be used to update and tune the refinery model to match government-collected aggregate energy consumption in the industry. A final report on this analysis should be available from LBNL by late 2012.

To date, the current analysis has not considered the ramifications of current trends in petroleum refining related to technology, crude oil quality, fuel specifications, and product slate. These trends will have a significant impact on the future path of the U.S. refining industry. Future challenges that will likely affect the industry include: lower gasoline-to-distillate product ratios due to ethanol blending into gasoline, vehicle hybridization, further dieselization of the U.S., and increased air travel; internationally agreed to marine SOx reductions requiring low-sulfur bunker fuels; refinery crude slate changes due to increased production of domestic shale oils, and increased imports of Canadian synthetic crude oils and dilbit blends; and further implementation of renewable and/or low-carbon fuel standards, which may result in the long term to the introduction of truly drop-in biofuels. Future sensitivity analyses will be needed to examine impacts of these potential changes, since many of these could have negative ramifications for improving efficiency and lowering emissions, while some may be positive. The role of CO_2 capture and sequestration in petroleum refining will also need to be examined more completely if CO_2 emissions are to be drastically reduced over the next fifty years.

Acknowledgements
The authors would like to express their appreciation to the U.S. Environmental Protection Agency for providing financial support for this project.

Any conclusions, comments or opinions expressed in this paper are solely those of the authors and do not represent any official positions held by LBNL, EPA, or the U.S. Government. Information contained herein has been based on the best data available to the authors at the time of preparation.

Copies of the slides used for this *2012 AIChE Spring Annual Meeting* presentation may be obtained by contacting John Marano at *marano@zoominternet.net*

References

1. Stockle, M.; Bullen, T., "Integrate Refinery Carbon Dioxide Plans Plant-Wide," *Hydrocarbon Processing*, pp.106-111, Nov. 2008.
2. Spoor, R.M., "Low-Carbon Refinery: Dream or Reality," *Hydrocarbon Processing*, pp.113-117, Nov. 2008.
3. Gunaseelan, P.; Bueler, C.; Chan, W.R., "Greenhouse Gas Emissions: Characterization and Management," *Hydrocarbon Processing*, pp.57-70, Sept. 2009.
4. Clarke, S.C., "CO_2 Management – A Refiners Perspective," Foster Wheeler Energy, Ltd., available at http://www.fwc.com/publications/tech_papers/oil_gas/CO2.pdf
5. Worrell, E.; and Galitsky, C., *Energy Efficiency Improvement and Cost Saving Opportunities for Petroleum Refineries - An ENERGY STAR® Guide for Energy and Plant Managers*, Ernest Orlando Lawrence Berkeley National Laboratory, Report No. LBNL-56183, Feb. 2005.
6. White, D.C., "Trading Silicon for Carbon: How to Reduce Energy Usage through Automation," *Hydrocarbon Processing*, pp.61-70, June 2010.
7. Beaubien, R., "Efficiency Improvements Cut GHG Emissions, Help Profits," *Oil & Gas Journal*, pp.20-26, Sept. 28, 2009.
8. Tice, M., "Carbon and Energy Management: Innovative Approaches for a Responsible Future," *National Petrochemical & Refiners Association Annual Meeting*, San Diego, CA, Paper No. AM-08-37, March 9-11, 2008.
9. *Energy and Environmental Profile of the U.S. Petroleum Refining Industry*, prepared by Energetics Inc. for the U.S. DOE Office of Energy Efficiency and Renewable Energy, Industrial Technologies Program, Nov. 2007.

10. *Energy Bandwidth for Petroleum Refining Processes*, prepared by Energetics Inc. for the U.S. DOE Office of Energy Efficiency and Renewable Energy, Industrial Technologies Program, Oct. 2006.
11. Glasgow, I.M.; Polcar, S.; Davis, E.; Nguyen, T.; Price, J.; Stuecheli, C; Palmer, R.E., "How will Carbon Emissions Regulations Revise Energy Conservation Economics?," *Hydrocarbon Processing*, pp.35-41, June 2010.
12. Rossiter, A.P., "Improve Energy Efficiency via Heat Integration," *Chemical Engineering Progress*, pp.33-42, Dec. 2010.
13. Carbonetto, B.; Pecchi, P., "Going 'Green' with FCC Expander Technology," *Chemical Engineering Progress*, pp.79-83, Jan. 2011.
14. Marano, J.J.; Rogers, S.; Spath, P.L.; Mann, M.K., "Life-Cycle Assessment of Biomass-Derived Refinery Feedstocks for Reducing CO_2 Emissions," *Proceedings of the Third Biomass Conference of the Americas*, Montreal, Quebec, August 25-28, 1997, pp.325-337.
15. Hirshfeld, D.; Kolb, J., in *Comparison of North American and Imported Crude Oil Lifecycle GHG Emissions, Final Report*, prepared by Tiax LLC and MathPro Inc. for the Alberta Energy Research Institute, Tiax Report No. D5595, July 6, 2009.
16. Bredeson, L., "Quick Overview of Complexity Issues in Refinery Modeling," *Energy Information Administration Technical Workshop*, Sept. 30, 2009, available at http://www.eia.gov/oiaf/emdworkshop/pdf/lbredeson_lfmm.pdf
17. Marano, J.J., "GHG Profiles for Refining & Upgrading Fossil Distillate Fuels," *9th International Conference on Greenhouse Gas Control Technologies*, Washington, DC, November 16-20, 2008.

PROFITABILITY THROUGH LEAN SIX SIGMA
Save Energy - Save Environment

Author
Bhanumurthy, Maddala V,
Tel No.: 00966-43978540 /00966-506285745
FaxNo.:00966-43978118
BHANUMMV@ARAMCO.COM
MADDALA.BHANUMURTHY@ARAMCO.COM
Saudi Aramco/Yanbu Refinery,Yanbu
Co-Author
Mohammad A. Al-Malki
Tel No.: 00966-43978101

Biography:

Bhanumurthy, Maddala Joined Saudi Aramco as a Senior Process Engineer in 2007 and is currently working as Yanbu Refinery Energy Coordinator. He is a Chemical Engineering Graduate with MBA from Andhra University. He had more than 25 years refinery experience in reputed organizations like Hindustan Petroleum Corporation Limited, Qatar Petroleum. Before joining SA, he worked as a Deputy General Manager/FCC Head with Essar Oil Limited, India.

Mohammad A. Al-Malki is a Chemical Engineer graduated from King Fahad University of Petroleum and Mineral and joined Saudi Aramco in 2011 and currently working as Utilities Area Engineer.

Abstract Summary:

Energy performance is measured as Energy Intensity Index (EII) and it is ratio of Actual Energy/Standard Energy Consumption. Lower EII indicates lower energy consumption per unit of output and better performance. YR EII is 78.9 while World's best refinery EII is 73.5 One unit change in EII will result in a cost avoidance of 0.77 MM$. We can avoid up to $ 4.08 MM expenditure. Actual Energy Consumption is comprised of Fuel Gas (FG), Fuel Oil, and Electrical Consumption. Reduce energy consumption to improve EII. Standard energy consumption is calculated from unit throughputs. Reduction in throughputs increases EII. LSS project is to identify and reduce defects in energy system causing high indices which will enable to meet best performing refineries EII. LSS drives speed, quality, and cost simultaneously. Lean focuses on the elimination of waste while Six Sigma focuses on reducing process variation and then on improving the process capability through the methodology DMAIC (Define-Measure-Analyze-Improve-Control).

Project started with team selection, Project scope finalization and identification of CTQ (critical to quality) factors. Brainstorming conducted for identified defects causing process deviations and reducing process capability. For each identified defect, Fishbone created, data collected, Failure Mode Effect Analysis (FMEA), detailed process maps and Pareto charts developed. Data analyzed using, hypothesis tests, and regression analysis to draw conclusions and developed solutions for short and long term .

Short-term measures like steam optimization, Boiler load management, Flue gas oxygen optimization were implemented. The LSS project implementation improved EII to 77.3 by end of 2011, while it was 78.1 by end of 2010 from 78.9 in year 2010 beginning with a cost avoidance of $1.4MM/year.

Yanbu Refinery, SA

Saudi Aramco's Yanbu Refinery Department (YRD) is a 235 MBD hydro skimming refinery located in western province of Saudi Arabia. YRD is committed to reduce energy costs as part of Saudi Aramco Energy conservation policy and reduce environmental emissions. Yanbu refinery (YR) adopted Lean Six Sigma (LSS) approach with Management and Organizational Consulting department support in its aim to reduce energy costs, improve energy key performance indicator (KPI) and reduced emissions.

Energy Intensity Index (EII)

Energy Intensity Index (EII) is ratio of Actual Energy/Standard Energy Consumption. Lower EII indicates lower energy consumption per unit of output. YRD uses Solomon methodology to calculate EII. Actual Energy Consumption is comprised of Fuel Gas (FG), Fuel Oil and Electrical Consumption. Energy consumption is to be reduced to improve EII. Standard energy consumption is calculated from formulas using unit throughputs. Any Reduction in throughputs will increase EII.

LSS Approach

LSS project is an ideal way to improve profitability by energy optimization, reduce EII and meet best performing refineries KPI. LSS drives speed, quality, and cost simultaneously. Lean focuses on the elimination of waste while Six Sigma focuses on reducing process variation and then on improving the process capability through the methodology DMAIC (Define-Measure-Analyze-Improve-Control) as shown in Fig1. LSS compliments to engineering solution where a problem is converted into a mathematical problem to obtain mathematical solution which will be applied to resolve the original problem.

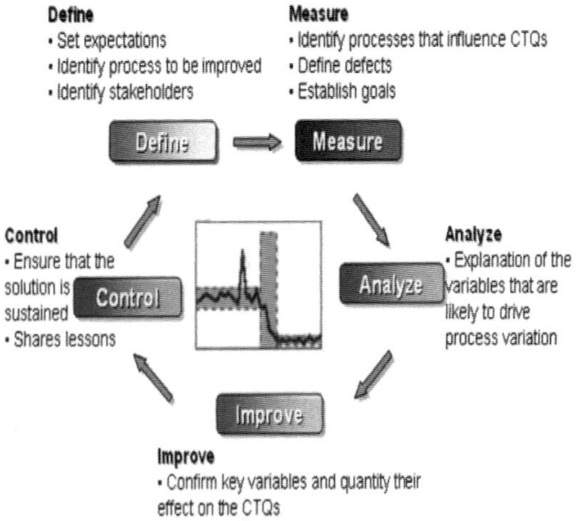

Fig1. DMAIC PROCESS

Features that set Six Sigma apart from other initiatives include a clear focus on achieving measurable and quantifiable financial returns, commitment to making decisions based on verifiable data, rather than assumptions and guesswork and an increased emphasis on strong management leadership and support. A special infrastructure of "Champion," "Master Black Belt and Green Belts", etc. will lead and implement the Six Sigma approach.

LSS Project

LSS project is to identify and reduce defects in energy system causing high indices which will enable to meet best performing refineries EII. Project started with selection of members from Process, Electrical, Mechanical Engineers, and Operations Foremen and supported by master black belt from Organization Consulting Department. Project scope clarified with problem and goal statements. Refinery EII is 78.9 while World's best refinery EII is 73.5 based on 2008 Solomon fuels study. One unit change in EII will result in a cost avoidance of 0.77 MM$. We can avoid up to $ 4.08 MM expenditure. To reduce EII by one unit, reduce total Fuel consumption by 97 BPD or Power by 65MW/Day.

Process boundaries identified using SIPOC (Supplier, Input, Process, Output, and Customer). Team identified CTQ (critical to quality) parameters as shown in figure and defects causing process deviations to meet customer requirements. For fuel gas system critical parameters impacting fuel gas composition include boilers load, boiler/heaters efficiencies, feed to heater inlet and outlet temperatures, stack temperatures, oxygen in stack flue gases etc, as shown in Fig2.

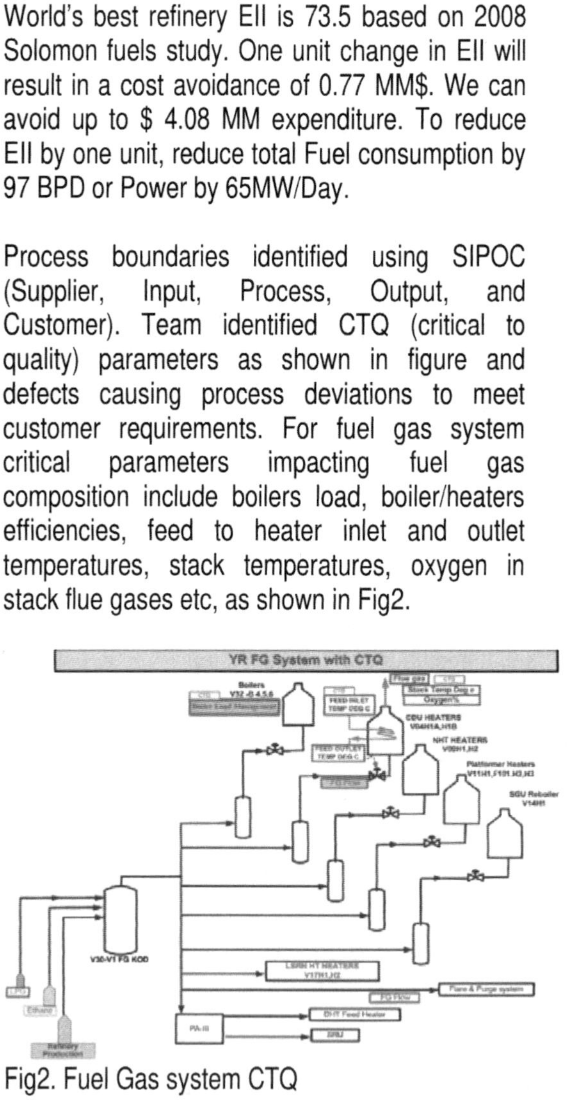

Fig2. Fuel Gas system CTQ

Pareto principle uses 80: 20 rule. This helps to identify major consumers/ areas (80%) to focus and minimize the defects in those areas for quick results. Pareto chart will be developed and used for all factors influencing Energy Consumption separately.

Pareto chart for factors influencing Standard Energy Consumption was prepared as shown in Fig 3. Refinery need to maintain high availability to maximize capacity utilization of the key units CDU, Diesel Hydrotreater and continuous catalytic regenerating reforming units. It also aids in maximizing refinery input by allowing blending operations.

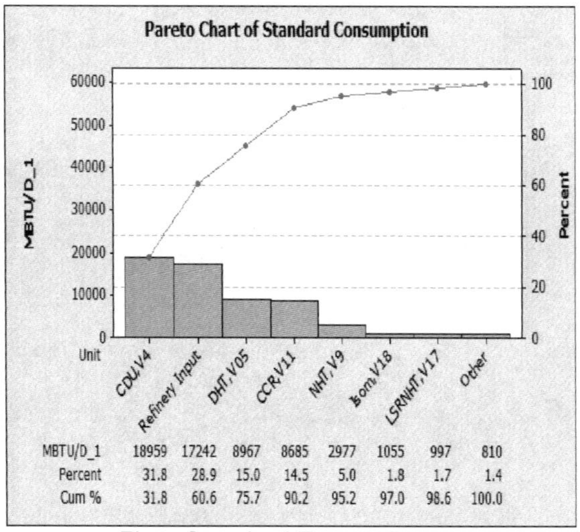

Fig3. Standard Energy Pareto

Data collected and created Pareto chart for factors influencing Actual Energy Consumption as shown in Fig4.

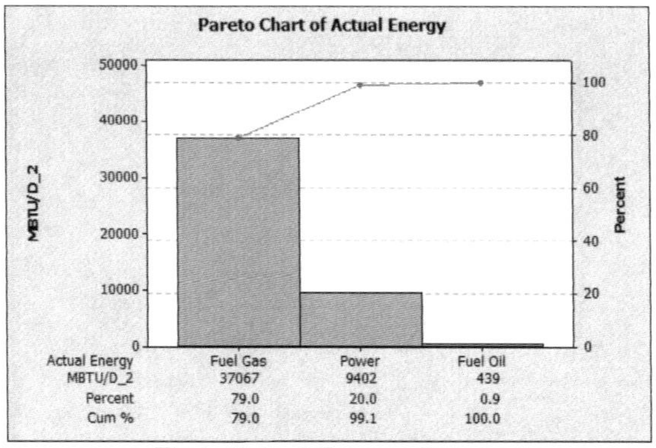

Fig4. Actual Energy Pareto

As can be seen from Pareto, refinery uses fuel oil (FO) only during emergencies as part of refinery environmental commitment and hence FO system is out of scope. Power consumption variation is mainly attributed to changes in ambient conditions as shown in fig 5 and will be excluded from further study as there is little control over ambient conditions.

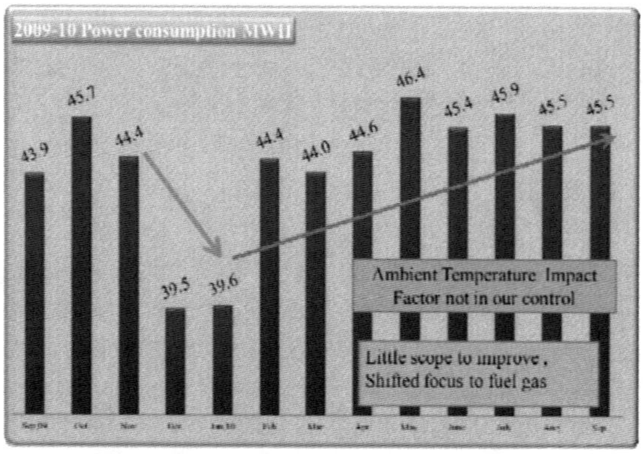

Fig5. Power Consumption Variation

Pareto chart for fuel gas (FG) Consumption prepared as shown in Fig6. CCR, Crude units and boilers consume 80% of fuel gas and will be focus areas for further study

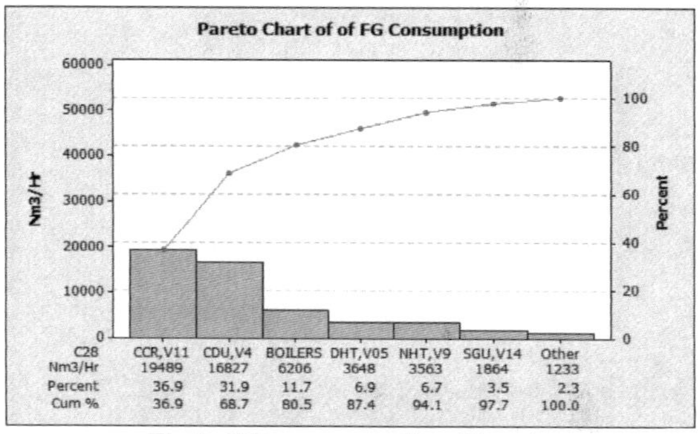

Fig6. FG consumption Pareto

Brainstorming was done for identifying defects for each actual energy factor and are represented in Fishbone chart as shown in Fig 7.

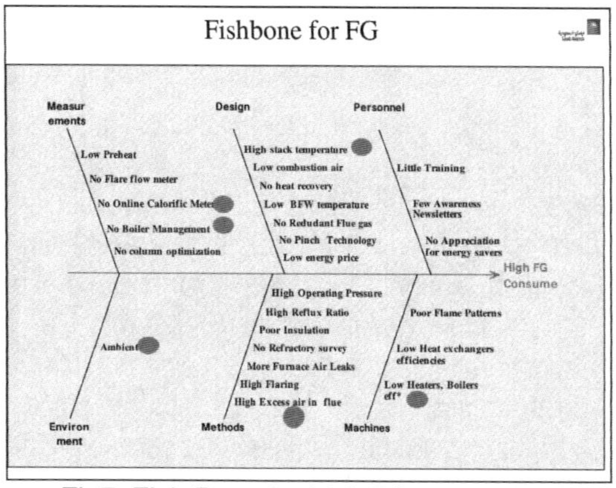

Fig7. Fish Bone for high FG consumption

Failure Mode Effect Analysis (FMEA) used for ranking the above factors based on severity, occurrence and determination. Detailed and specific process maps developed and analyzed for value and non-value added steps to identify opportunities.

Accordingly major fuel gas consuming heaters/boilers stack gases analyzed for oxygen. Prior to the project, heaters discharging flue gases to atmosphere with oxygen above 4% were 70% and were optimized to less than 40 % initially and finally to 10%. Accordingly the major fuel gas consumers, crude and platformer Heaters, oxygen in stack gases optimized. By optimizing these heaters a saving of $ 0.7MM/Year is realized.

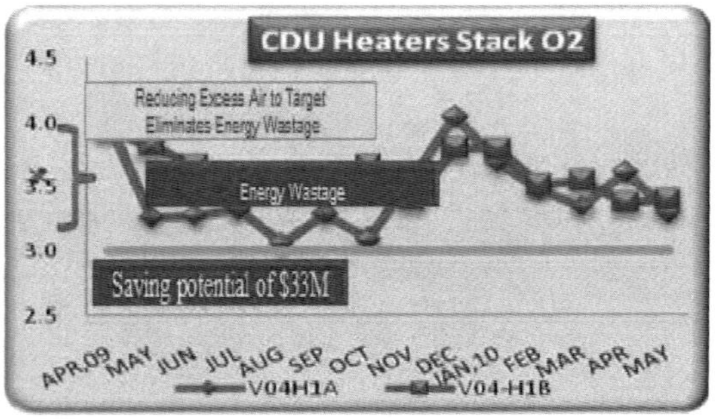

Fig8. Crude Heater flue gas oxygen

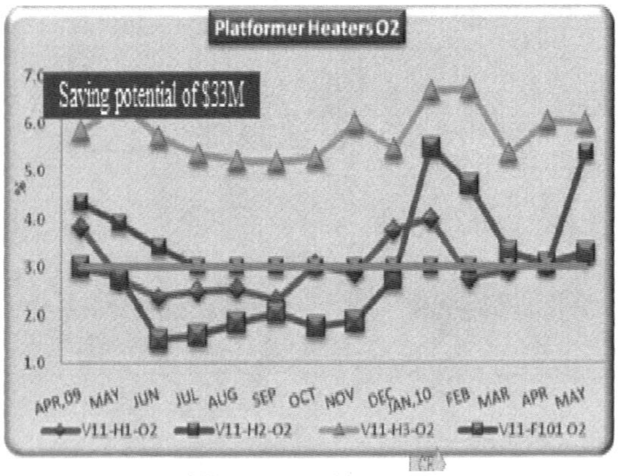

Fig9. CCR heaters flue gas oxygen

Flue gases generated after combustion are vented through stack and flue gas temperatures indicate amount of energy lost through stack gases. Flue gases exit temperatures can be reduced to 188-200 deg C, considering sox dew point impact, before venting through stack to minimize energy loss. Lean process recommends identifying and eliminating any wastage of energy in the refinery. Accordingly, Crude/CCR Heaters/Boilers, the three major FG consumers in the refinery stack flue gas temperatures data analyzed. Of the three major FG consumers, boilers and platformer Heaters, have waste heat recovery equipment on flue gases before being vented to atmosphere while Crude heaters consuming 31% of FG have no waste heat recovery equipment and discharge hot flue gases to atmosphere at above 480 deg C. Techno economical option will be evaluated and implemented to minimize energy loss through crude heaters.

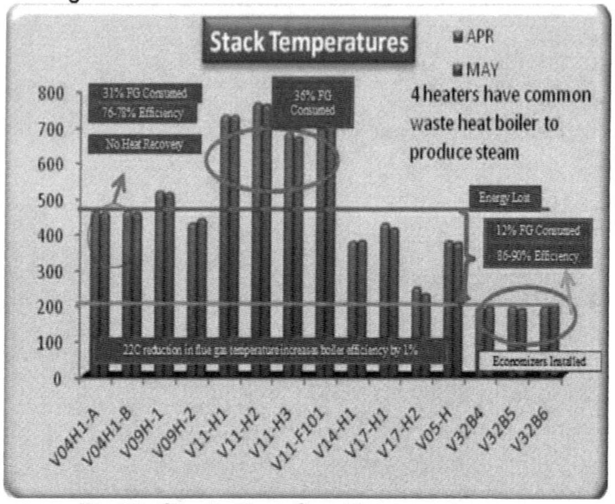

Fig10. Stack Flue Gas Temperatures

Refinery has three boilers with two boilers online always while the third one remains on hot standby condition to generate 150 psig(10.5 Kg/cm2) steam. Boiler loads analyzed as one of the major FG consumer utilizing hypothesis tests, and regression analysis to draw conclusions. It can be noted from the Fig 11 Boiler 5 operated at higher load than the other boilers while it had the lowest efficiency as compared to other two boilers.

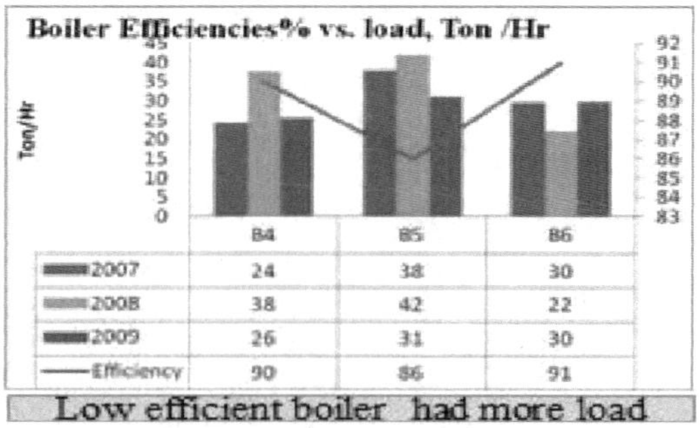

Fig11. Boilers Load & Efficiency

Six sigma tools were used to analyze data including boiler loads and fuel gas flow. In the first phase, fuel gas vs steam flow data analyzed using correlation and regression analysis. Strong positive Correlation is shown between fuel gas vs steam flow indicating high accuracy of the data as shown below.

Boiler FG versus steam flow Correlations:

Pearson correlation = 0.873

Then regression analysis performed between FG and steam flow for each boiler.

Regression Analysis:

Boiler 6 FG = 115 + 0.0605 steam flow

Boiler 5 FG = 696 + 0.0497 steam flow

From the regression analysis, boiler FG consumption fig12 was developed. It was obvious from the chart that Boiler#5 is consuming higher FG than Boiler#6 at normal operating loads.

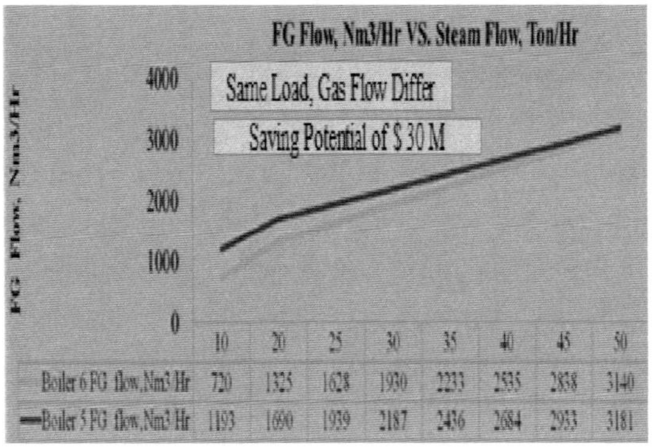

Fig12. Boiler FG consumption

It was recommended to operate high efficient Boiler#6&4 at higher loads than Boiler#5 to optimize boilers FG consumption. The regression analysis also allowed detecting meter errors and rectifying them early.

It was observed that boilers load is on increasing trend as shown in Fig13.

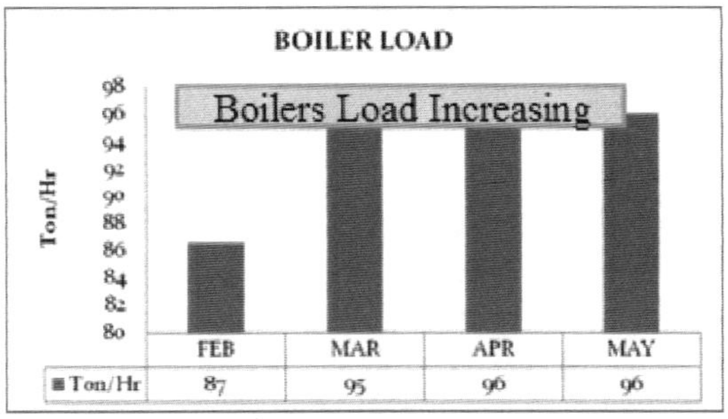

Fig13. Boilers load

Pareto chart developed for steam consumers in the refinery as shown below in Fig14.

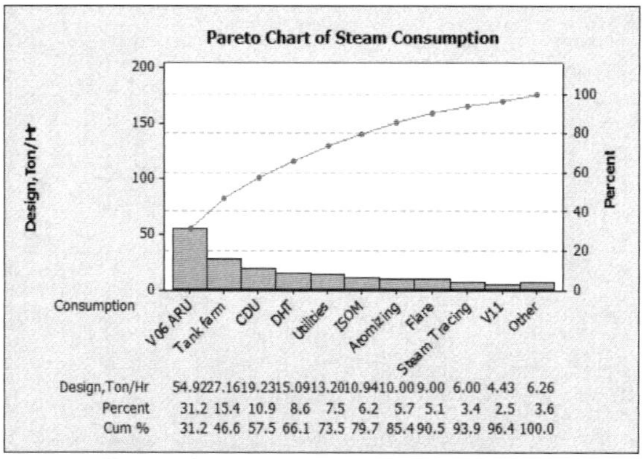

Fig14. Steam consumption Pareto

Amine regeneration unit (ARU) is the major consumer as can be seen From Fig 14. Accordingly ARU operations were optimized with the help of Area Engineers and Operations support as shown in Fig 15.

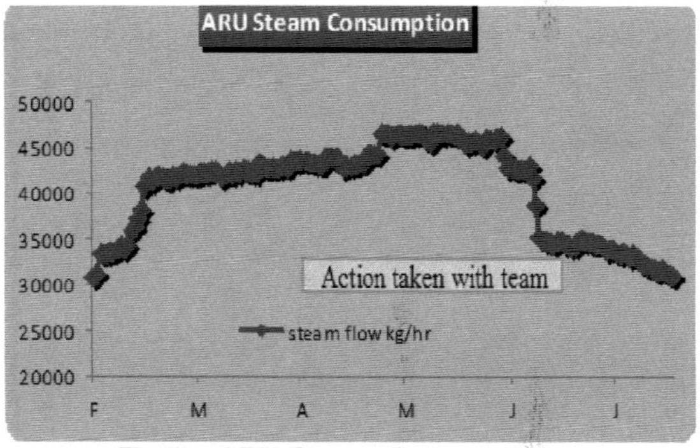

Fig15. ARU Steam Optimization

The Pareto chart can be also used to trouble shooting and restoring operations quickly during any partial steam failures/load shedding.

Various Process maps were developed during the project. Condensate system process map as shown in Fig16 also identified that refinery had excess 50 psig Low pressure steam (3.5 Kg/cm2) and air coolers are used as required to condense the excess steam to avoid venting. LSS project recommended shifting some steam users from medium pressure to low pressure steam. This will help to reduce steam generation from boilers using fuel gas firing and power consumption towards fin fans used to condense the steam.

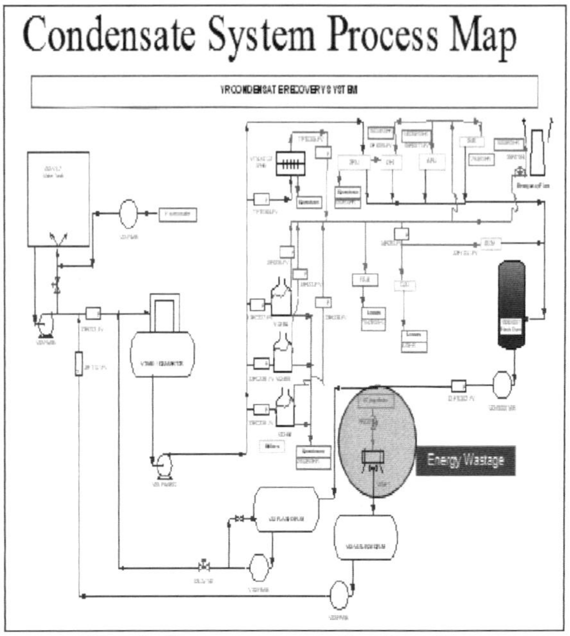

Fig16. Steam consumption Pareto

Save Energy –Save Environment;
Reduction of energy usage also reduces Harmful emissions of Sox,Nox,Co2 as shown in Fig 17.

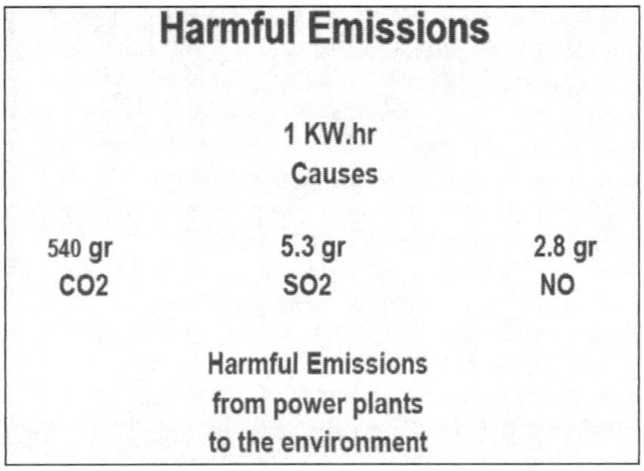

Fig17. Harmful emissions

Recommendations:
Project identified several short and long term recommendations to optimize energy consumption and reduce emissions. Short-term measures implemented include steam optimization, Boiler load management, Flue gas oxygen optimization, inaccurate flow meters identification and correction. Long-term measures include shifting users from Medium to low pressure steam and waste heat recovery from flues gases. Summary of key recommendations is shown in Table 1.

#	Root Cause	Data Output analysis	Recommendation
1	No Boiler management	Correlation proved	Increase load on high efficient boiler and reduce load on low efficient boiler
2	High steam consumption	Pareto chart helped in identifying consumer	Monitor and optimize steam consumption
3	Low condensate recovery	Lean	Enhance condensate recovery
4	High number of defective traps	Lean	Reduce defective traps percentage by forming special team
5	High oxygen in Heater Flue gases (At Upper limit of range 3-5%)	Relation	1. Reduce oxygen in flue gas to Minimum Target levels 2. Redundant analyzer like CO analyzer 3. Auto draft control
6	High Heater Flue gas temperature	Lean	Install heat recovery equipment for CDU heaters
7	YRD is long on LP steam and using fin-fans to condense LP steam.	Lean	Switch consumers from medium to low pressure steam
8	Fuel gas quality variation	Correlations affected	Install Online analyzer to measure and monitor fuel gas heating value

Table 1.LSS Project recommendations

Project Summary
Lean six sigma compliments to engineering solution with speed, quality, and cost simultaneously. Identify defects in major focua areas will yield quick results. Based on project recommendations, Refinery implemented several short-term measures like steam optimization, Boiler load management, Flue gases oxygen optimization. Long-term measures recommended include shifting users from Medium to low pressure steam and waste heat recovery from flues gases. Equipment efficiency, plant reliability play critical role and require continuous monitoring and enhancement. The LSS project implementation improved EII to 77.3 by end of 2011, while it was 78.1 by end of 2010 from 78.9 in year beginning with a cost avoidance of $1.4MM/year. Energy conservation reduces costs and emissions.

Gratitude
Authors wishes to record their sincere regards for the YRD management and OCD of Saudi Aramco for extending their support for successful completion and implementation of project and presenting the paper. Authors also wishes to record sincere appreciation to the 2012 AIChE Spring meeting organizers.

AIChE Paper Number 247037

REDUCING CO_2 EMISSIONS FROM REFINERIES THROUGH AMINE CHEMISTRY UPGRADE

Michelle Valenzuela

Technical Service and Development Manager

Dow Oil & Gas

Ana Eisenhauer

Refinery Strategic Marketing Manager

Dow Oil & Gas

Linda Pirtle

Senior Technical Services and Development Specialist

Dow Oil & Gas

Prepared for Presentation at the 2012 Spring National Meeting

April 3, 2012

Houston, TX

REDUCING CO_2 EMISSIONS FROM REFINERIES THROUGH AMINE CHEMISTRY UPGRADE

Michelle Valenzuela

Technical Service and Development Manager

Dow Oil & Gas

Ana Eisenhauer

Refinery Strategic Marketing Manager

Dow Oil & Gas

Linda Pirtle

Senior Technical Services and Development Specialist

Dow Oil & Gas

Regional initiatives are underway to reduce and limit the emissions of greenhouse gases. With CO_2 emissions making up over 80% of the greenhouse gases, cap-and-trade programs will focus on those industries that consume the most energy. Refineries are among the top energy consumers and are seeking opportunities to reduce usage. With tightening margins, energy management programs will not only help refineries meet CO_2 emission regulations, but can also provide a competitive advantage.

With the trend towards heavier and higher-sulfur containing crudes, refineries are increasing processing capabilities, which can include capital-intensive projects and additional energy consumption. Energy conservation plans should include optimization of these processes. One area to consider includes the acid gas removal systems in refineries.

Through the selection and use of optimal solvents and implementation of energy efficiency techniques, which require minimal capital investment and expenditures, refineries can reduce energy usage, overall CO_2 emissions, and total cost in acid gas systems. This paper will discuss these approaches and share case studies detailing the implementation and results.

Introduction

Despite the lack of a global greenhouse gas (GHG) emission reduction agreement, regional schemes are being developed to limit CO_2 and other GHG emissions to the atmosphere. In the United States, the state of California is moving forward with regulations that will cap emissions from installations producing over 25,000 tons CO_2e/year. Equipment such as boilers and fired-heaters, commonly used in refineries for energy generation, are included in the law and will be obligated to reduce CO_2 emissions.

To reduce CO_2 emissions, refineries often rely on energy efficiency improvements. Three types of energy efficiency projects include:

1) Equipment Efficiency — The replacement of aging assets with high-efficiency technology

2) Operational Efficiency — The use of energy management and operating strategies to optimize energy-intensive assets

3) "Chemical," or Solvent Efficiency — The use of improved chemicals to reduce overall energy requirements of equipment

In reducing total energy consumption, it is also important to consider regulations and overall plant emissions. The trend towards higher sulfur-containing crude products affects energy consumption of refineries. To meet stringent regulations in the global market, refineries must increase processing capabilities to accommodate a heavier, sour crude slate. This includes upgrading hydrotreating capacity to address the additional sulfur load and adding acid gas removal systems to treat the resulting H_2S. The result is overall increased energy demand and CO_2 emissions. The question then becomes, "Which is more important: addressing energy consumption and CO_2 emissions, or environmental regulations and overall plant emissions?" The answer is "both."

Specific to acid gas removal systems, it is possible to reduce overall energy requirements while maintaining environmental compliance with the use of chemical and operational efficiency. Both require minimal capital investment, which is important for an industry already struggling with tightening margins.

Chemical efficiency includes upgrading the chemistry of the solvent to increase capacity and reduce required processing energy (i.e. reboiler duties). Often refineries design and operate acid gas removal systems with the sole objective of meeting permitted sulfur emission limits. Monoethanolamine (MEA) and diethanolamine (DEA) are amine solvents commonly used for removal of H_2S from sour gas streams. However, these solvents degrade to form corrosive products. As a result, refineries maintain low solvent concentrations to reduce degradation products and lessen the corrosion potential. Less amine in the system requires increased circulation rates and more aggressive stripping of the lean amine to meet stringent H_2S and SO_2 regulations. These practices result in excessive energy consumption and, subsequently, excessive CO_2 emissions.

Operational efficiency requires an understanding of the system performance. With the use of mass and energy balances, a plan is developed to optimize appropriate operating conditions to reduce energy consumption while maintaining environmental performance.

This paper will discuss these two approaches in more detail and will share field experiences and learnings from actual implementation of these practices.

Environmental Drivers for Energy Efficiency

Greenhouse gases, which include carbon dioxide, methane, and nitrous oxide, contribute to the global warming effect by trapping radiation on the earth's surface. Global and regional programs have been initiated to reduce the emission of GHG.

The United Nations Framework Convention on Climate Change (UNFCCC) is an international treaty established in 1992 with the intent of stabilizing GHG emissions as a global response to climate change. To further enforce efforts towards emission reduction, the UNFCCC adopted the Kyoto Protocol in 2007, legally binding developed countries to GHG reduction targets through 2012.

The United States signed the Kyoto Protocol but never ratified it. Subsequent legislation to enact a cap-and-trade program to control GHG in the U.S. also failed in 2009. In the absence of a market-based program, the United States Environmental Protection Agency (EPA) issued the Mandatory Reporting of Greenhouse Gases Rule (40 CFR Part 98, Greenhouse Gas Reporting program), requiring facilities that emit 25,000 metric tons or more per year of GHGs to submit annual reports to the EPA. The purpose of the rule is to collect accurate and timely GHG data to inform future policy decisions.

The EPA is currently working on rules to limit the emissions of GHG from large sources under the Clean Air Act but has yet to implement or enforce any regulations. Although a federal mandate has not been set for GHG emission control, several regional initiatives have been established. In 2007, a collaboration of seven U.S. states and four Canadian provinces formed the Western Climate initiative (WCI) to identify, evaluate, and implement policies to address climate change.

Through WCI, states and provinces can implement regional cap-and-trade programs with the intent of integrating the programs into national or federal programs as they are developed. Of the WCI partners, California is the only state that has enacted regulation to implement a GHG emissions cap-and-trade program.

On October 2011, the California Air Resource Board adopted the final cap-and-trade Resolution 11-32. The regulation covers 360 businesses representing 600 facilities and is divided into two phases: the first, beginning in 2013, includes all major industrial sources along with electricity utilities; the second, starting in 2015, brings in distributors of transportation fuels, natural gas and other fuels producing over 25,000 tons CO_2e/year. CO_2 emissions are calculated based on the consumption of fossil fuel by a stationary combustion facility (boiler, generator, fired heater, etc). California refineries will be required to comply with this resolution starting in January 2013. To reduce CO_2 emissions, refineries will need to focus on energy efficiency opportunities.

Opportunities for Energy Management

Although the trend towards energy efficiency seems relatively recent with the latest global and domestic implementation of GHG initiatives, refineries have long been accustomed to energy management and optimization programs as a means of staying competitive. As energy prices increased in the 1970s-1980s, refineries were forced to evaluate plant energy demand to find opportunities to reduce fuel, natural gas, steam, and electricity usage. Since the initiation of energy management in the 1970s, refineries have reduced energy usage by 30%. However, the increasing use of sour crudes, in combination with stricter control and accountability for overall emissions, has led the refining industry to revisit energy goals and strategies.

Refineries must also consider ultra-low-sulfur diesel (ULSD) requirements and environmental regulations when considering energy management plans and goals. Additional hydrotreating capability to remove more sulfur, especially from sour crudes, will result in more sulfur-containing waste or "acid" gases. Further processing of the acid gases is required before final disposition: emission, incineration, combustion, etc. Typically, refineries use amine-treating systems to process waste gases in refineries.

As with the other processing systems in the refinery, amine systems can require a significant amount of energy depending on asset age and condition, acid gas loads, solvent used, and operating parameters. This paper will include energy-reduction techniques for amine systems that use products, technologies, and services in the AMINE MANAGEMENT^{SM} Program from Dow.

Refinery Amine Systems

Refinery amine systems are required to effectively remove acid gases from a variety of gas and/or liquid streams to targeted levels. Most U.S. refineries use amine treating systems (Figure 1) consisting of the following equipment:

- <u>Absorber:</u> Used to "scrub" or remove acid gases from the inlet stream with an amine solvent

- Rich Amine Flash Tank: Used to separate co-absorbed components (i.e. hydrocarbons) from the rich amine
- Lean/Rich Cross Exchanger: Used to heat the rich amine solvent by exchange with the hot, lean amine from the regenerator
- Regenerator (including reboiler): Used to steam strip the amine to remove acid gases before recycling to absorber as lean amine
- Condenser: Used to cool the acid gas stripped from the rich amine; condensed water returns to top of regenerator
- Amine Cooler: Used to cool the lean amine before entering the absorber

Figure 1: Typical Refinery Amine System

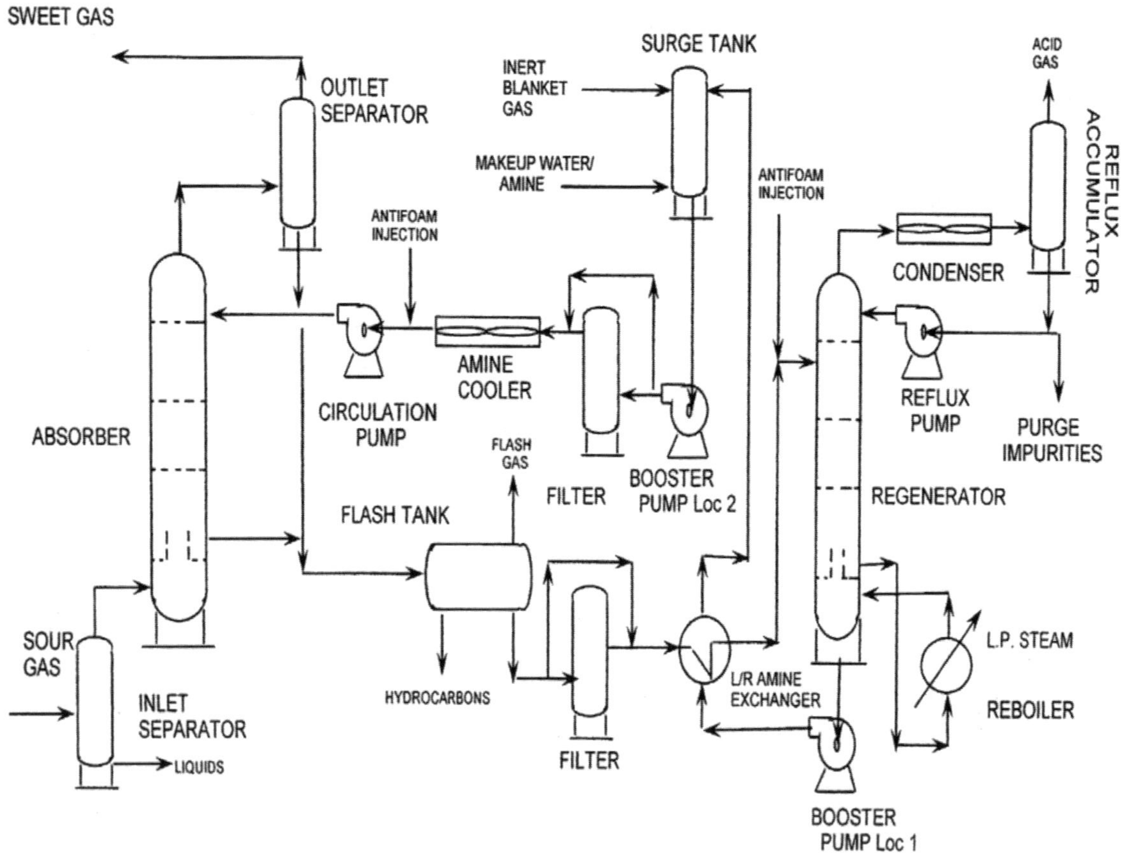

Solvent Selection

The development of amine solvents for acid gas treating has evolved from traditional generic amines, monoethanolamine (MEA), diethanolamine (DEA), and diglycolamine (DGA), to more selective gas treating amines that include diisopropanolamine (DIPA) and methyldiethanolamine (MDEA). The increased selectivity of the amine allows for preferential H_2S removal and CO_2 "slipping" into the treated gas stream, which is especially important for improved sulfur plant operations. MDEA, a tertiary amine, has become the most common selective solvent due to its resistance to degradation, low tendency for corrosion, higher acid gas loading capacity, and lower energy consumption.

The latest advances in amine technology have included the introduction of formulated specialty solvents developed to overcome the weaknesses inherent with generic amines. Proprietary additives are combined with the amine offering the greatest potential for system optimization. Initially, formulated solvents were based on MDEA to provide increased H_2S selectivity and energy savings. However, additional formulations and products have been developed to provide enhanced performance benefits in specific refinery applications. Performance solvents, for example, are now available to achieve the following objectives:

- Treat acid gas to meet <10ppm H_2S specifications in low pressure (i.e. tail gas units) applications while minimizing steam usage
- Reduce solvent losses in LPG service
- Reduce corrosion rates while increasing capacity
- Reduce energy consumption and, in turn, CO_2 emissions from utility generation

Table 1 below compares the overall performance of the gas treating solvents.

Table 1: Typical Amine Solvents Used in Refinery Acid Gas Treating Systems

Amine	Acid Gas Selectivity	Increasing Energy Savings
Formulated Solvent	Very high H_2S Removal	↑
MDEA	High H_2S Removal	
DIPA	Some H_2S Removal	
DGA	Total Acid Gas Removal	
DEA	Total Acid Gas Removal	
MEA	Total Acid Gas Removal	

The energy savings realized with formulated solvents stem from a reduction in sensible heat, a lower heat of reaction, and a reduction in necessary stripping steam. Tertiary amine-based solvents can operate at higher concentrations (up to 50wt% solution) without increased concern for corrosion. Therefore, circulation rates can be decreased, and the sensible heat requirement in the regenerator reboiler is reduced. The heat of reaction for the formulated solvents and tertiary amine is lower than primary and secondary amines. This reduces the amount of energy required to regenerate the solvent. Furthermore, formulated solvents can tolerate higher acid gas loading, thus reducing the stripping steam requirement and demand on the reboiler.

Identifying the optimal solvent is critical when reducing energy consumption. Many factors should be considered to ensure compliance with environmental requirements as well. Nevertheless, even with the best solvent selection, operation of the amine unit is equally important. To maximize the performance potential of the amine solvent, the operation of the amine unit should be optimized.

Amine System Performance and Management

A proactive approach to managing amine systems is required to provide both economic and performance benefits. The AMINE MANAGEMENT^{SM} Program from Dow is a comprehensive service program developed to provide sustainable results that reduce overall energy costs for amine treating systems while also meeting the rapidly changing demands of refinery unit operations.

As part of the program, an initial survey of the refinery amine system(s) is performed to gather key operating data (e.g. flows, temperatures, etc.), as well as equipment data (e.g. absorber and regenerator design data, reboiler duties, etc.). Amine samples are also collected and analyzed. The information collected is used to develop a fundamental understanding of current performance and identify opportunities for improvements. For instance, very low lean loadings (based on the amine analysis) and high overhead regenerator temperatures (based on the operating data collection) could indicate over-circulating or over-stripping of the solvent. This can result in unnecessary energy usage and increased amine losses, which can be remedied by reducing circulation or steam rates.

To complete the amine unit evaluation, technical tools are used to outline an improvement plan to meet refinery objectives. These tools include Dow proprietary simulation platforms, routine analytical analysis, pilot plant capabilities, heat stable amine salt management, and extensive knowledge and experience with more than 800 customers worldwide. An improvement plan is then presented to the refinery, implemented, validated, and sustained through routine visits and system reviews with Dow engineers.

The Use of Simulation Tools

Dow proprietary simulation platforms are used to develop operational solutions to meet energy reduction and/or other targets. State of the art simulation tools provide design and operating information to help achieve the following:

- Optimization of amine flow requirements and treatment results
- Gas treating solvent selection (including formulated solvents)
- Exploration of process limitations
- Process Flow Diagram
- Material balance with stream data tables
- Exchanger duties and other design information

The simulation output provides details on the current system performance, expected benefits from optimized performance, and any limitations of the amine system. An example of a temperature profile output is shown in Figure 2. Under ideal liquid-to-gas (L/G) ratios, the temperature profile of the absorber tower is shown with the "temperature bulge," or exotherm, at the bottom of the tower. However, in an effort to aggressively reduce steam usage, it is possible to reduce the L/G ratio such that the temperature bulge shifts to the middle or top of the absorber tower. This can compromise the treating capability and H_2S removal of the absorber system. When optimizing systems to reduce energy usage, it is important to understand limits of the process. An upset in the amine unit affecting a downstream unit or causing process downtime can quickly minimize the value of a poorly implemented energy savings plan.

Figure 2: Amine Absorber Temperature Profile with Normal L/G

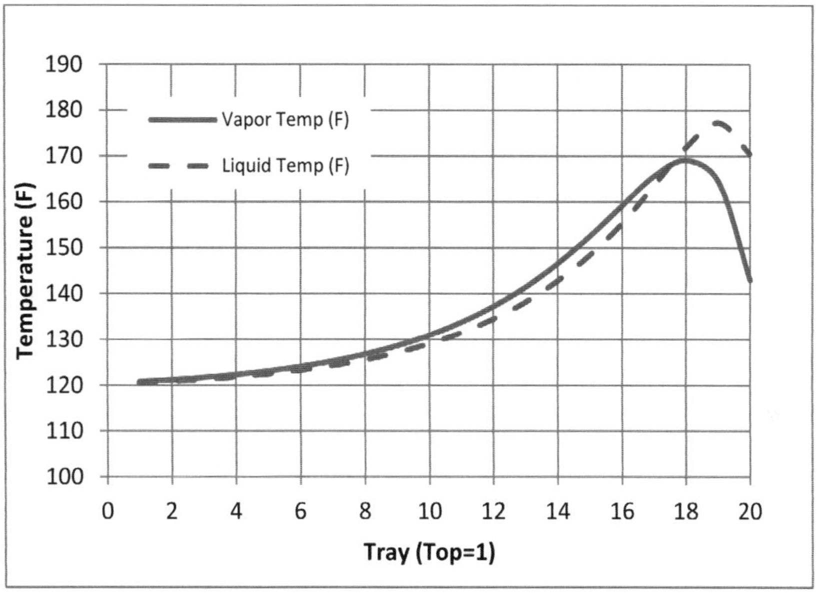

The Use of Pilot Plant Capabilities

In conjunction with the use of simulation tools, pilot-plant results are beneficial in modeling simulation output or plant performance, developing customer specific formulations, and validating operational improvements. As part of the AMINE MANAGEMENT^{SM} Program, pilot plant capabilities have helped:

- Demonstrate lower energy and CO_2 removal with solvent conversion
- Quantify organic sulfur removal performance for new gas treating applications
- Validate CO_2 slip and improve H_2S removal in low pressure applications (i.e. Tail Gas Units)

- Screen new additives and formulations for better acid gas removal with less energy usage

Solvent Hygiene

Often, energy management programs focus on equipment or process optimization. However, in amine systems, routine solvent analysis is important in identifying opportunities for energy savings. Monitoring amine solvent health is required to maintain optimal system reliability and operational performance. Ideally, routine sample analysis should include identification and quantification of all the components in the solvent. This would include amine concentration, acid gas content, individual degradation components, specific contaminants, and water content.

Contaminants in the system, such as formate, thiosulfate, thiocyanate, and other organic and inorganic acids, bind the amine to form Heat Stable Amine Salts (HSAS), which reduce functionality. High HSAS levels reduce system capacity, which can require increased steam usage and/or higher amine circulation rate to maintain system performance. Thus, it is important to monitor and maintain low levels of HSAS in the system.

Acid Gas loadings in the lean amine can serve as an indicator of stripping performance. In refinery systems, H_2S is present and measured as acid gas loading in the amine solvent. The absence of H_2S in the lean amine sample could indicate the amine is over-stripped. Operating data would then be used to confirm regenerator conditions and steam rates could be reduced.

As part of the AMINE MANAGEMENT^{SM} Program, a routine solvent analysis program is developed to monitor and trend solvent health as a means of determining system performance, anticipating and/or troubleshooting problems, and identifying opportunities for system improvement.

Case Studies

The following case studies will show how the implementation of the AMINE MANAGEMENT Program led to significant cost savings for each refinery. In all cases, solvent selection coupled with an optimization plan led to better treating performance, reduced steam consumption, and lower CO_2 emissions. The refineries below continue to use the AMINE MANAGEMENT Program from Dow to proactively identify opportunities for optimization and improved operational performance.

Case Study #1

A large-size, U.S. refinery completed a capacity expansion project to increase heavy crude processing capability. The expansion was successful but added strain on the utilities and downstream assets, increasing energy costs and limiting amine treating capacity. The shift in the crude slate increased the sulfur content and resulting acid gas volumes. Although the refinery was equipped with a complex amine treating system consisting of three regenerator trains and 18 gas and liquid contactors, the current solvent and system conditions were unable to maintain reliable acid gas removal from the process streams. A comprehensive, on-site evaluation of the refinery operations was conducted using the AMINE MANAGEMENT℠ Program, including detailed simulations of the existing amine systems and a comprehensive review of maintenance and operating procedures. The assessment showed that the refinery could achieve the following objectives:

- Address increased acid gas volumes without additional capital expenditure
- Decrease operating costs by reducing energy use
- Increase system reliability by reducing corrosion

The AMINE MANAGEMENT Program proposal recommended a solvent upgrade from the current solvent, DEA, to a high capacity, UCARSOL™ Specialty Solvent formulated for primary amine treating systems in refineries. Recommended operating concentration for DEA is less than 30% due to degradation that increases corrosion; however, the UCARSOL Solvent can target a much higher operating concentration, increasing the capacity available to remove acid gas. With the upgrade to the UCARSOL Solvent, the refinery increased solvent concentration from <30 wt% to an optimal target, allowing for reduced circulation rates and steam usage with the added benefit of reduced corrosion potential.

In addition to the solvent upgrade, process and solvent conditions were optimized to minimize environmental exceedances and improve system reliability. Amine loading targets and steam-to-lean amine (lb/gal) ratios were established to prevent over-stripping. The steam-to-lean ratio is a quick means of determining the amount of steam used to regenerate, or strip, the amine and should not be used as a replacement for reflux ratio. Steam-to-lean ratios were reduced by 20% indicating less steam usage. The lean and rich amine loadings were set at optimal targets allowing for reduced reboiler duty (or increased treating capacity).

When optimizing the system, targets and/or limits should account for occasional process swings (e.g., increase in H_2S content in inlet gas, increasing gas content, equipment fouling, etc). One of the regeneration systems at the refinery was able to aggressively reduce the steam-to-lean ratio by 25%, without affecting treating performance at existing conditions. However, to ensure safe operation and avoid H_2S and downstream SO_2 exceedances, the ratio was set at a conservative target.

The overall steam usage by the amine system regenerators is shown over a four-year period (2002-2006) in Figure 3. The solvent upgrade and optimization resulted in a significant energy reduction – 175,000 MMBTU/y – which translates to approximately $1.7 million/year in cost savings and a reduction of approximately 10,000 tons/year CO_2e emissions. Sulfur Recovery Plant energy usage was reduced by nine percent, and better utilization of the new solvent helped reduce amine losses by 17 percent.

Figure 3: Steam Usage for Refinery Amine Regenerators

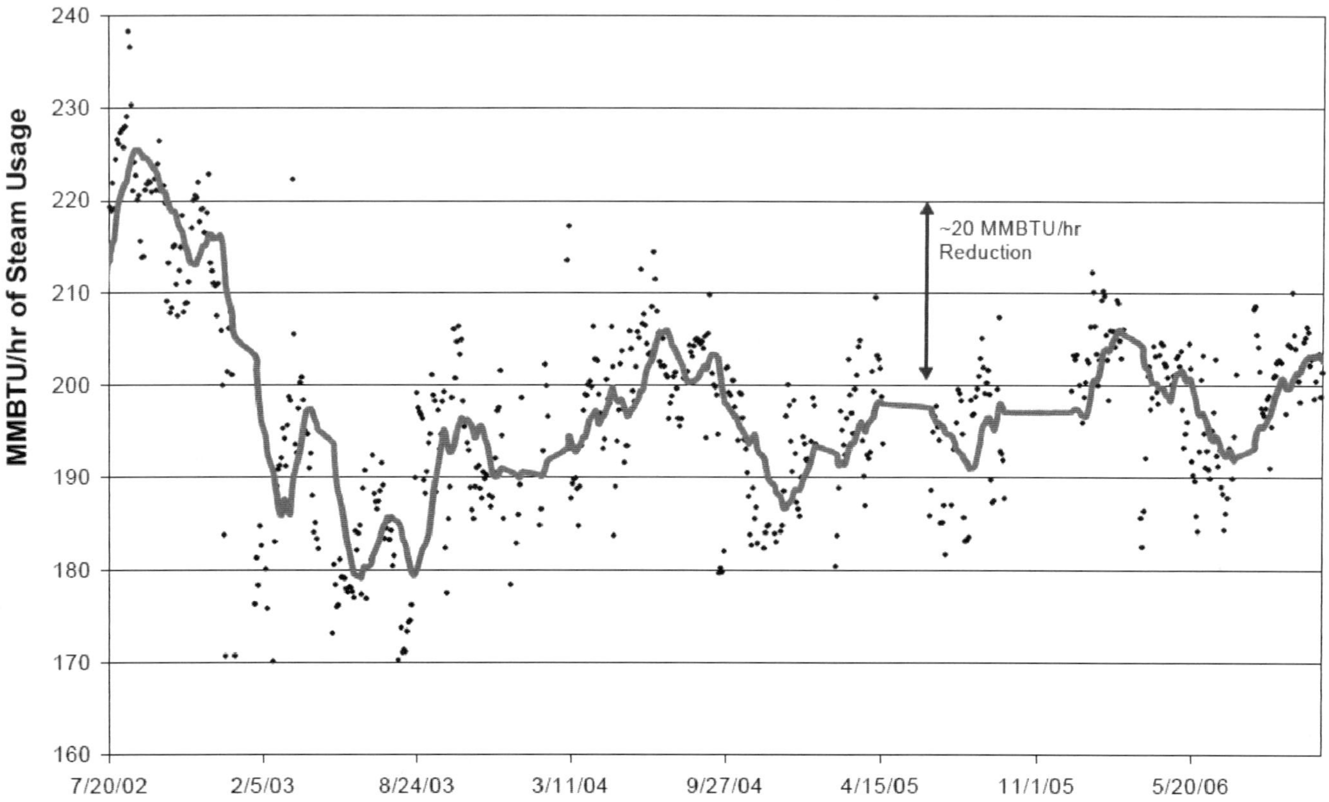

Case Study #2

This case study will show how a systematic approach using the AMINE MANAGEMENT[SM] Program from Dow was able to address the refinery needs. Initially, refinery objectives were focused on identifying a new solvent to enhance processing flexibility; however, through the years, opportunities to reduce solvent losses, enhance treating performance, and reduce energy usage have resulted in over 350,000 MMBTU/y savings and reduced CO_2 emissions.

A mid-size refinery in the U.S. used two generic solvents to treat process gas in the main and tail gas units. However, to increase system reliability and flexibility, the refinery moved to an integrated amine treating system requiring only one solvent for both primary and tail gas systems. The solvent would be required to meet treating needs for the contactors on the main system, including LPG liquid treaters, as well as provide CO_2 slipping and H_2S selectivity for the tail gas application. A comprehensive evaluation of the system was conducted using the AMINE MANAGEMENT Program to meet the following refinery objectives:

- Provide a single amine solvent to be used across the refinery
- Provide enhanced total sulfur removal
- Maintain adequate CO_2 slip, while removing meeting H_2S removal targets

A new specialty solvent was developed to meet the refinery needs. The refinery upgraded to a UCARSOL™ Solvent formulated for both primary and tail gas applications. Hesitant to increase solvent strength, the refinery operated at less than 30 wt% to minimize losses. However, with tools from the AMINE MANAGEMENT℠ Program, losses were tracked to reduce overall losses by 34% (as shown in Figure 4).

Figure 4: Refinery Amine Loss Reduction

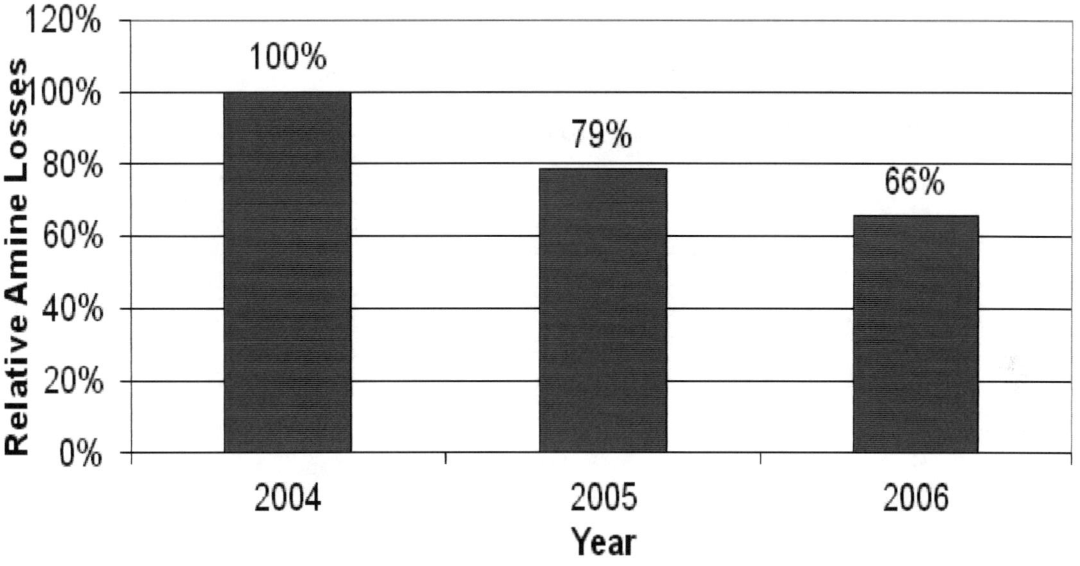

With losses optimized, another comprehensive evaluation by Dow engineers was performed and showed that increasing solvent concentration to greater than 40 wt% would result in a 26% decrease in energy costs, totaling a net savings of more than $1MM/year. Conversely, if energy usage was not a key objective, the concentration upgrade would increase overall treating capacity of the amine system by over 30%. Additional recommendations included reducing steam-to-lean ratios and adding improved process control logic to maintain optimal circulation rates.

The refinery increased the amine strength to 35wt% and implemented the optimization recommendations, which resulted in a reduced amine circulation rate. The lower rate led to a 30% decrease in regenerator duty (shown in Figure 5), and the energy reduction was approximately 40 MMBTU/h. These improvements translate to 350,000 MMBTU/y energy savings, 20,000 tons/y CO_2e emission reduction, and $1.1MM/year in total cost savings.

Figure 5: Refinery Trend in Total Regenerator Reboiler Duty

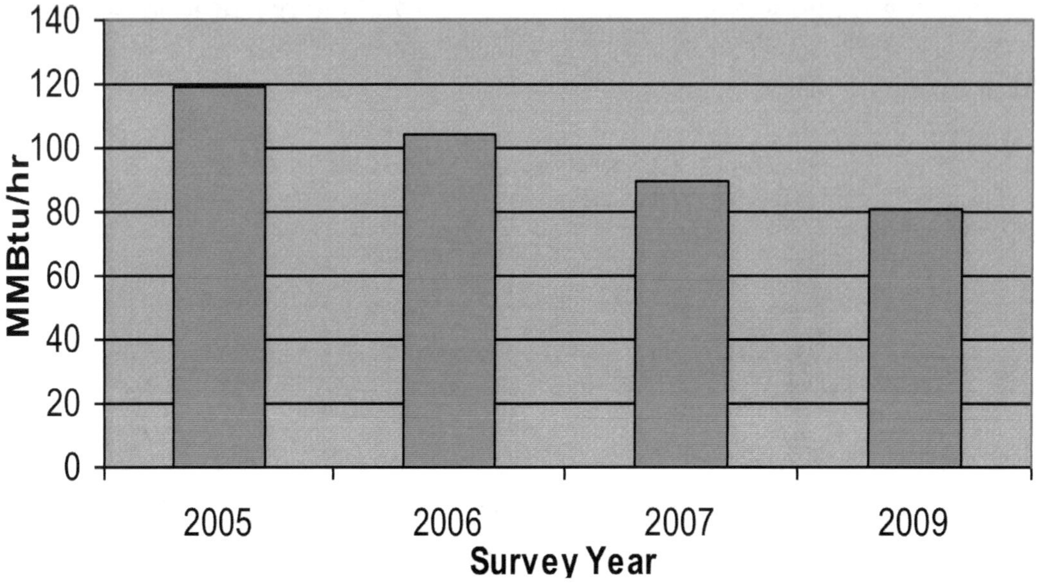

Case Study #3

With the flexibility to process large volumes of heavy crude, a large, U.S. refinery needed to achieve the following objectives to become more competitive:

- Increase gas treating capacity
- Improve system reliability
- Reduce total operating costs

The existing amine treating system used MEA. Although MEA is aggressive in total acid gas removal, the primary amine readily degrades, thus increasing the corrosion potential of the solvent. To minimize corrosion, the amine requires frequent and costly reclamation to remove degradation by-products. Operating concentrations are limited to 20wt%, and regeneration for MEA requires more stripping steam due to a higher heat of reaction as compared to formulated solvents. The use of MEA limited long-term capacity goals for the refinery.

An evaluation of the amine system was performed using the AMINE MANAGEMENT℠ Program. The evaluation determined that a solvent upgrade and optimization would help the refinery increase treating capacity by 48%, reduce amine losses by 50%, and reduce energy consumption by approximately 500,000 MMBTU/y. A two-step approach was developed and implemented to help the refinery achieve outlined objectives.

<u>Phase 1: Solvent Upgrade</u> The refinery upgraded from MEA to a UCARSOL™ Solvent designed to minimize energy usage and meet acid gas removal requirements in refinery applications. With the formulated solvent, the operating concentration was increased to an optimal range, and steam-to-lean ratios were reduced by 20%. Phase 1 of the optimization plan resulted in an immediate circulation rate reduction and an energy savings of over 400,000 MMBTU/y, translating to 22,000 tons CO_2e/y reduced emissions. Furthermore, the upgrade to the UCARSOL™ Solvent occurred with no shutdown or incidents.

A routine analytical plan was implemented to ensure optimal solvent and system performance. Solvent losses were also tracked, monitored, and improved. Overall, solvent health improved, solvent losses were reduced by 59%, and corrosion decreased.

Phase 2: Optimization The next step was to focus on optimization of the amine treating system. The individual gas contactors were evaluated using simulation techniques from the AMINE MANAGEMENT℠ Program. Proprietary equations were developed and implemented to optimize circulation rates, and rich loadings were increased to target an optimal range. This resulted in an additional 15% reduction in circulation rates and 300,000 MMBTU/y in energy savings, with a further reduction of 18,000 tons CO_2e/y.

The two-phase approach led to 700,000 MMBTU/y in energy savings, a reduction of 40,000 tons CO_2e/y emissions, a 59% decrease in solvent losses, and improved system performance with a total cost savings of over $3MM/year.

Summary

Energy conservation as a means of reducing CO_2 emissions and total operating costs is a target objective for refineries. As crude slates shift towards heavier, sour crudes, it is necessary to consider cost-efficient means of increasing processing flexibility while heeding environmental regulatory requirements. Specific to acid gas removal systems, this paper reviewed the implementation of AMINE MANAGEMENT℠ Program tools and techniques to reduce energy consumption and CO_2 emissions, resulting in a reduction in total operating costs.

In the case studies discussed, upgrading from generic solvents to formulated solvents improved overall system performance and increased gas treating capacity, meeting refinery objectives to address process expansion and flexibility needs. With the use of proprietary simulation tools, optimal solvents were selected based on the refineries' needs, and solvent upgrades were safely implemented with no environmental exceedances. Operating limitations were identified and targets for optimization were set specific to acid gas loadings and stripping steam ratios. Routine analytical plans were established to monitor solvent performance and proactively address system issues. Additional capacity gains and energy savings were realized with proprietary equations for individual absorbers and improved system process logic. The successful implementation of these optimization techniques, coupled with the solvent upgrade, led to 1,225,000 MMBTU/y in energy savings, a reduction of 70,000 tons CO2e/year emissions, and nearly $6 MM in operating cost savings combined for the three case studies discussed.

References

Ackman, R, Pirtle, L, "Amine Treating: A Review of the Technology", AICHE Spring National Meeting, April 2012.

"Annual Energy Outlook 2011", Available at http://www.eia.gov/oiaf/aeo/tablebrowser/, February 12, 2012.

Burns, D, Gregory, A, "Waste Minimization and Compliance Benefits To Be Gained From A Good Amine Management Program", NPRA, October 1995.

Echt, W, Wendt, C, "Reduce Sulfur Emissions From Claus Sulfur Recovery Unit Tail Gas Treaters", AICHE Spring National Meeting, March 1993.

"Energy Consumption", Available at www.need.org/needpdf/infobook_activities/IntInfo, February 12, 2012.

"Energy Efficieny Roadmap for Petroleum Refiners of California", California Energy Commission, Prepared by Energetics Incorporated, April 2004.

GPSA Engineering Data Book, Eleventh Edition (1998), Gas Processors Suppliers Association.

"Greenhouse Gas Emissions", Available at http://epa.gov/climatechange/emissions/index.html, February 12, 2012.

Gunn, Matt, "Amine System Upgrade to UCARSOL™ LE-801", Brimstone Symposium, September 2007.

"Kyoto Protocol", Available at http://unfccc.int/kyoto_protocol/items/2830.php, February 12, 2012.

"Manufacturing Energy and Carbon Footprint", Available at http://www1.eere.energy.gov/industry/rd/footprints.html, February 12, 2012.

"Petroleum Refining Industry Profile", Available at http://www1.eere.energy.gov/manufacturing/industries_technologies/petroleumrefining_profile.html, February 12, 2012.

"Refinery CO_2 Management Strategies", Available at http://www.hydrocarbonpublishing.com/ReportP/energy09.php, February 12, 2012.

"Western Climate Initiative", Available at http://www.westernclimateinitiative.org/index.php, February 12, 2012.

(86d) Carbon Capture Ready Heavy Oil Extraction and Upgrading

Eloy Flores III, Chemical Engineering Department, Division of Chemistry and Chemical Engineering, Southwest Research Institute, San Antonio, TX

Carbon dioxide emissions from heavy oil production are typically higher than that of producing conventional crudes. Ultimately, carbon dioxide emissions can make heavy oil production less competitive if regulated. Carbon dioxide emission regulations will affect the overall economics of heavy oil production and upgrading. The additional carbon dioxide gas emissions from heavy oil production, when compared to conventional crude oils, come from the additional energy that is typically needed for heavy oil extraction processes. A heavy oil upgrading process test facility, currently in operation at Southwest Research Institute (SwRI), has the capability to extract and upgrade heavy oil while producing a capture ready concentrated stream of carbon dioxide gas. The heavy oil upgrading process being demonstrated is a thermal cracking process, which is capable of directly providing the energy needed for extraction. Extraction will be done by steam assist gravity drain (SAGD). This process utilizes a fluidized bed process similar to that of a fluid catalytic cracking unit (FCC). Using oxygen combustion in the regeneration step, it is possible to generate a concentrated stream of capture ready carbon dioxide.

This presentation will examine and describe an alternative heavy oil upgrading process, currently being demonstrated, for carbon dioxide capture. It will also evaluate similar traditional heavy oil upgrading processes from literature that utilize oxygen combustion techniques.

(86e) Move to Atomic and Renewable Energy in An Oil Exporting Region?

Ahmed Khogeer, KAUST & Thuwal development Projects, Saudi Aramco, Thuwal, Saudi Arabia and **Hisham Bamufleh**, Chemical Engineering, King AbdulAziz University, Jeddah, Saudi Arabia

Move to Renewable energy in an oil exporting region?

With its abundant resources of oil and gas, and having 25 % of the world oil reserves and low cost of production, it was not the forecasted that Arab gulf countries will consider energy resources other than what they are exporting to the outer world. However, for many reasons, these countries are considering nuclear and renewable energies particularly solar for many residential and industrial uses. In this paper we will cover the reasons, economics, feasibility and some major announced projects in the region.

(86f) Process Integration for Dimethyl Ether Synthesis Via Catalytic Dehydration of Methanol

Mansour Emtir, Process Engineering, Harouge oil operations, Tripoli, Libya, Aymen Elbabour, Chemical Engineering Department, Academy of Graduate Studies, Tripoli, Libya and Elmahboub Edreder, Rsearch & Development, Libyan petroleum institute, Tripoli, Libya

Distillation is the most widely used separation technique in the process industry. Its main disadvantage is the high consumption of thermal energy: the heat requirements of the distillation columns often represent most part of the total energy cost of the plant. The energy saving in distillation is then an essential field of the chemical engineering research since the early 1970's; appropriate integration of the distillation columns with the overall process often results in a significant reduction of the energy consumption in many cases.

In this work the concept of the integration with the overall process has been applied by two integrated design approaches:

- Heat-integrated design; where the distillation system of the process will be heat-integrated with the background process (i.e. integration between the distillation columns and the heat exchangers).
- Reactive distillation design; where the reaction and distillation will be accomplished in the same unit operation.

The process of dimethyl ether (DME) synthesis by the catalytic dehydration of methanol has been considered as a case study.

Heat-integrated design of DME process showed a significant total annual cost saving of 41% compared to the conventional process, and the reactive DME distillation column design showed highest saving in total annual cost by 81%.

In addition to the cost saving of the two proposed designs, also the flue gas emissions have been reduced using process integration methodology, where 51% of the total emissions has been reduced by the heat-integrated design and 86% by the reactive distillation column design.

(87a) The Role of the Intermolecular Potential on the Dynamics of C2H4 Confined in Cylindrical Nanopores

Fernando J. A. L. Cruz[1], Erich A. Müller[2]and José P. B. Mota[1], (1)Requimte/CQFB, Universidade Nova de Lisboa, Caparica, Portugal, (2)Dept. Chem. Eng., Imperial College London, London, United Kingdom

I. Introduction

The confinement of fluids in nanoporous solids can be accompanied by striking effects on the molecular dynamics,[1]which do not necessarily possess an analogy in the bulk phase. Upon confinement, molecules interact with the solid walls to an extent that markedly depends on the chemical nature of the solid and its corresponding pore size. Bhatia*et al.*[2]studied the transport of CH_4adsorbed into armchair single-walled carbon nanotubes (SWCNTs), and concluded that the diffusion coefficient decreases with adsorbate density. The dynamics of $CH_4$and C_2H_6inside a zig-zag SWCNT have been studied by Krishna*et al.*[3]The diffusivity of gases and liquids confined in carbon nanotubes is far from being thoroughly understood and, furthermore, results are sometimes conflicting. We aim herein to understand the effect of the level of detail used in the description of the fluid's energetics, upon the effective observed dynamics. Much can be gained or lost in terms of computational effort and physical detail by coarse-graining (CG) molecular models; Depa*et al.*showed that by CG the intermolecular potential of polymers,[4]the bulk fluid would exhibit faster dynamics than the corresponding atomistic representations, and related this to the enhanced softness of the CG potential. It is the purpose of this work to study the influence of the intermolecular potential on the dynamical behavior of C_2H_4, in a bulk phase, and also confined inside a (16,0) SWCNT ($D_{eff}\approx$ 9.1 Å). This issue has been partially addressed by Mao and Sinnott[5]and Cruz and Müller.[6]We employ molecular dynamics (MD) simulations with five distinct intermolecular potential models, whose level of CG corresponds to different degrees of detailing the C_2H_4molecule; including a fully atomistic representation of ethylene (AA-OPLS), validated by comparison against bulk experimental data,[7]but also simpler models with varying degrees of CG ranging down to a single isotropic Lennard-Jones sphere (1CLJ). The capabilities and drawbacks of each of these different molecular descriptors are explored in a wide range of densities ($\rho\leq$ 15.751 mol/L) and temperatures (220 $\leq T$(K) $\leq$ 340).

II. Results and Discussion

The effect of confinement on the self-diffusion coefficient,D, of ethylene, is observed to be dependent on the particular potential model employed, in a manner which is different than in bulk, where the calculated$D$values are rather independent of the potential used. As $C_2H_4$is confined, differences between molecular potentials are enhanced resulting in discrepancies in the observed self-diffusivities. It is generally argued that coarse-graining intermolecular potentials results in increased self-diffusivities. In the present case, if the AA-OPLS model is used as benchmark, it becomes clear that the effect of CG is unpredictable, and an appropriate and accurate parameterization may faithfully reproduce self-diffusion data. Furthermore, the existence of explicit electrostatic details in the potential seems to induce short-range order in the confined fluid, resulting in a slowing down of molecular mobility. The radial distribution function for the simple 1CLJ potential exhibits a sharp peak at$r$= 4.73 Å and a second (less intense) long-range peak located at$r$= 8.73 Å, indicating that molecules are densely packed together with almost no free volume between them ($\sigma\approx$ 4.6 Å). When the fluid model is increasingly refined, a different picture is obtained. If the C_2H_4molecule is described with the TraPPE potential, not only does the initial$g(r)$peak becomes much broader, with an half-width length of$\Delta r\approx$ 1.95 Å, but it also appears closer at$r$= 4.48 Å; the second peak observed in the 1CLJ model is no longer clearly observable. If electrostatics are now explicitly added to the fluid molecules, via atomic partial charges (AA-OPLS), a further refinement of the initial peak takes place.

The r= 4.48 Å peak of the TraPPE potential essentially gets split into two contributions, r= 4.03 Å and r= 4.78 Å, suggesting the insurgence of a very short-range ordered arrangement of molecules in the AA-OPLS model, that is absent in the other potentials.

In spite of the marked anisotropy introduced by the solid, both the bulk fluid and anisotropic systems data are observed to roughly collapse onto a same master curve, in this work described as a simple power-law dependency of the self-diffusion coefficient with molecular density, $D = D_0\rho^{\lambda}$ ($-1.269 \leq \lambda \leq -0.943$). The isochors show that the confined fluid response to temperature is similar, either above or bellow the critical value, T_c=282.35 K, and in all cases obeys the Arrhenius law, $D = A\exp(-E_a/RT)$. At constant density, it was observed[7] that the self-diffusion coefficients are remarkably dependent of the particular potential model, D(1CLJ) < D(AA-OPLS) < D(TraPPE). From the Arrhenius fits, it becomes clear that for each potential studied, regardless of its own specificities and degree of molecular detail, the activation energy (E_a/R) decreases monotonically with increasing molecular density. Because molecules in the dense fluid are highly packed, with little free volume between them, exchange of momentum via molecular collisions is highly efficient, thus dramatically increasing the self-diffusional process. We are unaware of experimental work reporting self-diffusion data for confined ethylene. Nonetheless, in order to probe the simultaneous effect of confinement and T on the fluid dynamics, we compare our results with bulk experimental data[8] using the AA-OPLS model as benchmark. It then becomes clear that, when the solid is saturated with fluid (15.024 mol/L), the D coefficients show a similar response to applied temperature, both in bulk and under confinement. The activation energies obtained from the Arrhenius plots, $(E_a/R)^{bulk}$= 214.12 K and $(E_a/R)^{SWCNT}$= 242.85 K, suggest that confinement exerts a major influence on the absolute values of D, but not so much on their intrinsic dependence on temperature. To a reasonable approximation, the self-diffusivity differences between the bulk and confined phases are roughly constant over the entire temperature domain, $(D^{bulk}/D^{SWCNT})^{[220K\text{-}340K]}$= 2.52 ± 0.13. A totally different picture results when density is lowered, and the existence of free volume becomes a relevant parameter. For the isochor at 5.331 mol/L, free volume inside the solid is more than 50 % of total available volume, and therefore a sharp discrepancy can be observed in the slope of the corresponding bulk system (5 mol/L), leading to activation energies that are quite different between the bulk and confined fluid, $(E_a/R)^{bulk}$= 301.64 K and $(E_a/R)^{SWCNT}$= 533.96 K. For those lower densities, the self-diffusivities ratio increases dramatically to $(D^{bulk}/D^{SWCNT})^{[220K\text{-}340K]}$= 10.2 ± 0.5. One can extrapolate the previous observations to a very dilute fluid, and postulate that as the free volume inside the solid increases, so do the differences in the corresponding activation energies between the bulk and confined fluid.

References

[1] Gelb *et al.*, Rep. Prog. Phys. **62**, 1573 (1999). Bhatia, Ads. Sci. Technol. **24**, 101 (2006). Coasne *et al.* Mol. Phys. **104**, 3491 (2006). Bhatia *et al.* Phys. Rev. Let. **100**, 236103 (2008). Gubbins *et al.*, Phys. Chem. Chem. Phys. **13**, 58 (2011).

[2] S. K. Bhatia, H. Chen, D. S. Sholl, Mol. Sim. **31**, 643 (2005).

[3] R. Krishna, J. M. van Baten, Fluid Phase Eq. **45**, 2084 (2006).

[4] P. Depa, C. Chen, J. K. Maranas, J. Chem. Phys. **134**, 014903 (2011).

[5] Z. Mao, S. B. Sinnott, Phys. Rev. Lett. **89**, 278301 (2002).

[6] F. J. A. L. Cruz, E. A. Müller, Adsorption **15**, 13 (2009).

[7] F. J. A. L. Cruz, E. A. Müller, J. P. B. Mota, RSC Advances **1**, 270 (2011).

[8] B. Arends, K. O. Prins, N. J. Trappeniers, Physica A **107**, 307 (1981). E. S. Baker, D. R. Brown, J. Jonas, J. Phys. Chem. **88**, 5425 (1984).

(87b) Evaluation of Molecular Dynamics Simulation of Polyaniline Nanostructure Film with the Experiments

Alireza Bahramian III, Department of chemical engineering, Hamedan university of technology, Hamedan, Iran

ABSTRACT: In order to develop predictive atomistic models in nanostructure films, there is an ongoing search for experimental work and molecular dynamics simulation. Accordingly, experiments work was performed on the nanostructure films of polyaniline that coated on TiO_2surface prepared with dip-coating method over a broad range of temperatures. The atomistic model of the polymers that performed in the Molecular dynamics (MD) simulation was generated using energy minimization, equilibrium molecular dynamics simulations, and a polymer-consistent (PC) force field functional form. The various morphological properties of the polymer such as the specific volume and their densities are computed. The results of simulations provide a reasonable compatibility with atomic force microscopy (AFM) and scanning electron microscopy (SEM) images due to the morphology and surface structure of Pani-Cl prepared films.

(87c) Predicting the Properties of CO2-Natural Gas Mixtures for CCSS Projects

Adel Elsharkawy, Petroleum Engineering, Kuwait University, Kuwait, Kuwait

There is mounting evidence that rapid global warming is caused by CO2 emission into atmosphere. There is a consensus among nations that CO2 emission to the atmosphere must be controlled via creating economically viable technologies for capturing, separation and long term storage of CO2 (CCSS). One of many possible proposed sequestrations is the injection of CO2 into depleted oil and gas fields. However for planning such CO2 sequestration projects, accurate estimation of CO2-natural gas properties is required.

In this study a large data bank containing physical properties of various mixtures of CO2-natural gas properties at various temperatures and pressures has been collected. The data bank has been used to assess the accuracy of the industry standard methods as well as the recently published ones. The study indicated that these methods have series errors when they applied at high pressures, high temperatures, or at high content of CO2.

A new method is proposed in this study to calculate compressibility factor and density of CO2-natural gas mixtures. It is capable of estimating the CO2- natural gas mixtures properties at various temperatures and pressures and for any level of CO2 content. The new method is useful for designing pipeline, compression and storage of CO2 in CCSS projects.

(87d) Catalytic Oxidation of Methyl Bromide Fumigation Emissions

Chia-Ying Chenand Joseph Pignatello, Department of Environmental Sciences, The Connecticut Agricultural Experiment Station, New Haven, CT

Methyl bromide (MeBr) is widely used for quarantine and pre-shipment (QPS) chamber fumigations. Typically, spent MeBr is released to the atmosphere where it is recognized to be a potent ozone depleting compound. While most agricultural uses of MeBr have been banned in the U. S. for this environmental concern, MeBr as a QPS fumigant has been granted a critical-use exemption under the Montreal Protocol because it is the only fumigant proven to achieve insect pest control at probit 9 mortality levels (> 99.9986%). Preservation of the current critical-use exemption in the face of continuing regulatory scrutiny will require methods to destroy or capture and reuse spent MeBr. One option for destroying organic vapors is combustion. The temperature of combustion can be greatly lowered by employing suitable nano-structured catalysts. A goal of this research is to design a catalytic system for the elimination of atmospheric emissions of methyl bromide following QPS fumigation. A number of catalysts have been synthesized by wet impregnation (e.g., Pd/CeO_2-Al_2O_3) and core-shell self-assembly (e.g., Pd@CeO_2-Al_2O_3) for testing and determination of products and their yields. The results of catalyst performance so far indicate that Pt/CeO_2-$Al_2O_3$and Pd/CeO_2-Al_2O_3nanoparticles give > 92 % and > 97 % conversion of methyl bromide, respectively, at $350^{o}C$ and are superior to the core-shell catalysts. Hydrobromic acid is a product and the reaction requires O_2but not moisture, indicating catalyzed oxidation, not hydrolysis. Catalyst activity is unchanged after at least three cycles. The results show promise in regard to the use of low temperature oxidation for the effective treatment of QPS emissions of MeBr.

(87e) On Evaluation, Prevention and Repair of Microbial Acid-Produced Attack of Concrete

Kal Renganathan Sharma, Department of Chemical Engineering, Prairie View A & M University, Prairie View, TX and Ramlingam Radhakrishnan, Civil & Environmental Engineering, Prairie View A & M University, Prairie View, TX

The damage due to corrosion of concrete in the United States is estimated to be*$276 billion*. Microbial Influenced Corrosion, MIC has been found to be responsible for 50% of the total cost of corrosion. The microorganisms responsible for the corrosion may be fungi, algae, sulfur oxidizing and nitrifying bacteria. Different mechanisms are involved in the formation of corroded layer in the concrete pillars in the water bodies in the interstate highways in Texas. The microorganisms may grow in the water phase and generate acid that may diffuse into the solid phase through an interface and react with cementicious portion in the concrete. By another mechanism, the microorganisms may directly grow in the solid phase using cementicious portion of the concrete as nutrients. Carbonation reactions between carbon dioxide and calcium hydroxide forming calcium carbonate may lower the alkalinity of the cement accelerating other microbial reactions that are favored in an acidic environment. 16 single stranded ribosomal riboxynucleic acid, 16S rRNA sequence distribution of the microorganisms responsible for the corroded concrete can be obtained using microarray sequence measurements (Sharma, 2008). The corrosion and organisms both are proposed to be characterized using sequence alignment algorithms and bioinformatics databases and SEM, scanning electron microscopy. Mathematical models that take into account the Michael Menten kinetics, Fick's diffusion and generalized Fick's diffusion and the liquid phase and solid phase reactions have been developed. Damped wave diffusion effects are also studied. Field corrosion rates are compared between experiment and theory. Simulation studies of biodeterioration on fresh concrete are presented. Remedial actions are provided to delay or avoid corrosion of concrete piers that serve as substructure in bridges.

The deterioration of concrete is by diverse mechanisms (1). A better understanding of the mechanisms that cause the damage to concrete structures is needed. After the damage, the effect of different mechanisms on the damage needs to be delineated from each other. The microorganisms that colonize the concrete needs to be identified and characterized using ribosomal riboxynucleic acid, rRNA*sequencing studies.*The reactions that participate in the corrosion needs to be mapped out. The mechanism can be one or combination of several such as: i) microorganism such as algae, fungi, nitrifying bacteria and sulfur oxidizing bacteria which may generate acids in the water phase and the acid will diffuse and attack the cementicious portion of the concrete by*leaching*; ii) chlorine ions in the salt present in water may diffuse and attack the steel portion of the concrete by a process called salt induced corrosion; iii) the carbon dioxide, CO_2, participates in the carbonication of the cementicious portion of the concrete decreasing the alkalinity of the concrete from 13 to a level where bacteria that thrive in a acid environment can act upon, thus microbial succession is not clear by a process called carbonation. By carbonation

CO_2+ $Ca(OH)_2$----------à $CaCO_3$

; iv) the microorganism may grown in the cementicious portion of the concrete and excrete the acid which in turns further attacks the cementicious portion of the concrete. The role of each of these mechanisms in the said locations of the concrete structures in the state of Texas needs to be*better evaluated.*

Who is impacted by the problem ?

The concrete*industry*is impacted by the problem. The size of the concrete industry is US $35 billion. In 2006, over 7 $km^3$of concrete are used every year. Concrete industry employs more than 2 million workers in US. Over 55,000 miles of highways in the nation are paved with concrete. Concrete word originates from the Latin*concretus*that means hardened. By a process of*hydration*concrete solidifies and hardens upon mixing with water. A stone like material is created after reactions of cement with water and bond formation between various components of the concrete. Among other things, concrete is used to make motorways/roads and bridges. Concrete is the most used man-made material.

Significance/Scope of the problem

Approximately US $20 billion are needed for bridge repairs in the US Interstate highways because of '*salt induced corrosion*'. Per a recent survey, damage due to corrosion in the United States is estimated at*$276 billion*. Surveys in United Kingdom, Japan, Australia and Germany place the cost of corrosion to be 1-5% of the gross domestic product, GDP. MIC has been found to be responsible for 50% of the total cost of corrosion. Corrosion of reinforced concrete and pre-stressed and other concrete structures is a multi-billion dollar problem in US and other countries. A parking garage collapsed in Minnesota and the damage was attributed to de-icing salts. Another spectacular failure was the collapse of the Berlin congress hall (32).

Per a 1997 report, amongst 581,862 bridges in and off the US federal-aid-system about 101,518 bridges were rated structurally deficient. The repair costs can range from US $78 - $112 billion. By 2011 the annual maintenance cost of bridges would be US $ 5.2 billion. The US department of commerce census bureau states that the dollar impact of corrosion on highway bridges is considerable. The annual direct cost of corrosion on highway bridges is considerable. The annual direct cost of corrosion on highway bridges is estimated to be US $6.43 - $10.15 billion. The break-up of the total cost is as follows: i) about US $ 3.79 billion is needed to replace structurally deficient bridges over the next decade; ii) US $ 1.07-2.93 billion is needed for maintenance and cost of capital for concrete deck bridges; iii) US $0.5 billion is needed in painting cost of corrosion.

Per the Federal Highway Administration about 30% of the nation's bridges are structurally deficient or functionally obsolete. Chloride induced corrosion of the reinforcing steel has been attributed as a cause of premature rehabilitation of the bridge structures. Better service life models are needed. This allows planners to determine the remaining time to first repair and subsequently rehabilitate a given bridge or set of brides and evaluate the effectiveness of the various protection methods.

(87f) Controlling the Selectivity of Styrene Oxidation Catalyzed by Atomically Precise Aun Nanoclusters

Yan Zhu, Low Carbon Conversion Center, Shanghai Advanced Research Institute, shanghai, China

Controlling the Selectivity of Styrene Oxidation Catalyzed by Atomically Precise Au_nNanoclusters

Yan Zhu[1,2]*,andYuhan Sun[1]

[1]Shanghai Advanced Research Institute, Chinese Acadamy of Sciences, Shanghai , China

[2]Department of Chemistry, Carnegie Mellon Univeristy, Pittsburgh, USA

E-mail:zhuy@sari.ac.cn

The major limitations of conventional nanocatalysts, such as the inherent size polydispersity and precise size control at the atomic level, preclude fundamental investigations on the precise structure-catalytic activity relationships. Therefore, it isofattain atomically precise metal nanoparticles and use such nanoparticles as well-defined catalysts. By solving their atomic structure of the nanoclusters, one will be able to precisely correlate the catalytic properties with the exact atomic structure of the nanoclusters and to learn what control the surface activation, surface active site structure and catalytic mechanism. By learning these fundamental principles, one will ultimately be able to design new types of highly active and highly selective nanocatalysts for a variety of catalytic processes.

Au_n(n=number of gold atom in cluster are ideally composed of an*exact* number of gold atoms, e.g.,*n*ranging from a dozen to hundreds) nanoclusters demonstrate the huge power of*atomically precise*Au_nnanoclusters for achieving super catalytic performance and atomically precise structure-property relationships, and achieving thecontrollable$_n$ nanoclusters are utilized to catalyze the partial oxidation of styrene as an example. The partial oxidation of styrene typically produces benzaldehyde and epoxide as major products. Herein, we found a strategy for controlling the catalytic selectivity by exploring the kinetic factor and the promotion effect of acetonitrile. High selectivity for benzaldeyde (~100%) under one set of conditions and for epoxide (>95%) under the other set of conditions has been respectively attained by tuning the site-specific catalytic properties of gold nanoclusters. Our present study not only provides new opportunities for unraveling gold nanocatalysis at the atomic level, but also benefit the future design of new catalysts for certain chemical processes to obtain the desired products with high selectivity.

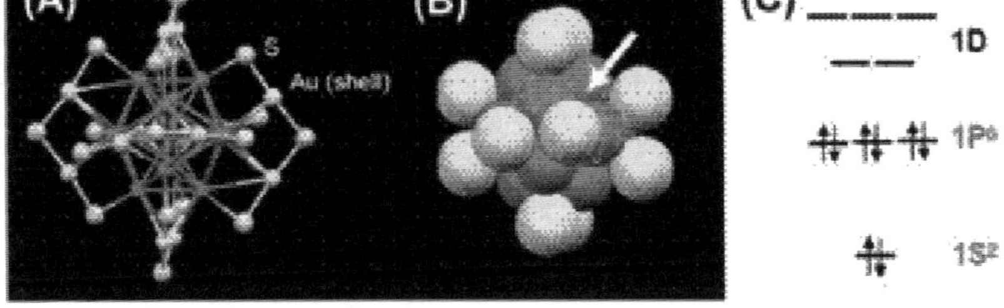

*Figure.*The geometric structure of $Au_{25}S_{18}$(the CH_2CH_2Ph moiety is omitted for clarity); (B) ball model; (C) electronic structures of $Au_{25}S_{18}$ nanoclusters. Color labels: pink for Au_{13}core atoms, blue for Au_{12}shell atoms, yellow for S.

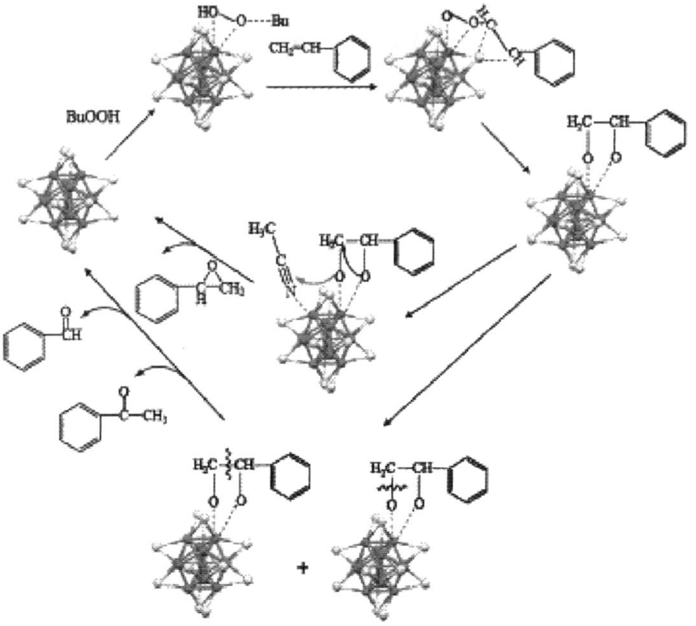

*Scheme.*Proposed mechanism for the epoxidation of styrene catalyzed by Aunanoclusters with TBHP as oxidant and CH_3CN as a promoter. (Magenta: Au_{13}core atoms, Cyan: Au_{12}shell atoms, the thiolate ligands are omitted for clarity.)

Acknowledgements

We thank financial support from CMUand

References

Zhu, M. Z.; Aikens, C. M.; Hollander, F. J.; Schatz, G. C.; Jin, R. C.*J. Am. Chem. Soc.*2008, 130, 5883.

Qian, H. F.;Eckenhoff, W. T.; Zhu, Y.; Pintauer, T.; Jin, R. C.J. Am. Chem. Soc.2010,

(106a) Synthesis of PVA Based Membranes Modified by Benzimidazole Group for Direct Methanol Fuel Cells

Pratima Gajbhiye, Chemical Engineering, Indian Institute of Technology, Kanpur, State of Uttar Pradesh, India, Jayant K. Singh, Chemical Engineering, Indian Institute of Technology, Kanpur, Ka, India and Anil Kumar, Department of Chemical Engineering, Indian Institute of Technology Kanpur, Kanpur, India

Abstract

In recent years, there has been a great effort to look for alternate membrane material and Polybenzimidazole (PBI) membranes have been synthesized where the conducting moiety of this polymer is benzimidazolium groups. The monomers used to synthesize these polymers are bis conjugated amines and diacids are expensive. In this work, we have synthesized grafted polymers of PVA as membrane materials where the grafts are either Poly Acrylic Acid, Poly maleic anhydride (PMA) or Poly Methacrylic acid (PMAA) and these materials were denoted by PVA-g-PAA, PVA-g-PMA, and PVA-g-PMAA. We then synthesized C1, C3 & C4 compounds (formula given at the bottom of Table S1) having bis conjugated amines whereas compound C2 having only one conjugated amine group. One of the steps in synthesizing C1 compound is nitration of phenyl ring, where there is a possibility of formation of two isomers out of which only one of them has strictly bis conjugated amines and can be polymerized with Isophthalic acid giving the formation of PBI. The other byproduct formed is an impurity and is minimized by carrying out the nitration step at low temperature but this cannot be completely removed from the product which is a mixture. Table S1 presents the summary of our experimental results on characterization of membrane, In the C1 to C4 compounds and in some cases (membranes 9, 10, 11), Sulfocuccinic acid were added to the casting polymer solution as additional crosslinking agent in order to achieve higher conductivity. We added the polymer, thus formed, to the casting solution, we first compared the results of the Fuel cell formed by adding C1 (C3 and C4) compound alone and heating at 180°C for some time. Table S1 has two sets of results, one on characterization (columns 1, 2, 3,4,5,6, and 7) of membranes while the other on Fuel cell (columns 8 to 11). For the latter columns 10 and 11 gave results on characterization on Fuel cell while columns 8 and 9 gave results when the Fuel cell was run on control mode for higher efficiency.

Experiments on membrane pore sizes are given in column 1 which lies between 8.88 to 9.52nm and they all are ultra filtration type. The values of the pore size are mostly governed by the nature of the graft chains, e.g. membranes 1,3,4,7 and 12 consist of PAA grafts and were found to have similar pore sizes. Similarly membranes 2,5,8,9 were of the similar pore size because the PVA have the same grafts of PMA. The next important thing is to determine the wall charges of these membranes and they were estimated by experimentally determining the rejection of Chromic (VI) acid solution through these membranes. Subsequently, a space charge model was proposed for the flow of the solution through their pores and a semi analytical solution was derived for the overall problem of separation of chromic acid solution. We observed that the high surface charge was because of the benzimidazolium groups present on the surface and they ranged from minimum of 465 to maximum of 525mV. The thermal degradation (TGA) experiments were carried out which showed that, through the modification of the polymer by benzimidazole groups, lead to stable membrane and the degradation temperature varies from 150°C for 10% weight loss and for approximately 80% weight reduction, a minimum of 300°C to 700°C.

The twelve membranes reported in Table S1 were first characterized completely in terms of pore size, wall charge, conductivity, IEC and thermal degradation. In fuel cell applications, higher temperatures give large current density and it is observed that some of the membranes (rows 1, 2, 5, 6, 7, 8 and 12) could easily be operated upto 150°C without the membrane degrading. However in our case, we insisted that the membrane of direct methanol fuel cell were operated at 90°C for which most of the fuel cell results for Nafion 117 are available. The catalysts used on the anode and cathode side were Pt-Ru and Pt loaded on carbon respectively, which were synthesized by following procedure given in literature. We find that in each case, the power density and the current density were higher and for some of our results (membranes 1,6,12) were atleast two to three times higher than that of Nafion 117 as seen in column 10 and 11.

It was recognized that the membrane material has hydrophilic PVA matrix and hydrophobic benzimidazolium groups at the surface formed by reaction of the C1 to C4 compounds with grafts (see Fig.S-1). When the reaction occurs (confirmed by IR and NMR), the hydrophobic benzimidazolium are formed which tends to separate as confirmed by the SEM results. In view of this we assumed that the pores of the membrane would have a thin hydrophobic layer with positive charge walls. This is confirmed by the contact angle measurements where the minimum contact angle is 51° and varied upto 89°. Even though most of the polymer matrix is PVA with its grafts, when the ions move through the pores (giving rise to current in Fuel cell), the pore walls get highly charged (as measured by column 2 of Table S1). Some of the membranes have Sulfo succinic acid SSA as part of the matrix (e.g. the membrane 9 to 11) which would tend to give sulfonate groups inside the PVA matrix as well as at the phase boundaries, this way giving rise to bipolar nature of the membrane (benzimidazolium group being strictly cationic in nature). In addition to this, some of the compounds such as C1 and C2 compounds have bonded sulfonate groups which would give the positive and negative charge separated by the molecular distance, once again producing a bipolar nature to the membrane. In remainder of the cases the charges would be induced on the phase boundaries as seen in Fig.S-1. The conductivity observed in our membranes is very high and could be attributed to the special bipolar structure of our polymers.

Table S1. Fuel cell and Membrane Characterization

Compound	Membrane	1	2	3	4	5	6	7	8	9	10	11

			MWCO (pore size, nm)	Wall Charge	Conductivity (S/cm)	IEC	Thermal Degradation temp (°C) 10% - 80%		Water Uptake	Contact Angle	Average Voltage Gain (V)	Average time cycle (min)	Power Density (mW/cm)	Current Density (mA/cm)
C1	1	PVA-g-PAA-C1	9.1	488	0.04	2.7	199.732	268.70	7.5	72	0.523	55.75	50.47	74.88
	2	PVA-g-PMA-C1	9.35	465	1.23×10^{-2}	1.8	142.659	447.6	10.691	74.917	0.568	42.896	0.305	3.81
	3	PVA-g-PAA-modified	9.52	488	1.51×10^{-2}	2.12	115.123	706.53 (28.93%)	13.665	75.581	0.778	69.85	37.37	64.44
C2	4	PVA-g-PAA-C2	9.52	525	6.59×10^{-2}	2.38	119.558	691.93	13.4	51.2	0.704	39.83	23.0020	50.55
	5	PVA-g-PMA-C2	9.36	519	1.04×10^{-2}	2.17	181.59	562.348	14.9	50.24	0.77	161.93	34.72	62.11
	6	PVA-g-PMAA-C2	8.99	482	1.16×10^{-2}	2.18	201.055	472.568	13.8	51.8	0.6066	147.81	76.72	92.33
C3	7	PVA-g-PAA-C3	9.0	512	2.51×10^{-2}	2.23	204.908	453.03	8.624	70.368	0.6883	82.55	36.397	75
	8	PVA-g-PMA-C3	9.5	510	3.51×10^{-2}	2.27	166.819	505.293	13.513	62.944	0.726	37.557	36.11	64.381
	9	PVA-g-PAA-SSA-C3	9.36	497	1.71×10^{-2}	2.51	154.914	813.769 (24.32%)	10.5	54.8	0.8096	99.423	23.45	57.778
	10	PVA-g-PMA-SSA-C3	9.83	480	2.28×10^{-2}	2.46	115.12	428.713	8.5	76.2	0.663	224.22	28.561	58.358
	11	PVA-g-PMAA-SSA-C3	8.88	525	3.66×10^{-2}	1.85	139.772	681.715 (30.78%)	7.2	89.2	0.77	18.36	20.832	48.11
C4	12	PVA-g-PAA-C4	9.5	495	2.56×10^{-2}	1.62	189.87	451.86	8.927	59.132	0.704	74.19	41.1	71
	13	Nafion 117			0.1								21.01	48.33

Compound C1, [Bis-3-Amino-4-{3-(triethylammoniumsulfonato) phenylamino}-phenyl sulfone] hydrochloride

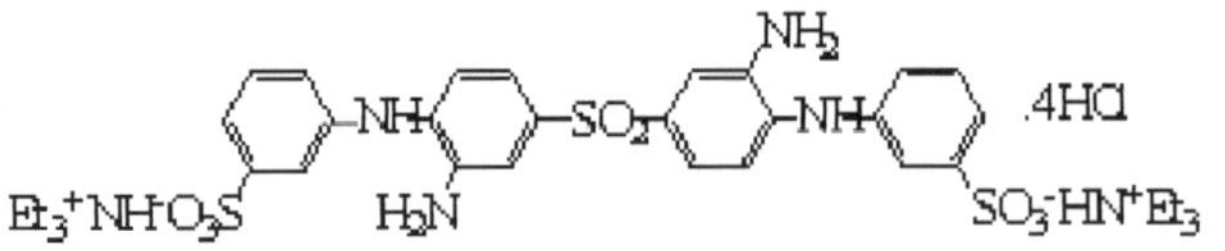

Compound C2, 3-Amino -4-[3-(triethyl ammonium sulfonato) phenyl amino] phenylene hydrochloride

Compound C3, Bis -4 [1, 2 diamino phenyl]-phenyl sulfone) hydrochloride

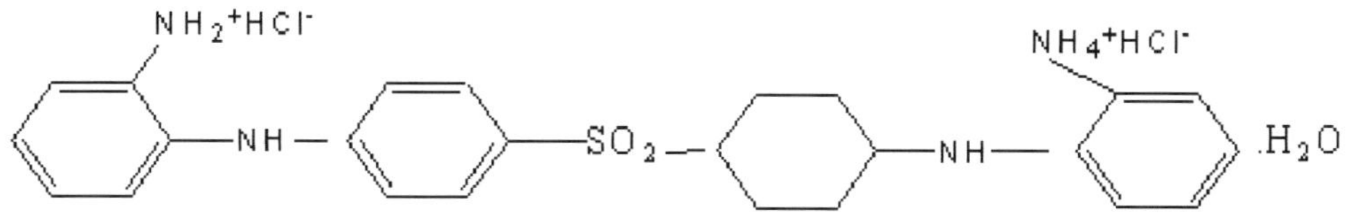

Compound C4, 1, 4 [1, 2 diamino phenyl] Phenylene hydrochloride

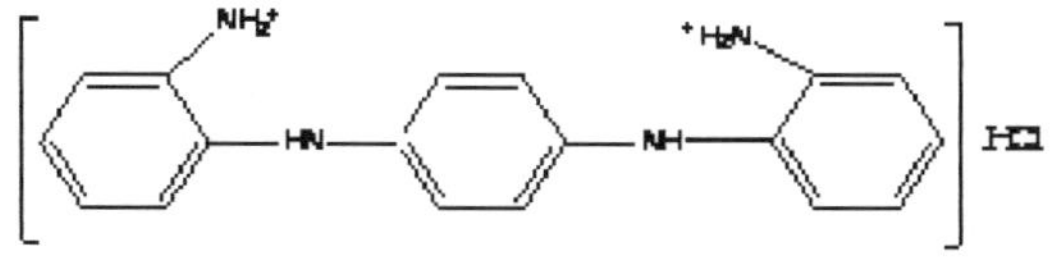

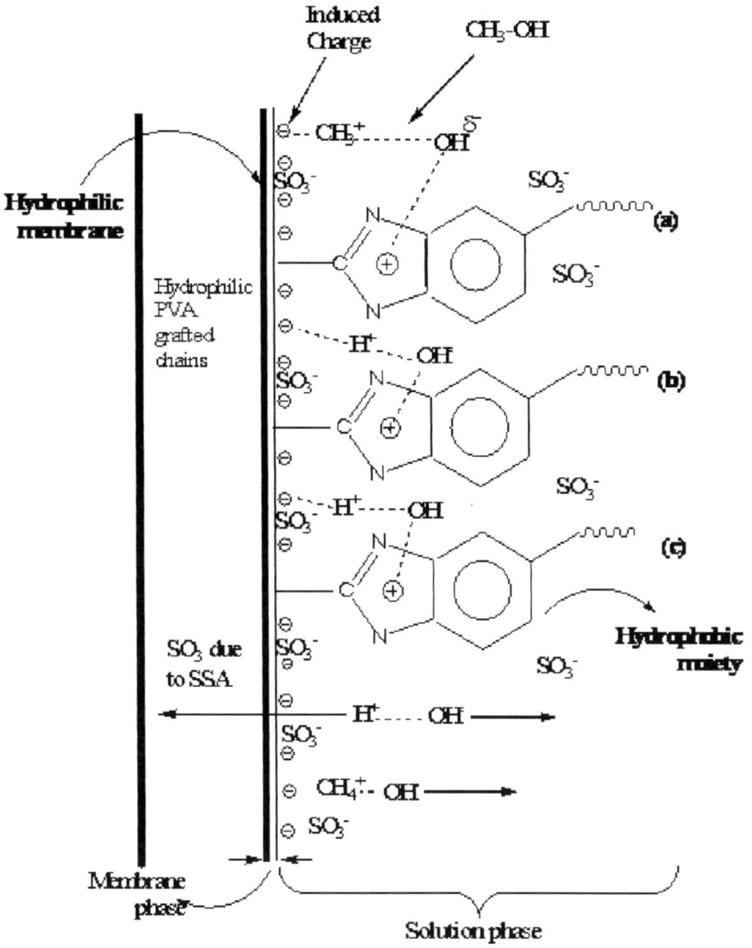

Figure S-1. The Bipolar nature of the membrane surface

(106b) Operation and Controls of Operating Direct Methanol Fuel Cells with PVA Based Membranes Modified by Benzimidazolium Groups

Pratima Gajbhiye, Chemical Engineering, Indian Institute of Technology, Kanpur, State of Uttar Pradesh, India, Pranjal Joshi, Chemical Engg, IIT Kanpur, Kanpur, India, Jayant K. Singh, Chemical Engineering, Indian Institute of Technology, Kanpur, Ka, India and A. Kumar, Department of Chemical Engineering, Indian Institute of Technology Kanpur, Kanpur, India

Operation and Control of Direct Methanol Fuel cell with PVA based membranes modified with benzimidazolium groups

Pratima Gajbhiye, Pranjal Joshi, Jayant K.Singh, Anil Kumar

Department of Chemical Engineering, Indian Institute of Technology, Kanpur 208016, India

Abstract

In recent years, in an effort to find alternate membrane material (other than Nafion 117), Polybenzimidazole (PBI) membranes have been synthesized where the conducting moiety of this polymer is benzimidazolium groups. The monomers used to synthesize these polymers are bis conjugated amines and diacids are expensive. In this work, we have synthesized grafted polymers of PVA as membrane materials where the grafts are either Poly Acrylic Acid, Poly maleic anhydride (PMA) or Poly Methacrylic acid (PMAA) and these materials were denoted by PVA-g-PAA, PVA-g-PMA, and PVA-g-PMAA. We then synthesized C1, C3 & C4 compounds (formula given at the bottom of Table S1) having bis conjugated amines whereas compound C2 having only one conjugated amine group. One of the steps in synthesizing C1 compound is nitration of phenyl ring, where there is a possibility of formation of two isomers out of which only one of them has strictly bis conjugated amines and can be polymerized with Isophthalic acid giving the formation of PBI. The other byproduct formed is an impurity and is minimized by carrying out the nitration step at low temperature but this cannot be completely removed from the product which is a mixture. Table S1 presents the summary of our experimental results on characterization of membrane, In the C1 to C4 compounds and in some cases (membranes 9, 10, 11), Sulfocuccinic acid were added to the casting polymer solution as additional crosslinking agent in order to achieve higher conductivity. We added the polymer, thus formed, to the casting solution, we first compared the results of the Fuel cell formed by adding C1 (C3 and C4) compound alone and heating at 180°C for some time. Table S1 has two sets of results, one on characterization (columns 1, 2, 3,4,5,6, and 7) of membranes while the other on Fuel cell (columns 8 to 11). For the latter columns 10 and 11 gave results on characterization on Fuel cell while columns 8 and 9 gave results when the Fuel cell was run on control mode for higher efficiency.

Experiments on membrane pore sizes are given in column 1 which lies between 8.88 to 9.52nm and they all are ultra filtration type. The values of the pore size are mostly governed by the nature of the graft chains, e.g. membranes 1,3,4,7 and 12 consist of PAA grafts and were found to have similar pore sizes. Similarly membranes 2,5,8,9 were of the similar pore size because the PVA have the same grafts of PMA. The next important thing is to determine the wall charges of these membranes and they were estimated by experimentally determining the rejection of Chromic (VI) acid solution through these membranes. Subsequently, a space charge model was proposed for the flow of the solution through their pores and a semi analytical solution was derived for the overall problem of separation of chromic acid solution. We observed that the high surface charge was because of the benzimidazolium groups present on the surface and they ranged from minimum of 465 to maximum of 525mV. The thermal degradation (TGA) experiments were carried out which showed that, through the modification of the polymer by benzimidazole groups, lead to stable membrane and the degradation temperature varies from 150°C for 10% weight loss and for approximately 80% weight reduction, a minimum of 300°C to 700°C.

The twelve membranes reported in Table S1 were first characterized completely in terms of pore size, wall charge, conductivity, IEC and thermal degradation. In fuel cell applications, higher temperatures give large current density and it is observed that some of the membranes (rows 1, 2, 5, 6, 7, 8 and 12) could easily be operated upto 150°C without the membrane degrading. However in our case, we insisted that the membrane of direct methanol fuel cell were operated at 90°C for which most of the fuel cell results for Nafion 117 are available. The catalysts used on the anode and cathode side were Pt-Ru and Pt loaded on carbon respectively, which were synthesized by following procedure given in literature. We find that in each case, the power density and the current density were higher and for some of our results (membranes 1,6,12) were atleast two to three times higher than that of Nafion 117 as seen in column 10 and 11. In normal Fuel cell operations, there is decay in the cathode catalyst which is further increased by the methanol crossover. The voltage of the Fuel cell increases, the average volatage gained is reported in column 8 and it is found that the voltage gain depends on the nature of the membrane which is not the same for all of them. In column 9, we have measured the average time cycle and we found that some of the membranes (membranes 6, 7, 9, 10) gave extremely good results.

It was recognized that the membrane material has hydrophilic PVA matrix and hydrophobic benzimidazolium groups at the surface formed by reaction of the C1 to C4 compounds with grafts (see Fig.S-1). When the reaction occurs (confirmed by IR and NMR), the hydrophobic benzimidazolium are formed which tends to separate as confirmed by the SEM results. In view of this we assumed that the pores of the membrane would have a thin hydrophobic layer with positive charge walls. This is confirmed by the contact angle measurements where the minimum contact angle is 51° and varied upto 89°. Even though most of the polymer matrix is PVA with its grafts, when the ions move through the pores (giving rise to current in Fuel cell), the pore walls get highly charged (as measured by column 2 of Table S1). Some of the membranes have Sulfo succinic acid SSA as part of the matrix (e.g. the membrane 9 to 11) which would tend to give sulfonate groups inside the PVA matrix as well as at the phase boundaries, this way giving rise to bipolar nature of the membrane (benzimidazolium group being strictly cationic in nature). In addition to this, some of the compounds such as C1 and C2 compounds have bonded sulfonate groups which would give the positive and negative charge separated by the molecular distance, once again producing a bipolar nature to the membrane. In remainder of the cases the charges would be induced on the phase boundaries as seen in Fig.S-1. The conductivity observed in our membranes is very high and could be attributed to the special bipolar structure of our polymers.

Table S1. Fuel cell and Membrane Characterization

Compound	**Membrane**		**1**	**2**	**3**	**4**	**5**		**6**	**7**	**8**	**9**	**10**	**11**
			MWCO (pore size, nm)	**Wall Charge**	**Conductivity (S/cm)**	**IEC**	**Thermal Degradation temp (°C) 10% - 80%**		**Water Uptake**	**Contact Angle**	**Average Voltage Gain (V)**	**Average time cycle (min)**	**Power Density (mW/ cm)**	**Current Density (mA/ cm)**
C1	1	PVA-g-PAA-C1	9.1	488	0.04	2.7	199.732	268.70	7.5	72	0.523	55.75	50.47	74.88
	2	PVA-g-PMA-C1	9.35	465	1.23 x10^{-2}	1.8	142.659	447.6	10.691	74.917	0.568	42.896	0.305	3.81
	3	PVA-g-PAA-modified	9.52	488	1.51 x10^{-2}	2.12	115.123	706.53 (28.93%)	13.665	75.581	0.778	69.85	37.37	64.44
C2	4	PVA-g-PAA-C2	9.52	525	6.59 x10^{-2}	2.38	119.558	691.93	13.4	51.2	0.704	39.83	23.0020	50.55
	5	PVA-g-PMA-C2	9.36	519	1.04 x10^{-2}	2.17	181.59	562.348	14.9	50.24	0.77	161.93	34.72	62.11
	6	PVA-g-PMAA-C2	8.99	482	1.16 x10^{-2}	2.18	201.055	472.568	13.8	51.8	0.6066	147.81	76.72	92.33
C3	7	PVA-g-PAA-C3	9.0	512	2.51 x10^{-2}	2.23	204.908	453.03	8.624	70.368	0.6883	82.55	36.397	75
	8	PVA-g-PMA-C3	9.5	510	3.51 x10^{-2}	2.27	166.819	505.293	13.513	62.944	0.726	37.557	36.11	64.381
	9	PVA-g-PAA-SSA-C3	9.36	497	1.71 x10^{-2}	2.51	154.914	813.769 (24.32%)	10.5	54.8	0.8096	99.423	23.45	57.778
	10	PVA-g-PMA-SSA-C3	9.83	480	2.28 x10^{-2}	2.46	115.12	428.713	8.5	76.2	0.663	224.22	28.561	58.358
	11	PVA-g-PMAA-	8.88	525	3.66 x10^{-2}	1.85	139.772	681.715	7.2	89.2	0.77	18.36	20.832	48.11

SOME FORMULAS/TEXT MAY BE MISSING DUE TO FORMATTING ISSUES IN THE ORIGINAL E-MEDIA VERSION

		SSA-C3						(30.78%)						
C4	12	PVA-g-PAA-C4	9.5	495	2.56×10^{-2}	1.62	189.87	451.86	8.927	59.132	0.704	74.19	41.1	71
	13	Nafion 117			0.1								21.01	48.33

Compound C1, [Bis-3-Amino-4-{3-(triethylammoniumsulfonato) phenylamino}-phenyl sulfone] hydrochloride

Compound C2, 3-Amino -4-[3-(triethyl ammonium sulfonato) phenyl amino] phenylene hydrochloride

Compound C3, Bis -4 [1, 2 diamino phenyl]-phenyl sulfone) hydrochloride

Compound C4, 1, 4 [1, 2 diamino phenyl] Phenylene hydrochloride

THE GRAPHIC ON THIS PAGE WAS NOT AVAILABLE ON THE E-MEDIA VERSION OF THIS PAPER.

Figure S-1. The bipolar nature of PVA-g-PAA-BASPAPS membrane which reduces the methanol crossover due to the molecular dissociation. In (a), the imidazolium group is shown to separate out from the PVA moiety of the chain into a separate phase. In (b), the water is shown to complex with the SO and the imidazolium group. As a result of this, interaction in (d) and (e), the water and the methanol is shown to split into anion and cation and the anion permeates through the membrane giving bipolar effect

(106e) Numerical Investigation in Heat Conduction Effect on Heat Exchange for Mini Cross-Flow Heat Exchanger (MCHE)

Hongyue Duan[1], Wei Mu[1], **Xigang Yuan**[1] and Lingai Luo[2], (1)State Key Laboratory of Chemical Engineering, Tianjin University, Tianjin, China, (2)LOCIE Chambery, Savoie University, Savoie, France

In a traditional heat exchanger, the effect of heat transfer within the wall material driven by the temperature gradient in axial direction of the fluid flow is usually very small because the area of the cross section of the tube wall is much smaller than that of the tube for heat exchange. However, in a mini cross-flow heat exchanger(MCHE), where the heat conducting medium is typically a cubic-shaped block of metal and the hot and cold streams are put into heat exchange via two sets of channels in the block, the heat conduction in the wall material in the direction parallel to the fluid flow may have a significant effect on the performance of the heat exchanger.

In this paper, a 2-channel-layer MCHE was simulated and a periodic boundary condition was applied so that the results could be applicable to entire MCHE. The proposed model and the mesh grid size were proved by the good agreement of the simulated overall heat transfer coefficient with that found in the experimental reported in the literature[1]. Then the effect of heat conduction on heat transfer performance in a MCHE was studied numerically. The simulation results suggest that the performance of the MCHE can be improved by increasing the wall thickness between the adjoining channels in the same layer or by decreasing the wall thickness between the adjacent layers. The heat conduction effect can also be weakened by optimizing the material conductivity or by increasing the diameter of the channels. New correlations were developed based on the simulated results to predict the value of the overall heat transfer coefficient as a function of the operating parameter such as Re in the channel and the design parameters of the MCHE such as the diameter of the channel, wall thickness and thermal conductivity.

[1]Luo L A, Fan Y L, Zhang W W*et al.*Integration of constructal distributors to a mini crossflow heat exchanger and their assembly configuration optimization[J]. Chemical Engineering Science, 2007, 62 (13): 3605-3619.

(106g) The Effects of Out Gassing Through Cylindrical Walls on the Transport Coefficients

Luma. H. Yaseen[1], **Nada B. Nakkash**[2] and M. N. Latif[1], (1)Al-Nahrain University, Baghdad, Iraq, (2)Chemical Engineering, Al-Nahrain University, Baghdad, Iraq

Abstract

This work concerns with studying the influence of blowing on friction factor and mass transfer coefficient when the flow is normal to a tube bank with outgassing porous cylinders. The friction factor and the mass transfer coefficients are measured for a range of Reynold's number and blowing rate and expressed in terms of various dimensionless groups.

The test rig consists of a rectangular duct (20.3 × 11.3 cm), five rows of three porous tubes in staggered arrangements fixed across the duct with 1.5pitches (both longitudinal and transversal). The outside and the inside diameters of tubes were 25.4 and 21.4 mm, respectively.

The friction factor measurements was studied for Reynold's number range of

$3000 < Re < 17{,}000$ and blowing rate up to 4×10^{-3}. The results show that the blowing intothe boundary layer for the cylinders increases the overall friction factor from 10%to 50% depending on the blowing rate. The mass transfer coefficient was measured for Reynold's number range of $5000 < Re < 16{,}000$ and blowing rate of $10^{-5} < m/G_\infty < 10^{-3}$.

A new definition of the blowing rate was considered and expressed by the mass transfer driving force. Two successful formulas were obtained to account the outgassing effect on the mass transfer coefficient. The mass transfer coefficient for the outgassing boundary layer was obtained first by determining the non-blowing coefficient and multiplying it with the correction factor .

The first model was modified to account the density variation associated with the boundary layer change:

$$St_m Sc^{2/3} = \frac{0.51\, Re^{-0.5}}{(1+\beta_m)^{4.7}} \left(\frac{\rho_s}{\rho_\infty}\right)^{0.4}$$

With the coefficient of determination R2 = 0.93.

The second model depends on the flat plate correction factor:

$$St_m Sc^{2/3} = 0.5\, Re^{-0.5} \left(\frac{Ln(1+\beta_m)}{\beta_m}\right)^{9.94}$$

With the coefficient of determination R2 = 0.92.

These two models are found to be in agreement with the previous literature and a reduction of about 80% in the mass transfer coefficient has been obtained due to the outgassing.

(106h) Down the Hubbert Curve

Ian Sutton, Sutton Technical Books, Houston, TX

Down the Hubbert Curve

In March 1956 M. King Hubbert, who was at that time Chief Geologist with the Shell Development Company published a seminal paper with the title "Nuclear Energy and the Fossil Fuels" (Hubbert 1956). In this lengthy and authoritative paper, which was presented before the American Petroleum Institute in San Antonio, Texas, Dr. Hubbert showed how key fossil fuel resources in the United States, particularly coal and oil, were of finite size and that their output would reach a maximum, and then decline. Dr. King's forecasts for the production of oil in the United States were accurate; he predicted that oil production would peak in the early 1970s, and that is what has happened. (The widely-used term "Peak Oil" is frequently accredited to Dr. Colin Campbell (2001)).

In the fifty years since the publication of Dr. King's paper great advances have been made in finding and extracting oil from what were considered to be marginal fields. But Dr. King's overall conclusion is still valid. Production of oil in the United States remains below the 1970's peak. Although Dr. Hubbert's analysis was confined to the United States, subsequent analysts applied his insights to other parts of the world. For example, the book *Twilight in the Desert: The Coming Saudi Oil Shock and the World Economy* (Simmons 2005) by the late Matthew Simmons achieved wide recognition by showing that the concept of peak oil most likely applies to reserves in Saudi Arabia, and other nations such as Iran and Kuwait.

The purpose of this paper is to examine how oil production and consumption have changed in the last five years, and to examine whether or not the concept of Peak Oil remains valid. Key to this discussion is an understanding of the concept of Energy Returned on Energy Invested (EROEI). It is concluded that the concept of Peak Oil, as first postulated by Dr. King, remains valid and current, and the production of oil peaked in the period 2005-2011 and is now set to move down the right hand side of the Hubbert Curve.

The title of this session is *Challenges in Instituting Energy Savings Project in the Present Market Conditions*. If the production of oil and has indeed already passed a peak then the challenge is not to do with "instituting" energy savings; those savings are going to occur whether we like it or not. The challenge will be to determine how the process industries can restructure themselves given that the energy resources available (both as direct energy and as petrochemical feedstock) will be in decline and that there will be intense demand for those same resources from other areas of the economy.

(106j) Statistical Decision Making for In-Service Inspection of Underground Waste Storage Tanks

Emily Mitchell, Emory University, Atlanta, GA and **Stephen P. Harris**, Savannah River National Laboratory, Aiken, SC

Pitting is a localized form of corrosion that occurs on passivated metal surfaces exposed to aggressive chemical environments. The pitting process involves the breakdown of the passive oxide film and subsequent anodic dissolution of the metal at a local site, while the remainder of the surface remains protected by the film.

Deep pits may result in lowering the maximum level of waste that made be stored in the tank and eventually lead to loss of confinement. Pit depths, established by ultrasonic testing, are subject to variables that cause uncertainty, including the repeatability of metric location, time between measurements, different transducers, and different inspectors.

The generalized extreme value (GEV) distribution is fit to the data gathered from ultrasonic testing and statistical projections are made for the full tank wall interior. A comparison is made with results from other well known statistical distributions. The estimated parameters are used in statistical tolerance limits for bounding pit depth in a waste tank.

The tolerance limits give an indication of extreme values of pit depth for the current condition of the tank walls. The upper tolerance limit for pit depth will then be compared to action levels for pitting. If the upper tolerance limit for pit depth exceeds the action level, more frequent examinations or increased tank surface area inspections are conducted.

(106k) Thermal Inbreathing Requirements of Low Pressure Storage Tanks At Elevated Temperatures

Donald E. Brooks, Pressure Relief Analysis, Lloyd's Register Energy Americas, Inc., Houston, TX

ISO 28300 (API 2000) "Petroleum, petrochemical and natural gas industries - Venting of atmospheric and low pressure storage tanks" provides guidance on determining the required thermal inbreathing flow rates to protect atmospheric and low pressure storage tanks against vacuum due to weather changes (e.g. a decrease in atmospheric temperature)." For heated, un-insulated tanks with vapor space temperatures above 58.9 °C (120 "F), ISO 28300 recommends an engineering review of thermal inbreathing requirements.

This paper describes one approach to evaluating thermal inbreathing requirements for un-insulated storage tanks with vapor space temperatures exceeding 58.8 °C (120 "F)." The tank is assumed to be exposed to the cooling effects of a sudden rain. A simplified unsteady state heat transfer analysis is performed to determine the tank wall and vapor space temperatures versus time. Initial gas mass and pressure are assumed to remain constant in the tank. Gas volume in the tank is calculated as a function of vapor space temperature using the ideal gas law. The rate of volume change is calculated for each time increment. The rate of volume change goes through a maximum value. This maximum value is selected as the thermal inbreathing requirement of the tank.

The calculation methodology described in this paper can be programmed in a spreadsheet for ease of use.

Dynamic Operational Risk Assessment with Bayesian Networks

Shubharthi Barua, Xiaodan Gao, M. Sam Mannan
Mary Kay O'Connor Process Safety Center
Texas A&M University, Jack E. Brown Engineering Bldg., 3122 TAMU, Room 244, College Station, TX 77843
shubharthi.barua@tamu.edu, xiaodan_gao08@mail.che.tamu.edu, mannan@tamu.edu

Extended Abstract Prepared for the Poster Presentation at the
8th Global Congress on Process Safety
Houston, TX
April 1 – 4, 2012

UNPUBLISHED

Dynamic Operational Risk Assessment with Bayesian Networks

Shubharthi Barua, Xiaodan Gao, M. Sam Mannan

Mary Kay O'Connor Process Safety Center
Texas A&M University, Jack E. Brown Engineering Bldg., 3122 TAMU Room 244, College Station, TX-77843
shubharthi.barua@tamu.edu, xiaodan_gao08@mail.che.tamu.edu, mannan@tamu.edu

Keywords: risk assessment, Bayesian network, dynamic Bayesian network, fault tree, event tree

Abstract

Oil/gas and petrochemical plants are complicated and dynamic in nature. Dynamic characteristics include ageing of equipment/components, season changes, stochastic processes, operator response times, inspection and testing time intervals, and timing of safety system operations, all of which are time dependent criteria that can influence dynamic processes. Conventional risk assessment methodologies can quantify dynamic changes in processes with limited capacity. Several methods are proposed for dynamic risk assessment such as C-DET, D-DET, DORA, etc., but all of them have deficiencies of some kind. Bayesian networks, a graphical inference technique, have some advantageous characteristics for risk analysis over other techniques. This research focuses on providing a framework for dynamic risk analysis based on Bayesian networks. A case study will be provided to illustrate this model.

1.0 Introduction

Oil/gas, chemical, and petrochemical process plants consist of numerous equipment and unit operations, thousands of control loops, and exhibit dynamic behavior. Different hazards and several types of risk are associated with these process facilities. For example, large process areas have the potential to stock a large amount of hazardous chemicals, which can result in the process area being highly congested. It is reported that process risk, i.e., risk in process facilities due to fires and explosions, greatly contributes to the total risk of offshore installations. Also, reactivity, toxic, and mechanical hazards have a significant effect on the total risk of these facilities. Therefore, it is very important to identify hazards, perform risk assessments, and take proper initiatives to minimize/remove hazards and risks; else a catastrophic accident may result.

From case histories, it has been observed that catastrophic accidents have a significant effect on people, environment, and society. Catastrophic accidents such as the Flixborough disaster, the Bhopal incident, and the Piper Alpha disaster caused fatalities and unbearable economic loss. In the past few years, several disastrous accidents in refineries, power plants, and offshore platforms have significantly affected people's perception, and consequently raised concern to emphasize process safety. A series of explosions in 2005 resulting from insufficient process safety and lack of risk reduction measures at the BP Texas City refinery accident resulted in 15 fatalities, 180 injuries, and $1.5 billion in losses [1]. The Kleen Energy power plant explosion in

Connecticut in 2010 is considered to be one of the worst industrial explosions in the USA [2]. In 2010, a fire and explosion resulting from a blowout at the Macondo well resulted in 11 deaths and 17 injuries. Also the continuous spill from the wellhead for 87 days had disastrous effects on the environment and wildlife surrounding the Gulf of Mexico. It has been reported that most of the accidents could be prevented or their severity could have been minimized by effective risk assessment and process safety management. But existing risk assessment methodologies cannot properly quantify various time-dependent effects. Therefore, the objective of this research is to develop a risk assessment methodology with the capability to capture variations due to different time-dependent effects.

2.0 Literature Review

N. Siu [3] summarized available methodologies for risk assessment. For example, the fault tree and event tree methods are two most common methodologies for risk assessment. But both techniques have limited ability to address different time-dependent effects such as equipment or component ageing, stochastic processes, operator response times, and inspection and testing time intervals. N. Siu [3] also described other methodology limitations while applying dynamic system analyses. Therefore, for dynamic systems, conventional techniques are not able to provide accurate results because of their inadequate ability to describe the variation of operational risks as time-dependent deviations, or the changes occurring in the process. Continuous event trees [4], discrete event trees [5], and Markov theory [6] methods have been proposed for dynamic reliability and safety analysis. But all of the proposed techniques have some deficiencies. For example, the Markov theory based model does not consider the effect of inspection on system-state transitions. X. Yang et al. [1] proposed the Dynamic Operational Risk Assessment (DORA) technique, which is able to take inspection and testing time intervals and repair time into account for evaluating operational failure.

Recently, Bayesian networks are receiving wide acceptance in reliability assessment, fault diagnosis, safety, risk, and maintenance analysis. Several authors have [7,8] compared some commonly used methods for accident analysis with Bayesian networks, and described several advantages of Bayesian networks over these methods. N. Khakzad et al. [2] illustrated parallelism between Bayesian networks and fault trees, and discussed the potential of Bayesian networks as a tool for process safety and accident analysis. H. Pasman and W. Rogers [9] described an approach to incorporate Bayesian networks into layer of protection analysis to make it more effective. These studies demonstrate the potential advantages of Bayesian networks application in risk assessment. This research provides a methodology for dynamic risk assessment based on Bayesian networks.

3.0 Methodology and Expected Result

3.01 Bayesian Networks

A Bayesian network, which is a probabilistic graphical model, describes causal influence relations among variables. A set of random variables and their conditional dependencies are represented by Bayesian networks via a directed acyclic graph (DAG). It has the ability to represent dependency among events clearly, accommodate multi mode and continuous random variables, and incorporate information i.e. generic, system specific and expert judgment to support optimum decision making.

A simple Bayesian network is shown in Figure 1. In a binary network, variables are represented by nodes and causal relationships among different nodes are shown by an arc. Conditional probability tables or defined probabilistic relations among nodes represent how one variable is linked with one or many variables. The nodes that influence another variable and have unconditional probability are called parent or root nodes. Nodes that are conditionally dependent on their direct parents are called intermediate nodes. The top node is defined as a leaf node, which is similar to the top event of a fault tree and initiating event for an event tree.

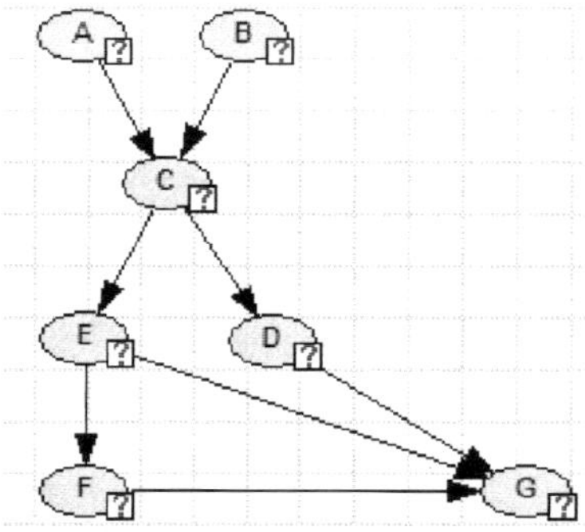

Figure 1: A simple Bayesian network

Let $N = (G, P)$ be a Bayesian network, where

$G = (V, E)$ is a directed acyclic graph; V (random variables) represents nodes; and E represents edges between pairs of nodes of DAG.

P represents probability distribution over V and $V = \{X_1, X_2, \dots, X_n\}$ is either discrete or continuous random variables [10]. These random variables are assigned to the nodes and the edges.

Bayesian networks can be represented by the joint probability distribution P(V);

$$P(V) = \sum_{X \in V} P\{X|pa(X)\} = P(X_1, X_2, \dots, X_N) = \sum_{i=1}^{n} P\{X_i|pa(X_i)\}$$

Here $pa(X_i)$ = parent nodes of X_i.

A main advantage of the application of Bayesian networks in risk analysis is the ability to update prior data using Bayes' theorem by incorporating new information as demonstrated by N. Khakzad et al. [2]. Also, Bayesian networks have an advantage of handling different types of uncertainty.

3.02 Dynamic Bayesian Networks

A general Bayesian network is a static model, i.e., the joint probability distribution is usually a representation of a fixed point or an interval of time [11]. A dynamic Bayesian network is a Bayesian network that has the ability to describe the evolution of joint probability distribution over time. Dean and Kanazawa [12] proposed modeling time in a discrete manner to represent the progression of time in a dynamic Bayesian network. In a dynamic Bayesian network, arcs links node/nodes from previous time slice to the node/nodes of the next time slice to represent temporal dependencies among them.

3.03 Methodology

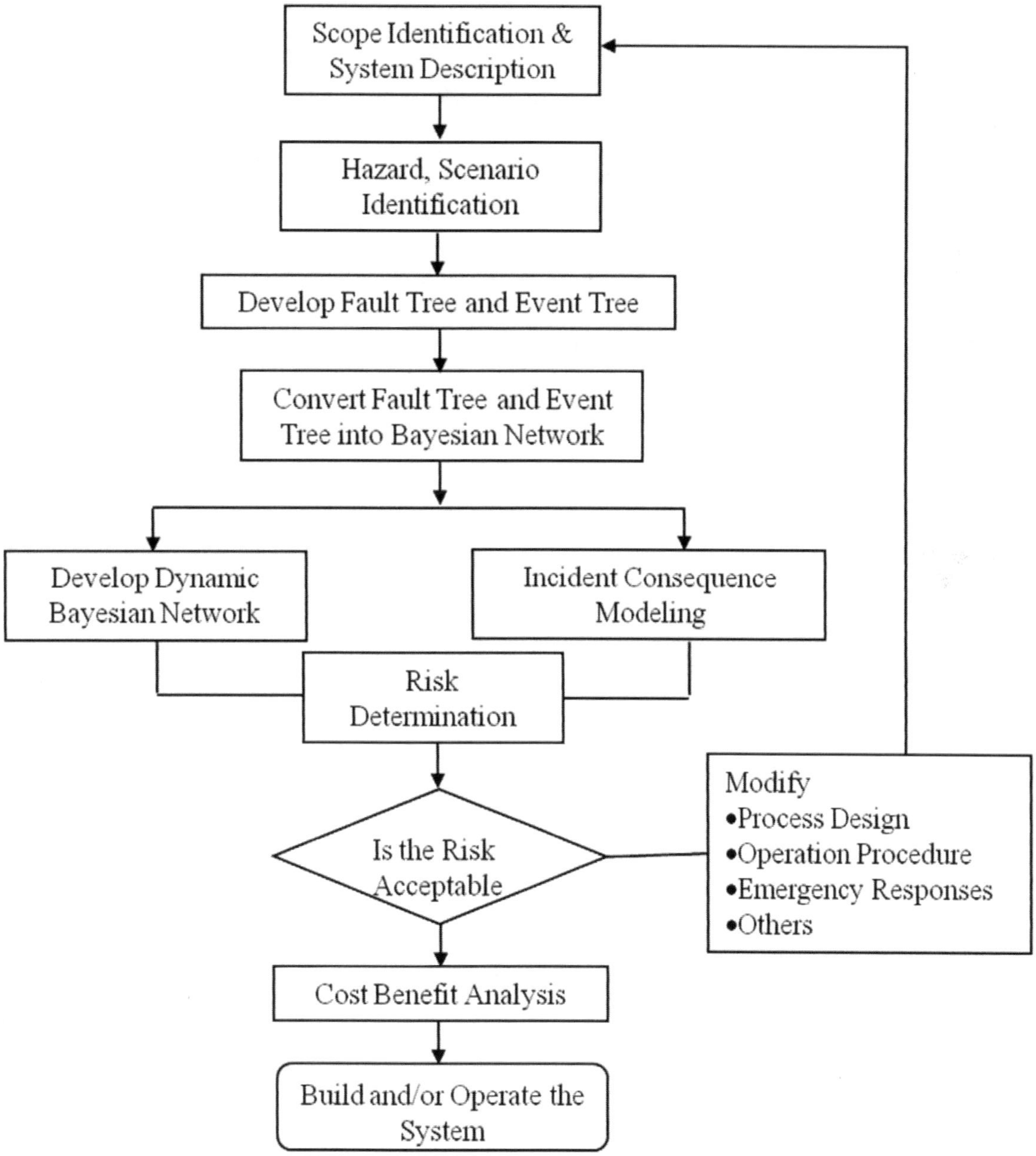

Figure 2: Framework for the dynamic Bayesian network based risk assessment

The objective of this research is to propose a dynamic risk assessment method based on dynamic Bayesian networks. According to the framework, a fault tree and event tree are developed for a system. The objective for following step is to convert the developed fault and event tree into a Bayesian network. Bobbio et al. [13] provided a methodology for conversion of fault trees into Bayesian networks. N. Khakzad et al. [2] showed equivalency between fault trees and Bayesian networks, and described the advantages of Bayesian networks for process safety analysis. Bearfield and March [14] proposed another methodology for generalizing event trees into Bayesian networks. The method has the ability to generalize simple event trees into Bayesian networks in a relatively easy manner, but the method becomes complicated for large event trees. Therefore, in the first phase, a methodology is proposed to convert fault trees and event trees into a static Bayesian network. In the second phase, a dynamic Bayesian network is developed from the static Bayesian network. The dynamic Bayesian network provides probability of a system at different time-slices for different states of the system. The probabilities are then combined with consequence modeling levels to calculate risk. GeNIe, which was developed at the Decision Systems Laboratory at the University of Pittsburgh, is used to create the model [15].

4.0 Conclusions

This article briefly describes a framework based on Bayesian network for dynamic risk assessment for the oil/gas, chemical, and petrochemical industries. This research illustrates the capability of Bayesian networks to represent all component interactions in complex systems, and to update all prior system probabilities in the presence of new information. The advantages of modeling flexibility and efficient computational mechanisms using dynamic Bayesian networks are also discussed.

4. References

[1] Yang, X. and Mannan, M.S., "The development and application of dynamic operational risk assessment in oil/gas and chemical process industry", Reliability Engineering and System Safety 95 (2010) 806-815

[2] Khakzad, N., Khan, F. and Amyotte,P., "Safety analysis in process facilities: Comparison of fault tree and Bayesian network approaches", Reliability Engineering and System Safety 96(2011) 925-932

[3] Siu,N. "Risk assessment for dynamic systems: an overview", Reliability Engineering and System Safety 43 (1994) 43-73

[4] Devooght, J. and Smidts,C. "Probabilistic reactor dynamics-I: the theory of continuous event trees", Nuclear Science and Engineering 1992; 111:229-40

[5] Amendola, A. "Accident sequence dynamic simulation versus event trees", Reliability Engineering and System Safety 1988; 22:3-25

[6] Howard, R. "Dynamic probability systems", Vols. I and II, New York, 1971

[7] Sklet, S. "Comparison of some selected methods for accident investigation", Journal of Hazardous Materials 2004; 111:29-37

[8] Nivolianitou, Z.S., Leopoulos, V.N. and Konstantinidou, M., "Comparison of techniques for accident scenario analysis in hazardous systems", Journal of Loss Prevention in the Process Industries 2004; 17"467-75

[9] Pasman, H. and Rogers, W.; "BBN, a tool to make LOPA more effective, QRA more transparent and flexible, and therefore to make safety more definable", 14th Annual Sysposium, Mary Kay O'Connor Process Safety Center, College Station, Texas, October 2011

[10] Donohue, S.K. and Dugan, J.B.; "Modeling the "Good enough to release" decision usin V&V preference structure and Bayesian belief networks", Reliability and Maintainability Symposium, 2003 Annual, 2003

[11] McNaught, K.R. and Zagorecki, A. "Using Dynamic Bayesian networks for prognostic modeling to inform maintenance decision making", IEEM 2009, Hond Kong, January 2010

[12] Dean, T. and Kanazawa, K., "A model for reasoning about persistence and causation", Computational Intelligence 1989; 5:142-150

[13] Bobbio A, Portinale L, Minichino M and Ciancamerla E., "Improving the analysis of dependable systems by mapping fault trees into Bayesian networks", Reliability Engineering and System Safety 71(3): 249-260

[14] Bearfield G and Marsh W, "Generalising event trees using Bayesian networks with a case study of train derailment", Available online at http://www.agenarisk.com/resources/white_papers/Generalising_Event_Trees.pdf , last modified: 23 February 2010, accessed: 7 January 2012

[15] Decision Systems Laboratory, University of Pittsburgh, GeNIe (Graphical Network Interface and SMILE (Structural Modeling, Inference, and Learning Enginer), Version 2.0 software, 2010. Available online at http://genie.sis.pitt.edu

(137b) Comparing the Performance of Compact Heat Exchange Reactor to CSTR for Fischer Tropsch Synthesis

Hani Gadalla[1], Zhijun Jia[1], Steven J. Vallee[1], Burtron H. Davis[2]and Dennis Sparks[2], (1)Chart Energy & Chemicals, La Crosse, WI, (2)Center for Applied Energy Research, University of Kentucky, Lexington, KY

A commercial Fischer Tropsch catalyst was studied in two Compact Heat Exchange Reactors (CHER) with different channel sizes, and in a continuous stir tank reactor (CSTR). The CHERs were constructed by Chart Energy and Chemicals with one reactor having a 5mm channels and the other having a 6mm channels, while maintaining the same catalyst volumeThe catalyst was crushed and sieved retaining particles ranging in size from 75-125μm. Testing was performed at the University of Kentucky Center for Applied Energy Research in Lexington, KY. The CO conversion, selectivity and hydrocarbon productivity were measured for each reactor over a wide range of operating conditions, as a function of temperature, pressure, H2:CO feed ratio, and flow rate. Both reactor types showed excellent temperature control, but the performances varied at different operating conditions. The larger channel CHER had a higher conversion than the 5mm channel, potentially due to slightly higher reactor bed temperature. The two CHERs approached and even exceeded the productivity of the CSTR for liquid hydrocarbons per gram of catalyst. At higher flow rates the methane selectivity increases at a lower rate for either of the CHERs than for the CSTR.

(130a) Climate Change and Greenhouse Gas Regulation Update

Mary Ellen Ternes, McAfee & Taft, Oklahoma City, OK

This presentation will review the most recent regulatory and litigation developments arising in the area of climate change and greenhouse gas emissions. Anticipated topics include the status of challenges to EPA greenhouse gas regulation, CAA permitting and GHG reporting enforcement developments and the status of greenhouse gas litigation.

(130b) Adaptation for Chemical Engineers: Incorporating Resource Shifts Into Siting and Planning

Margaret Peloso, Vinson & Elkins LLP, Washington, DC

From shifting shorelines to changes in the amount and seasonality of rainfall, climate change will dramatically alter patterns of natural resource availabiltiy. Understanding these resource challenges and incorporating them into project design will be critical to project success and sustainability. This presentation will review these challenges and discuss examples of how these challenges are being addressed in particular projects.

AUTHOR INDEX

AUTHOR INDEX